Computing
PROSODY

Springer
New York
Berlin
Heidelberg
Barcelona
Budapest
Hong Kong
London
Milan
Paris
Santa Clara
Singapore
Tokyo

Yoshinori Sagisaka • Nick Campbell • Norio Higuchi
Editors

Computing PROSODY

Computational Models for Processing Spontaneous Speech

With 75 Illustrations

Springer

Yoshinori Sagisaka
Nick Campbell
Norio Higuchi
ATR Interpreting Telecommunications
Research Labs
2-2, Hikaridai, Seika-cho, Soraku-gun
Kyoto, 619-02 Japan

Library of Congress Cataloging-in-Publication Data
Computing prosody : computational models for processing spontaneous speech / [edited by] Yoshinori Sagisaka, Nick Campbell, Norio Higuchi
p. cm.
"A collection of papers from the Spring '95 Workshop on Computational Approaches to Processing the Prosody of Spontaneous Speech ... Kyoto, Japan"—Pref.
Includes bibliographical references and indexes.
ISBN-13:978-1-4612-7476-6 e-ISBN-13:978-1-4612-2258-3
DOI: 10.1007/978-1-4612-2258-3

1. Prosodic analysis (Linguistics)—Data processing—Congresses. 2. Speech processing systems—Congresses. 3. Japanese language—Prosodic analysis—Data processing—Congresses. I. Sagisaka, Y. (Yoshinori) II. Campbell, Nick. III. Higuchi, Norio. IV. Workshop on Computational Approaches to Processing the Prosody of Spontaneous Speech (1995 : Kyoto, Japan)
P224.C66 1996
414'.6—dc20 96-18416

Printed on acid-free paper.

Softcover reprint of the hardcover 1st edition 1997

Production managed by Robert Wexler; manufacturing supervised by Jeffrey Taub.
Camera-ready copy prepared using the authors LaTeX files.
9 8 7 6 5 4 3 2 1
ISBN-13:978-1-4612-7476-6 Springer-Verlag New York Berlin Heidelberg SPIN 10539768

Preface

This book presents a collection of papers from the Spring 1995 Workshop on Computational Approaches to Processing the Prosody of Spontaneous Speech, hosted by the ATR Interpreting Telecommunications Research Laboratories in Kyoto, Japan. The workshop brought together leading researchers in the fields of speech and signal processing, electrical engineering, psychology, and linguistics, to discuss aspects of spontaneous speech prosody and to suggest approaches to its computational analysis and modelling.

The book is divided into four sections. Part I gives an overview and theoretical background to the nature of spontaneous speech, differentiating it from the lab-speech that has been the focus of so many earlier analyses. Part II focuses on the prosodic features of discourse and the structure of the spoken message, Part III on the generation and modelling of prosody for computer speech synthesis. Part IV discusses how prosodic information can be used in the context of automatic speech recognition. Each section of the book starts with an invited overview paper to situate the chapters in the context of current research.

We feel that this collection of papers offers interesting insights into the scope and nature of the problems concerned with the computational analysis and modelling of real spontaneous speech, and expect that these works will not only form the basis of further developments in each field but also merge to form an integrated computational model of prosody for a better understanding of human processing of the complex interactions of the speech chain.

Kyoto, Japan
February, 1996

Yoshinori Sagisaka
Nick Campbell
Norio Higuchi

Acknowledgment
The editors are particularly grateful to the many reviewers who gave so much of their time to help improve the contributions, and to the invited experts who contributed the Introductions to each section. We would also like to take this opportunity to express our thanks to the management of ATR ITL for providing the facilities for the workshop, and to M. Nishimura, Y. Shibata, T. Minami, and A. W. Black for their assistance with the technical details concerning production of the book.

Participants at the Spring 1995 Workshop on Computational Approaches to Processing the Prosody of Spontaneous Speech

Contents

III Prosody in Speech Synthesis 155

Contributors

Gérard Bailly. Institut de la Communication Parlée 46 avenue Félix Viallet, 38031, Grenoble, Cedex 1, France

Anton Batliner. Institut für Deutsche Philologie, Universität München Schellingstrasse 3, 80799 München, Germany

Mary E. Beckman. The Ohio State University, Dept. of Linguistics 222 Oxley Hall, 1712 Neil Ave., Columbus, OH 43210-1298 USA

Alan W. Black. ATR Interpreting Telecommunications Research Labs., Dept. 2 2-2, Hikaridai, Seika-cho, Soraku-gun, Kyoto 619-02, Japan

Gösta Bruce. Lund University, Dept. of Linguistics Helgonabacken 12 223 62, Lund, Sweden

Nick Campbell. ATR Interpreting Telecommunications Research Labs., Dept. 2 2-2, Hikaridai, Seika-cho, Soraku-gun, Kyoto 619-02, Japan

Anne Cutler. Max-Planck-Inst. for Psycholinguistics Wundtlaan 1 6525 XD, Nijmegen, The Netherlands

Shigeru Fujio. ATR Interpreting Telecommunications Research Labs., Dept. 2 2-2, Hikaridai, Seika-cho, Soraku-gun, Kyoto 619-02, Japan

Hiroya Fujisaki. Science University of Tokyo, Dept. of Applied Electonics 2641, Yamazaki, Noda, Chiba 278, Japan

Sadaoki Furui. NTT Human Interface Labs., Furui Research Labs. 3-9-1 Midori-cho, Musashino, Tokyo 180, Japan

Björn Granström. KTH Dept. of Speech Communication and Music Acoustics Box 70014, S-10044, Stockholm, Sweden

Kiell Gustafson. KTH Dept. of Speech Communication and Music Acoustics Box 70014, S-10044, Stockholm, Sweden

Wolfgang Hess. Institut für Kommunikationsforschung und Phonetik (IKP) Universität Bonn, Poppeldorfer Allee 47 D-53115, Bonn, Germany

Norio Higuchi. ATR Interpreting Telecommunications Research Labs., Dept. 2 2-2, Hikaridai, Seika-cho, Soraku-gun, Kyoto 619-02, Japan

Toshio Hirai. ATR Interpreting Telecommunications Research Labs., Dept. 2 2-2, Hikaridai, Seika-cho, Soraku-gun, Kyoto 619-02, Japan

Keikichi Hirose. University of Tokyo, Dept. of Information and Communication Engineering 7-3-1, Hongo, Bunkyo-ku, Tokyo, Japan

Merle Horne. Lund University, Dept. of Linguistics and Phonetics Helgonabacken 12, S-223 62, Lund, Sweden

David House. Lund University, Dept. of Linguistics and Phonetics Helgonabacken 12, S-223 62, Lund, Sweden

Andrew Hunt. Sun Microsystems Laboratories, 2 Elizabeth Drive, Chelmsford, MA 01824

Hiroaki Kato. ATR Human Information Processing Research Labs., Dept. 1 2-2, Hikaridai, Seika-cho, Soraku-gun, Kyoto 619-02, Japan

Andreas Kiessling. Lehrstuhl für Mustererkennung, Universität Martenstrasse 3, 90158 Erlangen-Nürnberg, Germany

Klaus Kohler. Institut fur Phonetik und digitale Sprachverarbeitung (IPDS) Christian-Albrechts-Universitat, D-24098, Kiel, Germany

Ralf Kompe. Lehrstuhl für Mustererkennung, Universität Martenstrasse 3, 90158 Erlangen-Nürnberg, Germany

D. R. Ladd. University of Edinburgh, Dept. of Linguistics George Square, Edinburgh, Scotland, United Kindom

Kikuo Maekawa. The National Language Research Institute, Dept. of Language Behavior 3-9-14, Nishigaoka, Kita-ku, Tokyo 115, Japan

Mitsuru Nakai. Japan Advanced Institute of Science and Technology 1-1 Asahidai, Tatsunokuchi, Ishikawa 932-12, Japan

Shin'ya Nakajima. NTT Human Interface Labs., Speech and Acoustics Lab. 3-9-11, Midori-cho, Musashino-shi, Tokyo 180, Japan

Christine Nakatani. AT&T Labs-Research, Murray Hill, NJ 07974-0636 and Harvard University, Aiken Computation Laboratory 33 Oxford St., Cambridge, MA 02138 USA

Mari Ostendorf. Boston University Electrical, Computer and Systems Engineering 44 Cummington St., Boston, MA 02215 USA

Anton Petzold. Institut für Kommunikationsforschung und Phonetik, Universität Bonn Poppeldorfer Allee 47 D-53115, Bonn, Germany

Matthias Reyelt. Institut für Nachrichtentechnik, Technische Universität Schleininzstrasse 23, 38092 Braunschweig, Germany

Ken Ross. Boston University Electrical, Computer and Systems Engineering 44 Cummington St., Boston, MA 02215 USA

Yoshinori Sagisaka. ATR Interpreting Telecommunications Research Labs., Dept. 1 2-2, Hikaridai, Seika-cho, Soraku-gun, Kyoto 619-02, Japan

Jan van Santen. AT&T Bell Labs 2D-452, 600 Mountain Ave., Murray Hill, NJ 07974-0636 USA

Hiroshi Shimodaira. Japan Advanced Institute of Science and Technology 15, Asahidai, Tatsuguchi-cho, Nomi-gun, Ishikawa 923-12, Japan

Harald Singer. ATR Interpreting Telecommunications Research Labs., Dept. 1 2-2, Hikaridai, Seika-cho, Soraku-gun, Kyoto 619-02, Japan

Volker Strom. Institut für Kommunikationsforschung und Phonetik, Universität Bonn Poppeldorfer Allee 47 D-53115, Bonn, Germany

Jacques Terken. Institute for Perception Research P.O. Box 513, 5600 MB Eindhoven, The Netherlands

Paul Touati. Lund University, Dept of Linguistics and Phonetics Helgonabacken 12, S-223 62, Lund, Sweden

Hajime Tsukada. NTT Human Interface Labs., Speech and Acoustics Lab. 3-9-11, Midori-cho, Musashino-shi, Tokyo 180, Japan

Minoru Tsuzaki. ATR Human Information Processing Research Labs., Dept. 1 2-2, Hikaridai, Seika-cho, Soraku-gun, Kyoto 619-02, Japan

Part I

The Prosody of Spontaneous Speech

1
Introduction to Part I

D. R. Ladd

1.1 Naturalness and Spontaneous Speech

The topic touched on by the papers in this introductory section is the difference between spontaneous speech and careful speech produced in the laboratory. Beckman essays a panoramic taxonomy of kinds of spontaneous speech and discusses the ways in which the varieties she identifies differ with respect to prosody. Fujisaki suggests a methodology for making models. Bruce et al. present a useful summary of research on prosodic features of Swedish spontaneous speech carried out over the last several years by their group in Lund and by a cooperating group at KTH in Stockholm. These three contributions, both by what they say and by the very fact of their having been written, make it clear that spontaneous speech has quickly moved to a prominent place on the research agenda for linguists and phoneticians concerned with technological applications.

Many basic problems in speech synthesis and recognition have been solved, at least to an extent that makes limited but real applications possible. Yet far more fundamental work is required before we reach the manifest ultimate goal of speech technology research, namely the use of ordinary spoken language to interact with computers (or, in the case of interpretive telephony, to interact with other human beings with the aid of a computer intermediary). By far the biggest hurdle on our way to attaining this goal is our lack of knowledge about how the linguist's idealised descriptions of language structure and language sound patterns relate to the superficially disorderly way that language is put to use in real interactions between real people.

What is already clear, however, is that this is an extremely difficult area to study. It sometimes appears that the only way we can investigate what makes natural speech natural is by destroying its naturalness. This theme is echoed both by Beckman and by Bruce and his colleagues and occurs repeatedly throughout the book. To be sure, this is not a new observation. More than 30 years ago, Lehiste [Leh63, 353] suggested that:

> "... the linguist who wants to use experimental methods in his investigation is forced to trade a certain amount of naturalness for rigorousness in his experimental design.... The

> two requirements of naturalness and rigorousness appear to occupy the opposite ends of a continuous scale of gradual mutual exclusion. At one extreme we find complete naturalness coupled with complete lack of control; at the other end, complete control over the experimental design at the expense of naturalness and linguistic relevance."

While Lehiste's remarks show that the problem is not new, the scientific context now makes the dilemma more acute. In acquiring basic knowledge about speech and sound patterns, it has so far been possible to steer a reasonable course between the two incompatible desiderata Lehiste identifies. Now it may no longer be possible. If we want to know about speech, a compromise between naturalness and control is attainable; but if we want to know about *NATURAL* speech, naturalness is paramount, and new approaches to control may be necessary.

However, the quote from Lehiste implicitly draws attention to a source of understanding that is worth cultivating. Lehiste talks of "the linguist"—not "the speech scientist", not "the engineer", not even "the phonetician". Laudably, Lehiste assumes that linguists can and should work in the laboratory, and that the theoretical constructs of linguists are in principle relevant to describing the behavior of actual speakers. This point has sometimes been forgotten, both by linguists—whose academic culture to a great extent prizes theorising over applications—and by experimental phoneticians and engineers, who have often tended to belittle, dismiss, or ignore what linguists have to say. This has been especially true in the area of prosody, where a real split between "experimentalist" and "impressionistic" approaches was evident especially during the period from about 1950 to 1980 (see [Lad96, Chap. 1] for more discussion).

While the split between the two approaches is still with us, it has begun to narrow somewhat, notably with the appearance of Pierrehumbert's work on English intonation [Pie80, Pie81] and with the development of Laboratory Phonology (e.g., Kingston and Beckman [KB90]). The coming together of experimental methodology and serious theoretical work provides the setting for many of the papers brought together in this book on the prosody of spontaneous speech. In the 1950s, Fry [Fry55, Fry58] showed that the most consistent acoustic correlate of "stress" in English two-syllable utterances is pitch movement or pitch obtrusion, and for many years after that the "experimentalist" community took that finding to justify the assumption that pitch movement is the essential phonetic basis of stress. Ideas about intensity and force of articulation were discredited, and discussions within theoretical linguistics of fine distinctions of relative prominence (e.g., Chomsky and Halle [CH68]) were dismissed as empirically baseless (by, e.g., [VL72, Lie60]). But by the mid-1980s the "impressionistic" view began to seem more plausible. Evidence stubbornly persisted of cases where perceived stress could not be related to pitch movement (e.g., [Hus78]).

Direct evidence was provided for basic differences between Japanese and English, showing that pitch movement really is the essence of accent in Japanese, while intensity and duration play a key role in English [Bec86]. Findings like these, combined with the theoretical notion of phonological "association" between pitch features and segmental features [Gol76, Pie80], yield a clear distinction between "pitch accent" and "stress"—a distinction that is simply incoherent given the experimentalist understanding of the 1950s and 1960s. This new view has led to new empirical discoveries, such as the finding that (at least in Dutch) shallower spectral tilt is a reliable acoustic correlate of lexical stress, regardless of whether the stressed syllable also bears sentence-level pitch accent [SvH96]. In this context see also Maekawa's finding [Mae96, this volume] that in Japanese formant structure may signal emphasis as distinct from lexical accent. It seems likely that many further discoveries in this area remain to be made, and that they will inevitably lead to improvements in both synthesis and recognition technology. Some of these discoveries could already have been made if experimentalists and theorists had not ignored each other's views about stress during the 1960s and 1970s.

To return to spontaneous speech, then, I believe it is important for speech technology researchers to value the work of a wide variety of researchers—not only those researchers whose methods and approach they find congenial. Many linguists have studied "discourse" and "coherence" and similar phenomena making use of the linguist's traditional method of theorising on the basis of carefully chosen individual cases (see, e.g., [HH76]). There has recently been much discussion of "focus" and "attention" at the intersection of linguistics, artificial intelligence, and philosophy (e.g., the papers in Cohen, Morgan, and Pollack 1990 [CMP90]). For anyone whose preferred approach to studying natural speech is based on statistical analysis of data from corpora of recorded dialogue, some of this other work must appear speculative, unverifiable, even unscientific. But that is exactly the attitude that served as an obstacle to progress in understanding stress and accent in the 1960s and 1970s. The field of speech technology is too young, and the problem of natural conversational interaction too multi-faceted, for a single approach to yield the understanding required for successful applications. We must all pay attention to one another.

References

[Bec86] M. E. Beckman. *Stress and Non-Stress Accent.* Netherlands Phonetic Archives 7. Dordrecht: Foris Publications, 1986.

[CH68] N. Chomsky and M. Halle. *The Sound Pattern of English.* New York: Harper and Row, 1968.

[CMP90] P. R. Cohen, J. Morgan, and M. E. Pollack, editors. *Intentions in Communication*. Cambridge, MA: MIT Press, 1990.

[Fry55] D. B. Fry. Duration and intensity as physical correlates of linguistic stress. *J. Acoust. Soc. Am.*, 27:765–768, 1955.

[Fry58] D. B. Fry. Experiments in the perception of stress. *Language and Speech*, 1:126–152, 1958.

[Gol76] J. Goldsmith. *Autosegmental Phonology*. Ph.D. thesis, MIT, 1976; published 1979 by Garland Press, New York.

[HH76] M. A. K. Halliday and R. Hasan. *Cohesion in English*. London: Longman, 1976.

[Hus78] V. Huss. English word stress in the post-nuclear position. *Phonetica*, 35:86–105, 1978.

[KB90] J. Kingston and M. E. Beckman, editors. *Papers in Laboratory Phonology I: Between the Grammar and Physics of Speech*. Cambridge, UK: Cambridge University Press, 1990.

[Lad96] D. R. Ladd. *Intonational Phonology*. Cambridge, UK: Cambridge University Press, 1996.

[Leh63] I. Lehiste. Review of K. Hadding-Koch, Acoustico-phonetic studies in the intonation of Southern Swedish, Lund: Gleerup. *Language*, 39:352–360, 1963.

[Lie60] P. Lieberman. Some acoustic correlates of word stress in American English. *J. Acoust. Soc. Am.*, 32:451–454, 1960.

[Mae96] K. Maekawa. Effects of focus on vowel formant frequency in Japanese. in *Computing Prosody: Approaches to a Computational Analysis of the Prosody of Spontaneous Speech*. New York: Springer-Verlag, 1997.

[Pie80] J. B. Pierrehumbert. *The Phonology and Phonetics of English Intonation*. Ph.D. thesis, Massachusetts Institute of Technology, Distributed by the Indiana University Linguistics Club, 1980.

[Pie81] J. Pierrehumbert. Synthesizing intonation. *J. Acoust. Soc. Am.*, 70:985–995, 1981.

[SvH96] A. Sluijter and V. van Heuven. Spectral balance as an acoustic correlate of linguistic stress. *J. Acoust. Soc. Am.*, 1996.

[VL72] R. Vanderslice and P. Ladefoged. Binary suprasegmental features and transformational word-accentuation rules. *Language*, 48:819–838, 1972.

2
A Typology of Spontaneous Speech

Mary E. Beckman

ABSTRACT Building accurate computational models of the prosody of spontaneous speech is a daunting enterprise because speech produced without a carefully devised written script does not readily allow the explicit control and repeated observation that read "lab speech" corpora are designed to provide. The prosody of spontaneous speech is affected profoundly by the social and rhetorical context of the recording, and these contextual factors can themselves vary widely in ways beyond our current understanding and control, so that there are many types of spontaneous speech which differ substantially not just from lab speech but also from each other. This paper motivates the study of spontaneous speech by describing several important aspects of prosody and its function that cannot be studied fully in lab speech, either because the relevant phenomena do not occur at all in lab speech or occur in a limited range of types. It then lists and characterizes some kinds of spontaneous speech that have been successfully recorded and analysed by scientists working on some of these aspects of prosody or on related discourse phenomena.

2.1 Introduction

The purpose of this paper is not to describe a specific computational model of some phenomenon in the prosody of spontaneous speech, but to play the role of Linnaeus. I will delimit what is meant by "spontaneous speech" and the kinds of prosodic phenomena that could (or should) be modelled for it. The history of current prosodic models already delimits the object to some extent. All current successful models have been developed and tested in the context of cumulative large-scale analyses of "read speech"—corpora of utterances produced in good recording conditions in response to the prompts provided by written scripts. Our initial delimitation is thus a negative definition. The "spontaneous speech" which we want to model is speech that is not read to script. In order to substitute a more positive definition, it is useful to consider why we study the prosody of read speech and why it is necessary to look at any other kind of speech. In the next section, therefore, I will sketch an answer by describing several phenomena that have been of particular concern in modelling the prosody of English and several other languages, and discuss why an examination of these phenomena in read speech cannot serve our whole purpose. First, however, let me motivate the exercise more generally by considering why a typology of spontaneous speech is necessary at all.

A discussion of types is necessary because spontaneous speech is not homogeneous. Speech produced without a written script can be intended for many different communicative purposes, and an important part of a fluent speaker's competence is to know how to adjust the speech to the purpose. A mother calling out the names of her children to tell them to come in to dinner will not sound the same when she produces the same names in response to the questions of the new neighbor next door. If the mother is speaking English, she will sound different in part because she uses qualitatively different intonation contours. When we decide to expand the coverage of our model of some particular prosodic phenomenon to spontaneous speech, therefore, it is not enough to say that spontaneous speech differs from read speech. We must think carefully about how different types of spontaneous speech are likely to differ from read speech, about whether those differences will make the spontaneous speech a useful source of data for extending our knowledge beyond the range of prosodic phenomena or values on which our models are based.

Of course, read speech is not homogenous either. For example, a professional *karuta* caller reading a *Hyakunin issyuu* poem for a New Year's poetry-card contest does not sound like a high-school teacher reading the same poem in front of his students in Classical Japanese class (see [Hom91] for a description of some of the prosodic differences involved). However, when we define spontaneous speech in contrast to read speech, what we are thinking of is fairly homogeneous. Our synthesis models are based upon **lab speech**—multiply repeated productions of relatively small corpora of sentences designed by the experimenter to vary only in certain dimensions of interest for the prosodic model. Recognition models are necessarily based on larger corpora (and hence fewer repeated productions of each sentence type), but the utterances are often still characterizable as lab speech. The collection and analysis of lab speech has allowed us to isolate and understand many prosodic phenomena that we know to be important in generating natural-sounding speech synthesis, or that we can predict will be important for building robust spoken-language understanding systems.

2.2 Some Prosodic Phenomena

One such prosodic phenomenon is **pitch accent**, which in the English prosodic system means any one of an inventory of paradigmatically contrasting intonational morphemes which are distinguished syntagmatically from the other components of an intonation contour in that an accent is necessarily associated with one of the more prominent syllables in the utterance. Restating this from the point of view of another aspect of prosody, association with a pitch accent defines a particularly high level of rhythmic prominence, or "phrase stress" (see, e.g., [BE94, dJ95]). It is important to accurately model the placement of accents relative to the text, because the level of syllabic prominence defined by the association is closely related to the discourse phenomenon of **focus**, and different focus patterns can cause radically different interpretations of a sentence, as in the following example.

Suppose that a speaker of English is using a spoken language translation system to communicate by telephone with a Japanese travel agent. The client wants to go from Yokohama to Shirahama, and utters the sentence in (1). This can mean one of three very different things, depending on the focus in the part after the word *want*. With narrow focus on *seat*, as in (1a), the client is saying that he has already paid for the ticket from Yokohama to Shin-Osaka, and merely wants the agent to reserve a seat for him on a particular train for that first leg of the journey. In this rendition, the sentence says nothing about the second leg of the trip, and the travel agent would not be amiss in then asking whether to book a reserved seat on the Kuroshio express down the Kii Peninsula coast. With narrow focus as in (1b), by contrast, the client is telling the travel agent to get him a reserved seat ticket for the Shinkansen, but the cheaper ticket without a guaranteed seat for the trip between Shin-Osaka and Shirahama. Finally, with broad focus on the entire complex noun phrase, as in (1c), the client seems to be making the same request as in (1a), but this time implying that he has made other arrangements for the second leg of the trip.

(1) a. I only need [a seat]$_F$ on the Shinkansen.
b. I only need a seat on the [Shinkansen]$_F$.
c. I only need [a seat on the Shinkansen]$_F$.

Rooth [Roo92] and others working on the semantics of sentences like (1) have proposed a model of the interaction between focus domain and the interpretation of *only* that is explicit enough that it probably could be encoded into the language understanding component of the machine translation system, if the correct focus representation were provided in the input. Moreover, if the standard understanding of how speakers signal narrow focus is correct, then it should be possible to provide a representation of the focus domain by training the recognition module to recognize where pitch accents have been placed in the intonation contour. That is, in the narrow-focus readings of sentence (1) and similar sentences that provide the data for the semantic models, focus domain corresponds rather straightforwardly to placement of nuclear pitch accent (the last pitch accent in the intonation contour). The focus pattern of (1a) would be indicated if the client placed the nuclear accent on *seat*. The contrasting focus pattern in (1b) would be indicated if he put the nuclear accent on the second syllable (i.e., the normally stressed syllable) of *Shinkansen*, although here there is a potential for ambiguity with the broad focus of (1c). In order to render these different meanings with the postposition translating *only* after the appropriate noun phrase in the Japanese that the agent gets, the recognition component must parse the accent placement that signals the corresponding focus domain pattern.

This understanding of the ambiguity between narrow focus and broad focus interpretations of late accents is very old in the literature on the phonology and semantics of sentence stress in English. Ladd [Lad80] summarizes literature going back several decades before 1980. A crucial aspect of the understanding is that the relevant prosodic dimension relating intonation to focus is nuclear accent **placement**, independent of pitch accent type. For example, in interpreting the scope of *only* in the question in (2), the same relationship between accent placement and focus domain holds as in the statement in (1), even though here the accent is not the high pitch target before the steep fall of the declarative intonation pattern, but the low pitch target at the start of the rise at the end of a "yes–no question" intonation. This understanding is compatible with experimental studies of acoustic correlates of "neutral" versus "emphatic stress" patterns, such as [CEM86].

(2) Do you only need a seat on the Shinkansen?

However, there is a problem with the observations that underlie the standard understanding of the prosodic contrasts in (1), a problem having to do with the distribution of accent types in lab speech. If lab speech can be identified with any communicative function, that function is essentially the pedagogical one of recitation, as in reading class in elementary school. The speaker in front of the microphone is being asked to produce a set of utterances in order for the experimenter to examine the form of the utterances rather than their content. I suspect that all cultures with a pedagogical tradition of recitation have conventions about the prosodic patterns appropriate for such **citation form** productions. A convention for American English, for example, is that declarative productions have a "flat hat" contour. This is an intonation pattern involving a rise to high pitch on the stressed syllable of a content word early in the utterance, high level pitch across optionally many content words, and a steep fall from high pitch for a nuclear accent on a content word late in the utterance, placed to make the accentuation compatible with the broadest possible focus interpretation—in other words, the intonation pattern that in the ToBI transcription system [PBH94] would be transcribed as minimally H* H* L- L%. The work on focus domain and accent, then, compares such a "neutral" citation form production with productions of the same sentence in which the subject has been asked to "emphasize" one or another content word earlier in the utterance, with the word to be emphasized indicated by italics, underlying, capital letters, or the like. Or, if the experimenter is more sophisticated about the function of prosody in discourse contexts, the subject might be provided with short dialogues to induce the intended narrow focus by simulating "contrast" with something in the immediately preceding context of the target utterance.

This experimental paradigm for examining prosodic correlates of broad versus narrow focus has been extremely useful for building quite detailed computational models not just for the prosodic system of English, but also for a host of other languages, including Swedish [Bru77], French [Tou87], Mandarin Chinese [Shi88], and Japanese [PB88]. For the most part, varying "emphasis" in this way has proved an extremely useful means of quickly getting many repetitions of a few types varying in a well controlled way (although see Arvaniti's criticism of an application of this method to Greek [Arv90]). However, underlining or otherwise inducing narrow focus in a target sentence does not change the quasi-pedagogical communicative function of lab speech. Like the "neutral" pattern, the "non-neutral" productions with variable "emphasis" are also produced as recitation forms. In American English, therefore, the narrow-focus productions also will have the "flat-hat" pattern or some variant of it compatible with an early nuclear pitch accent placement. In other words, our models of the relationship between focus and intonation are based on a very limited distribution of prosodic forms. For English, our understanding of focus as a function of accent placement alone is based almost entirely on productions with a very limited distribution of accent types—essentially the H* accent in the flat-hat pattern, with some support from L* in yes–no questions. We really do not know what speakers will do in other situations, where narrow focus arises as a result of the interaction between other communicative purposes and a potentially more complex distribution of discourse functions such as given versus new information. Recently, in fact, there have been two suggestions that the relationship between focus domain and intonation is not simply a matter of accent placement alone.

First, Ladd [LVJ94] has interpreted some experimental data on prominence perception as suggesting that there are other, more subtle differences of pitch range between "neutral" and "emphatic" interpretations of late nuclear pitch accents. The results that Ladd cites for this suggestion are also compatible with the suggestion that a rising accent may be more conducive to a narrow focus interpretation than a simple peak accent. That is, using the ToBI transcription system, we might interpret these results on prominence perception as suggesting that the L+H* accent has an inherently greater prominence than the H* accent. This is in keeping with the meanings of the two English rising accents suggested by Hirschberg and her colleagues (see [WH85, PH90a]). That is, both L+H* and L*+H seem to differ from plain H* in explicitly evoking a semantic scale or a choice of value along some scale. One way in which narrow focus could occur naturally in communicative situations other than the recitation paradigm above is if the context calls for an explicit contrast between the accented entity and other values in the presupposition set, as in the hypothetical context for the sentence with *only* in example (1a) above. If Hirschberg and her colleagues are correct about the meaning of L+H*, this would be a natural place to expect narrow focus on *seat* to be signaled not just by the early placement of nuclear accent but also by the use of a L+H* to make salient the contrast between making a seat reservation and any other service that the agent might want to sell.

Second, the discussion of a third type of non-low pitch accent in [PH90a] further suggests that downstepped later accents are compatible with an early focus interpretation (see also [Aye95]). That is, in some contexts it seems equally felicitous to deaccent content words after the new information in a sentence or to produce a series of accents of the type that is transcribed as !H* in the ToBI system. This is reminiscent of Bruce's finding for Swedish that word accents are downstepped after the focal accent [Bru82].

To evaluate either of these two suggestions about a relationship between focus and accent type, it becomes important to ask what accent types speakers actually do produce spontaneously in situations such as (1b) versus (1c), where the progress of a conversation depends on whether the listener interprets a given accent placement as indicating broad focus or late narrow focus. To answer this question will take at least three steps, no one of which is easy. First, we need to devise some means of obtaining recordings of conversations in which we can insure that such situations occur with sufficient frequency to be able make reliable observations of the distribution of accent types for different kinds of discourse situation that might call for different kinds of broad and narrow focus. Second, we need to transcribe accent types in these recordings, which means we need to agree upon an inventory of accent types for English and train transcribers to label those types reliably. Finally, we need to develop an independent understanding of **discourse structure** to annotate the recorded utterances for the potentially relevant categories of old versus new information and the like, lest our analysis be so circular as to be useless in developing the more complete model of the distribution of accents and accent types. This is a formidable task, but it must be a fruitful one, because if we are successful in answering the question for English, we can then apply the lessons drawn from the exercise also in developing models for other languages with similar intonation systems, where pitch accents constitute a diverse inventory of pragmatically contrastive intonational morphemes—languages such as Dutch, German, and Italian (see [CtH81, GR88, Koh87, Ave90, GS95]).

I have spent this much time motivating why we want to look at spontaneous speech for a better modelling of how focus domain relates to accentuation in English because this is a good example of how lab speech has served us well and of how it fails to serve our needs completely.

The same characterization holds for many other phenomena which I will describe more briefly in the rest of this section.

It is conventional wisdom that prosody must be related somehow, directly or indirectly, to **syntactic structure**. There is an old literature, going back to [Leh73, OKDA73], and even earlier, showing that in lab speech productions of examples of bracketing ambiguities, such as (3), speakers can make differences in **prosodic phrasing, pitch range reset**, and the like to help the listener recover the intended syntactic structure.

(3) a. [[Old men] and [women]].
b. [Old [men and women]].

Most readers will be familiar with this literature on English, but there are related findings concerning prosodic disambiguation of syntactic bracketing ambiguities for many other languages, including Swedish [BGGH92, Str93], Italian and Spanish [AHP95], Mandarin Chinese [TA90], and Japanese ([UHI$^+$81, AT91, Ven94]). Studies on the time course of resolution of partial ambiguities (e.g., [VYB94]) suggest that these differences can be useful for human listeners even when complete recognition of the text would eventually resolve the ambiguity. Results of an experiment by Silverman and colleagues [SKS$^+$93] can be interpreted as evidence that such processing considerations play an important role in determining whether intelligible synthetic speech remains intelligible in difficult listening conditions, such as deciphering proper names over telephone lines.

Moreover, in languages such as Japanese and Korean, where prosodic (minor) phrasing and pitch range reset at (major) phrasal boundaries are the functional equivalent of pitch accent placement in English with respect to cueing focus domain, this aspect of modelling prosody and syntax goes well beyond syntactic bracketing ambiguities. For example, studies by Maekawa [Mae91] and by Jun and Oh [JO94] suggest that the prosodic correlates of the inherent focus patterns of WH questions will be important for recognizing WH words and distinguishing them from the corresponding indefinite pronouns. There is related work by Tsumaki [Tsu94] on focusing patterns of different Japanese adverbs.

However, most of these studies are based on lab speech, primarily on lab speech elicited in experiments where the speakers cannot help but be aware of the syntactic ambiguity involved. Lehiste [Leh73] suggests that speakers will not produce the disambiguating cues unless they are aware of the contrast, which means that modelling the prosodic cues might be less helpful in recognition than in synthesis. In order to see whether this pessimism is warranted, we need to examine ambiguous or partially ambiguous examples in other communicative contexts which do not draw the speaker's attention to the form of the utterances.

There also is a fairly old literature on prosodic cues to **discourse topic structure** (e.g., [Leh75]). The first summary models of these results proposed that **overall pitch range** is a major component, with expanded pitch range at the beginning of a new topic, and reduced pitch range at the end (e.g., [BCK80]). Some more recent experiments, on the other hand, suggest that the impression of pitch range expansion or suppression may be due in part to somewhat more local aspects of downtrend such as **final lowering** (e.g., [HP86]) or, in some languages, **downstep** (e.g., [ABG+95]). Also, it may be important to differentiate an overall expansion of the pitch range from an upward shift of the floor of the pitch range [Tou95]. In recognition, these prosodic correlates of discourse topic structure might be used to build a model of the progression of discourse segments in a conversation in progress, which in turn would be useful for such things as resolution of pronominal reference (see, e.g., [GS86]). In synthesis, conversely, it should be important to model these correlates of discourse topic structure in order to aid the listener in making such resolutions.

Anecdotal observation also suggests other smaller and more immediate applications. For example, we might predict that material outside of the main flow of a narrative, such as an aside or a parenthetical expansion, will be set off by being in a different (typically reduced) pitch range. In this same vein, we can predict that when expressions such as *now* and *well* are used as **discourse markers** to signal the boundaries between discourse segments and the intentional relationships between segments, they will be prosodically differentiated from their literal uses by the lack of accent, or by being produced with L* pitch accents [HL87]. In Japanese, another related use of reduced pitch range is to mark off material that is postposed after the predicate. This construction is essentially a topicalizing devise to focus on the VP in an SOV language (it occurs in Korean and Turkish as well), and it seems to be consistently marked by extremely reduced pitch and amplitude on the postposed material [Venditti, personal communication].

There is some important recent work supporting the standard understanding of how pitch range manipulation functions to signal discourse topic structure (e.g., [GH92, Swe93]). However, there is also work (e.g., [Aye94]) suggesting that the signaling of topic structure will be subordinated to the exigencies of using such pitch features for negotiating turns and such in more interactive forms of spontaneous speech. For other contradictory evidence, see [GS93].

Prosodic cues to such control of **interactive structure** are by definition something that cannot be found in lab speech. So also are **disfluencies** and **repairs**, when the speaker corrects slips of the tongue or begins again after a false start. Closely related to both of these is the so-called **filled pause**, floor-holding vocalizations such as *uhm* in English or *eetto* in Japanese. It has been suggested that both interactive structure and disfluent regions might be identified in recognition by cues from intonation and rhythm (e.g., [Sch83, Hin83]). All of these things will be crucially important to model in highly interactive applications such as data-base querying systems (e.g., [KKN+93, HN93, SL93]).

One last set of phenomena which are impossible (or at least extremely difficult) to study in lab speech are prosodic cues to the speaker's **emotional state** and prosodic phenomena related to **code-switching**. Both of these will become more important as our synthesis systems become good enough to try to simulate a wider range of interactive situations that occur in all speech communities, or the larger range of styles available in communities where there is diglossia or pronounced sociolectal variation.

That is, in modelling communication among bilingual or bidialectal speakers, we will need to know when and why a speaker has switched from one language or dialect to the other. Since code-switching can be a mark of distancing versus solidarity or formality versus informality, it is unlikely that we will be able to observe this phenomenon in lab speech. Sociolinguists have studied to some extent the syntactic conditions for felicitous switching, but if we want to model this register-shifting devise in speech synthesis, we will need to understand the prosodic conditions as well. Also, if there are prosodic differences among the dialects, as in Osaka Japanese (e.g., [Kor92]) or Montreal French (e.g., [CS85]), then we must model the prosodic differences accurately, lest we inadvertently synthesize an intimate speaking style when we want a more formal one.

Even in systems built for that rare thing—the dialectally homogeneous speech community—it would be unfortunate to inadvertently synthesize an angry or bored voice for the system's responses in applications such as data-base querying or hotel reservation systems, and it may be useful to be able to recognize anger or irritation in the client's utterances. The literature on vocal expression of emotion is not easy to interpret (see, e.g., the recent reviews in [MA93, Sch95]), but it seems clear that strongly felt emotion affects at least such global settings as overall speed, overall pitch range, and voice quality, and also that human listeners can interpret these prosodic effects to gauge the type and particularly the strength of the emotion [LST+85]. Since lab speech, by definition, is a communicative situation where the speaker is cooperating obligingly with the experimenter's purpose, it is difficult to study these effects in lab speech. In order to model the effects for speech synthesis, it has been useful to record professional actors who have been asked to simulate the emotions of interest. However, there is some evidence that enacted emotion may differ prosodically from the natural expression of emotion in exaggerating only the most salient effects [Sch86]. Thus, for recognition, we will need recordings of non-professional speakers in interactive situations where they might spontaneously feel anger or irritation.

2.3 Types of Spontaneous Speech Recordings

As the above discussion should make clear, not all types of spontaneous speech can be suitable for examining all of the phenomena we want to be able to model in spontaneous speech. In recording spontaneous speech, then, we need to think about how to tailor our elicitation technique to the phenomena that we want to study. Among the questions that we must consider are whether the technique is conducive to getting large enough discourses for the particular prosodic phenomena to occur often enough or in enough variety to allow the kinds of analysis we want to make, and whether the communicative situation will allow sufficient control over the linguistic content and discourse structure for us to observe the relationships between relevant aspects of these and the prosodic phenomena. Also, will the recording be good enough for analysis, and can we make the recording without impinging on the speaker's legal rights?

A frequent technique in sociolinguistic research (e.g., [CS85]) has been to elicit an **unstructured narrative** from the speaker in an informal interview, by prompting with open-ended questions about the speaker's background or the like. With a skilled interviewer or a lucky choice of prompt, this technique can produce long fluent stretches of monologue narrative to analyse for topic structure. This, in fact, was the type of data that Brown and her colleagues used in their seminal description of pitch range and topic structure [BCK80]. As a last formal task after reading a list of sentences and a story, the subjects were shown a very old photograph of part of Edinburgh and asked to figure out where the photograph was taken. This prompted the subjects to ask questions about the photograph, as they tried to work out the place, and then to continue on talking for about 10 or 15 min about changes they had seen in Edinburgh over the years and so on. The technique is useful because it is legal so long as the subject's consent is first obtained, yet the tape recorder can often be placed so unobtrusively that the subject seems to forget after a while that the narrative is being recorded. There is a strong disadvantage, however, of providing no control over the content of the speech, so that it is often extremely difficult to get any independent gauge of discourse topic structure, particularly if the narrative is long and rambling. Also, there is virtually no control over where and when the speaker will require back-channel encouragement to continue talking (cf. the discussion in [Aye94]).

Asking the subject to produce an **extended descriptive narrative** that retells the story of a movie or the like provides a better guarantee against uncontrolled interaction with the interviewer and also provides somewhat more control over the content of the narrative. This was the method Chafe [Cha80] used to elicit extended narratives for an analysis of cues to discourse topic structure. The more controlled content and the specific task of "description" make this sort of elicited narrative easier to analyse for an independent model of topic structure (cf. [PL93, Swe95]).

The same holds true of most forms of public **performance narrative**, such as the after-dinner speech analysed by Hirschberg and Litman [HL87] or the professional story-teller's performance used by Fowler and Housum [FH87]. Recording performances of such narratives is a very common device for eliciting connected discourse in fieldwork, particularly for languages with strong traditions of oral transmission (see, e.g., [Woo93]). In some forms of performance narrative, there is the added interest of trying to model the intentional structure of the narrative, the prosodic correlates of rhetorical devices used to persuade and so on. This is particularly true of political speeches (e.g., [Tou93b, Tou95]) and sermons. These forms of performance also will typically yield a greater range of emotion. However, there is also the danger that any particular type of performance will have its own established conventions about prosody that make it impossible to generalize results to productions by non-professionals. The political speeches of Martin Luther King, for example, were often rhetorically successful because he performed them with the stylized poetic rhythms of Southern Baptist sermons. Also, it is not always possible to get clear recordings of public speech, since authentic performance can depend on recording in the natural context of a meeting hall or around the campfire.

At the other end of the scale from all three of these types of narrative in terms of control of content and topic structure are experimental **instruction monologues**—very short discourses in which the subject is asked to instruct an imagined (or real, but silently non-interacting) listener to perform some task, such as constructing a model of a house front from a set of pieces [Ter84] or following a route between fixed start and endpoints on a map [Swe93]. This method provides clear recordings with extremely good control over the content words and syntactic structures that will occur, along with potentially clear models of the discourse structure provided by the task itself, without interference from cues to interactive structure.

For work on interactive structure and turn-taking, the type of data that corresponds to the unstructured narrative for topic structure is **overhead conversation**—i.e., a surreptitious recording of a casual exchange, typically of an exchange that has no purpose other than to pass the time at a party or in chatting with a friend over the telephone. Fox [Fox87] used several such recordings, provided by E. Schegloff, to examine the relationship between anaphora and discourse structure, and this has been a frequent tool in work on interactive structure in general. It has all the disadvantages of the unstructured narrative, with the addition of such problems for acoustic analysis as frequently overlaid voices and generally bad recording quality. Unless the conversation is over the telephone, many of the cues to interactive structure will be gestural (e.g., patterns of eye contact), and require simultaneous video recording and analysis. In many countries, also, it is illegal to make such surreptitious recordings.

The last problem is surmounted in what we might call **enacted conversation**—conversation recorded from speakers who have given prior permission or who have been recruited to chat for a tape recorder in a quasi-party setting (e.g., [EP88]). The same holds true of various kinds of **public conversation**, such as radio interviews (used, e.g., by Fletcher [Fle91]) or recordings of radio call-in programs (used by Hirschberg and Litman [HL87]). However, these forms of conversation, particularly the enacted conversation, do not overcome the other problems of overlaid voices and so on.

Eliciting interactive **instruction dialogues** in the laboratory is a convenient way to get around many of these difficulties. The speakers can be placed out of each other's view and asked to communicate over microphone and headphone so that they can be recorded on separate channels even when their voices are actually overlapping in time as the listener provides back-channeling or corrections. As in the comparable instruction monologues, the content of the dialogues can be more fixed, and the task itself provides some independent evidence of the discourse structure. The MAP task dialogues [ABB+91] and the TRAINS dialogues [NA93a] are recent examples of this sort of spontaneous speech. See [GS93] for an example of very short discourses combining the relevant control aspects of the instruction monologue and instruction dialogue design.

Instruction dialogues or instruction monologues are a very useful tool for observing particular target forms, since the task can be tailored specifically to getting multiple repetitions. In [BSRdJ90], for example, we used instruction dialogues to get spontaneous repetitions of phrases contrasting in word boundary placement by having the subject instruct the listener in building an arrangement of pictures of people, labelled with names such as *M. Malone* versus *Emma Lone*. The listener feigned mishearing because of supposed noise on the line, so as to force the speaker to repeat each name several times, in the context of different feigned contrasting names. Van Wijk and Kempen [vWK87] similarly had subjects give short descriptions of a scene that changed unpredictably to elicit multiple self-repairs in syntactically controlled material. However, there then comes the danger of designing a task so different from anything that the speaker would do in natural life that the paradigm can only elicit something close to the recitation performance of lab speech. That is, the simpler and more tightly controlled the task, the more likely it will be that the subject's attention is drawn to the fact that it is the *form* of the performance rather than the task itself that is of interest to the investigator.

This last problem can be overcome by recording subjects performing natural but simple tasks that they have initiated themselves. In particular, researchers with connections to telephone companies and the like often can get recordings of natural **database querying dialogues** that have many of the advantages of overheard conversation, but the better-defined task-specified structure of artificial instruction dialogues. For example, Silverman and colleagues [SBSP92] used recordings of telephone directory queries, and Kiessling and colleagues [KKN+93] used railway timetable queries. This dialogue form also has the advantage of direct applicability, since database querying is exactly the kind of task where speech synthesis and recognition technology are likely to be first successfully used. Whereas large companies can apparently afford to be less careful about getting the speaker's explicit written consent before recording such interactions, however, the ordinary speech scientist probably cannot. A compromise solution, then, is to devise a technique for eliciting database querying dialogues and similarly natural domain-specific interactions in the laboratory.

This is the motivation for the **Wizard of Oz** technique, where the experimenter asks the speaker to test a computer database querying system or the like, while simulating the computer's response. Task domains that have been used in this technique are querying airline listings (as in the ATIS project [M92]), and making travel arrangements using a simulated speech translation system [FLKP95]. As in using any other general elicitation technique, however, it is important to keep in mind the ultimate goal of the elicitation. For example, if the researcher wants to study the kinds of disfluency that occur in real-life database querying dialogues, then some technique must be devised to let the subject rehearse the goal before addressing the system. My impression in listening to many of the ATIS utterances is that they are disfluent in a way that real-life database querying would not be, because the speaker is having some difficulty remembering all relevant travel points and time constraints in solving several complicated trip "scenarios" in quick succession. Similarly, if the researcher wants to observe interactive structure of the sort that occurs in ordinary information exchanges among humans, the simulated response should not be so delayed that the conversation becomes an exchange of short monologues rather than a real dialogue.

In short, the researcher must carefully attend to many aspects of the elicitation paradigm in order to have any luck in getting spontaneous speech that will be useful for the research purpose. Thus this typology of spontaneous speech has been as much a typology of elicitation paradigms as it has been of spontaneous speech phenomena. Therefore, it was necessary to not talk about some of the spontaneous speech phenomena discussed in the preceding section. Obviously, I have not talked about how to elicit spontaneous speech for studying a wider range of emotions or for observing prosodic phenomena related to code-switching. As far as I know, no one yet has worked out an elicitation paradigm with sufficient relevant control to allow fruitful analysis of these phenomena for computational modelling. I think we are now at the stage where enough experts in the relevant different areas are aware of each others' work that we can begin to seriously hone elicitation paradigms for modelling the prosody of such phenomena as repair, discourse topic organization, and interactive structure. Linguists and computer scientists working on dialogue models know that they need to talk to phoneticians and speech engineers who work on intonation and rhythm, and vice versa. Perhaps in another decade or so, work on the other areas of interest in spontaneous speech prosody listed above will be at a similar state of hopeful beginning.

References

[ABB+91] A. H. Anderson, M. Bader, E. G. Bard, E. Boyle, G. Docherty, S. Garrod, S. Isard, J. Kowtko, J. McAllister, J. Miller, H. Thompson, and R. Weinert. The HCRC map task corpus. *Language and Speech*, 34:351–366, 1991.

[ABG+95] G. Ayers, G. Bruce, B. Granström, K. Gustafson, M. Horne, D. House, and P. Touati. Modelling intonation in dialogue. In *Proceedings of the 13th International Congress of Phonetic Sciences, Stockholm, Sweden*, Vol. 2, pp. 278–281, 1995.

[AHP95] C. Avesani, J. Hirschberg, and P. Prieto. The intonational disambiguation of potentially ambiguous utterances in English, Italian, and Spanish. In *Proceedings of the 13th International Congress of Phonetic Sciences, Stockholm, Sweden*, Vol. 1, pp. 174–177, 1995.

[Arv90] A. Arvaniti. Review of *Stress and prosodic structure in Greek: A phonological, physiological and perceptual study*, by A. Botinis. *Journal of Phonetics*, 18:65–69, 1990.

[AT91] J. Azuma and Y. Tsukuma. Role of F_0 and pause in disambiguating syntactically ambiguous Japanese sentences. In *Proceedings of the XIIème International Congress of Phonetic Sciences, Aix-en-Provence, France*, Vol. 3, pp. 274–277, 1991.

[Ave90] C. Avesani. A contribution to the synthesis of Italian intonation. In *Proceedings of the International Conference on Spoken Language Processing, Kobe, Japan*, Vol. 2, pp. 833–836, 1990.

[Aye94] G. Ayers. Discourse functions of pitch range in spontaneous and read speech. *OSU Working Papers in Linguistics*, 44:1–49, 1994.

[Aye95] G. M. Ayers. Nuclear accent types and prominence: some psycholinguistic experiments. In *Proceedings of the 13th International Congress of Phonetic Sciences, Stockholm, Sweden*, Vol. 3, pp. 660–663, 1995.

[BCK80] G. Brown, K. L. Currie, and J. Kenworthy. *Questions of Intonation*. Croom Helm, 1980.

[BE94] M. E. Beckman and J. Edwards. Articulatory evidence for differentiating stress categories. In P. A. Keating, editor, *Phonological Structure and Phonetic Form: Papers in Laboratory Phonology III*, pp. 7–33. Cambridge, UK: Cambridge University Press, 1994.

[BGGH92] G. Bruce, B. Granström, K. Gustafson, and D. House. Aspects of prosodic phrasing in Swedish. In *Proceedings of the International Conference on Spoken Language Processing, Kobe, Japan*, Vol. 1, pp. 109–112, 1992.

[Bru77] G. Bruce. *Swedish Word Accents in Sentence Perspective.* Lund: Gleerup, 1977.

[Bru82] G. Bruce. Developing the Swedish intonational model. *Working Papers, Lund University*, 22:51–116, 1982.

[BSRdJ90] M. E. Beckman, M. G. Swora, J. Rauschenberg, and K. de Jong. Stress shift, stress clash, and polysyllabic shortening in a prosodically annotated discourse. In *Proceedings of the International Conference on Spoken Language Processing, Kobe, Japan*, Vol. 1, pp. 5–8, 1990.

[CEM86] W. E. Cooper, S. J. Eady, and P. R. Mueller. Acoustical aspects of contrastive stress in question-answer contexts. *J. Acoust. Soc. Am.*, 77:2142–2156, 1986.

[Cha80] W. L. Chafe. *The Pear Stories: Cognitive, Cultural and Linguistic Aspects of Narrative Production.* Norwood, NJ: Ablex, 1980.

[CS85] H. J. Cedergren and L. Simoneau. La chute des voyelles hautes en français de Montréal: Às-tu entendu la belle syncope?. In M. Lemieux and H. J. Cedergren, editors, *Les Tendences Dynamiques du Français Parlé á Montreál*, pp. 57–144. Montreal: Office de la Langue Française, 1985.

[CtH81] R. Collier and J. 't Hart. *Cursus Nederlandse Intonatie.* Leuven: Acco, 1981.

[dJ95] K. de Jong. The supraglottal articulation of prominence in English: Linguistic stress as localized hyperarticulation. *J. Acoust. Soc. Am.*, 97:491–504, 1995.

[EP88] J. Esser and A. Polomski. *Comparing Reading and Speaking Intonation.* Amsterdam: Rodopi, 1988.

[FH87] C. A. Fowler and J. Housum. Talkers' signalling of 'new' and 'old' words in speech, and listeners' perception and use of the distinction. *Journal of Memory & Language*, 26:489–504, 1987.

[Fle91] J. Fletcher. Rhythm and lengthening in French. *Journal of Phonetics*, 19:193–212, 1991.

[FLKP95] L. Fais, K. Loken-Kim, and Y-D. Park. Speakers' responses to requests for repetition in a multimedia language processing environment. *Proceedings of the International Conference on Cooperative Multimodal Communication*, pp. 129–144, 1995.

[Fox87] B. A. Fox. *Discourse Structure and Anaphora: Written and Conversational English.* Cambridge, UK: Cambridge University Press, 1987.

[GH92] B. Grosz and J. Hirschberg. Some intonational characteristics of discourse structure. In *Proceedings of the International Conference on Spoken Language Processing, Banff, Canada*, Vol. 1, pp. 429–432, 1992.

[GR88] C. Gussenhoven and A. C. M. Rietveld. Fundamental frequency declination in Dutch: Testing three hypotheses. *Journal of Phonetics*, 16:355–369, 1988.

[GS86] B. Grosz and C. Sidner. Attention, intentions, and the structure of discourse. *Computational Linguistics*, 12:175–204, 1986.

[GS93] R. Geluykens and M. Swerts. Local and global prosodic cues to discourse organization in dialogues. *Working Papers 41, Proc. ESCA Workshop on Prosody, Lund University, Sweden*, pp. 108–111, 1993.

[GS95] M. Grice and M. Savino. Low tone versus 'sag' in Bari Italian intonation: A perceptual experiment. In *Proceedings of the 13th International Congress of Phonetic Sciences, Stockholm, Sweden*, Vol. 4, pp. 658–661, 1995.

[Hin83] D. Hindle. Deterministic parsing of syntactic nonfluencies. *Proceedings of the 21st Annual Meeting of the Association for Computational Linguistics*, pp. 123–128, 1983.

[HL87] J. Hirschberg and D. Litman. Now let's talk about now: Identifying cue phrases intonationally. *Proceedings of the 25th Annual Meeting of the Association for Computational Linguistics*, pp. 163–171, 1987.

[HN93] J. Hirschberg and C. Nakatani. A speech-first model for repair identification in spoken language systems. *Proceedings of the European Conference on Speech Communication and Technology, Berlin, Germany*, pp. 1173–1176, 1993.

[Hom91] Y. Homma. The rhythm of Tanka, short Japanese poems; read in prose style and in contest style. In *Proceedings of the XIIème International Congress of Phonetic Sciences, Aix-en-Provence, France*, Vol. 2, pp. 314–317, 1991.

[HP86] J. Hirschberg and J. Pierrehumbert. The intonational structuring of discourse. *Proceedings of the 24th Annual Meeting of the Association for Computational Linguistics*, pp. 136–144, 1986.

[JO94] S. A. Jun and M. Oh. A prosodic analysis of three sentence types with 'wh' words in Korean. In *Proceedings of the International Conference on Spoken Language Processing, Yokohama, Japan*, Vol. 1, pp. 323–326, 1994.

[KKN+93] A. Kießling, R. Kompe, H. Niemann, E. Nöth, and A. Batliner. Roger, sorry, I'm still listening: Dialog guiding signals in information retrieval dialogs. *Working Papers 41, Proceedings of the ESCA Workshop on Prosody, Lund University, Sweden*, pp. 140–143, 1993.

[Koh87] K. J. Kohler. Categorical pitch perception. In *Proceedings of the 11th International Congress of Phonetic Sciences, Tallin, Estonia*, Vol. 5, pp. 331–333, 1987.

[Kor92] S. Kori. Nihongo bun'ontyoo no kenkyuu kadai. Paper presented at the International Symposium on Prosody, 1992.

[Lad80] D. R. Ladd. *The Structure of Intonational Meaning.* Bloomington: Indiana University Press, 1980.

[Leh73] I. Lehiste. Phonetic disambiguation of syntactic ambiguity. *Glossa*, 7:107–122, 1973.

[Leh75] I. Lehiste. The phonetic structure of paragraphs. In A. Cohen and S. Nooteboom, editors, *Structure and Process in Speech Perception*, pp. 195–203. Heidelberg: Springer, 1975.

[LST+85] D. R. Ladd, K. E. A. Silverman, F. Tolkmitt, G. Bergmann, and K. R. Scherer. Evidence for the independent function of intonation contour type, voice quality, and F_0 range in signaling speaker affect. *J. Acoust. Soc. Am.*, 78:435–444, 1985.

[LVJ94] D. R. Ladd, J. Verhoeven, and K. Jacobs. Influence of adjacent pitch accents on each other's perceived prominence, two contradictory effects. *Journal of Phonetics*, 22:87–99, 1994.

[M92] L. Hirschman. MADCOW. Multi-site data collection for a spoken language corpus. *Proceedings DARPA Speech and Natural Language Workshop*, pp. 7–14, 1992.

[MA93] I. R. Murray and J. L. Arnott. Toward the simulation of emotion in synthetic speech: A review of the literature on human vocal emotion. *J. Acoust. Soc. Am.*, 93:1097–1108, 1993.

[Mae91] K. Maekawa. Perception of intonational characteristics of wh and non-wh questions in Tokyo Japanese. In *Proceedings of the XIIème International Congress of Phonetic Sciences, Aix-en-Provence, France*, Vol. 4, pp. 202–205, 1991.

[NA93a] S. Nakajima and J. Allen. A study on prosody and discourse structure in cooperative dialogues. *Phonetica*, 50:197–210, 1993.

[OKDA73] M. H. O'Malley, D. R. Kloker, and D. Dara-Abrams. Recovering parentheses from spoken algebraic expressions. *IEEE Trans. Audio and Electroacoustics*, AU-21:217–220, 1973.

[PB88] J. B. Pierrehumbert and M. E. Beckman. *Japanese Tone Structure*. Cambridge, MA: MIT Press, 1988.

[PBH94] J. Pitrelli, M. E. Beckman, and J. Hirschberg. Evaluation of prosodic transcription labelling reliability in the ToBI framework. In *Proceedings of the International Conference on Spoken Language Processing, Yokohama, Japan*, Vol. 1, pp. 123–126, 1994.

[PH90a] J. Pierrehumbert and J. Hirschberg. The meaning of intonation contours in the interpretation of discourse. In P. R. Cohen, J. Morgan, and M. E. Pollack, editors, *Intentions in Communication*, pp. 271–311. Cambridge, MA: MIT Press, 1990.

[PL93] R. Passonneau and D. Litman. Feasibility of automated discourse segmentation. *Proceedings of the 31st Annual Meeting of the Association for Computational Linguistics*, pp. 148–163, 1993.

[Roo92] M. Rooth. A theory of focus interpretation. *Natural Language Semantics*, 1:75–116, 1992.

[SBSP92] K. E. A. Silverman, E. Blaauw, J. Spitz, and J. Pitrelli. Towards using prosody in speech recognition/understanding systems: differences between read and spontaneous speech. *Proceedings DARPA Speech and Natural Language Workshop*, pp. 435–440, 1992.

[Sch83] D. Schaffer. The role of intonation as a cue in turn taking in conversation. *Journal of Phonetics*, 11:243–344, 1983.

[Sch86] K. R. Scherer. Vocal affect expression: a review and model for future research. *Psychological Bulletin*, 99:143–165, 1986.

[Sch95] K. R. Scherer. How emotion is expressed in speech and singing. In *Proceedings of the 13th International Congress of Phonetic Sciences, Stockholm, Sweden*, Vol. 3, pp. 90–96, 1995.

[Shi88] C-L. Shih. Tone and intonation in Mandarin. *Working Papers, Cornell Phonetics Laboratory*, 3:83–109, 1988.

[SKS+93] K. E. A. Silverman, A. Kalyanswamy, J. Silverman, S. Basson, and D. Yashchin. Synthesizer intelligibility in the context of a name-and-address information service. *Proceedings of the European Conference on Speech Communication and Technology, Berlin, Germany*, pp. 2169–2172, 1993.

[SL93] E. E. Shriberg and R. J. Lickley. Intonation of clause-internal filled pauses. *Phonetica*, 50:172–179, 1993.

[Str93] E. Strangert. Perceived pauses, silent intervals, and syntactic boundaries. *PHONUM*, 1:35–38, 1993.

[Swe93] M. Swerts. On the prosodic prediction of discourse finality. *Working Papers 41, Proceedings of the ESCA Workshop on Prosody, Lund University, Sweden*, pp. 96–99, 1993.

[Swe95] M. Swerts. Combining statistical and phonetic analyses of spontaneous discourse segmentation. In *Proceedings of the 13th International Congress of Phonetic Sciences, Stockholm, Sweden*, Vol. 4, pp. 208–211, 1995.

[TA90] Y. Tsukuma and J. Azuma. Prosodic features determining the comprehension of syntactically ambiguous sentences in Mandarin Chinese. In *Proceedings of the International Conference on Spoken Language Processing, Kobe, Japan*, Vol. 1, pp. 505–508, 1990.

[Ter84] J. M. B. Terken. The distribution of pitch accents in instructions as a function of discourse structure. *Language & Speech*, 27:269–289, 1984.

[Tou87] P. Touati. *Structure Prosodiques du Suëdois et du Francais.* Lund: Lund University Press, 1987.

[Tou93b] P. Touati. Prosodic aspects of political rhetoric. *Working Papers 41, Proceedings of the ESCA Workshop on Prosody, Lund University, Sweden*, pp. 168–171, 1993.

[Tou95] P. Touati. Pitch range and register in French political speech. In *Proceedings of the 13th International Congress of Phonetic Sciences, Stockholm, Sweden*, Vol. 4, pp. 244–247, 1995.

[Tsu94] J. Tsumaki. Intonational properties of adverbs in Tokyo Japanese. In *Proceedings of the International Conference on Spoken Language Processing, Yokohama, Japan*, Vol. 4, pp. 1727–1730, 1994.

[UHI+81] T. Uyeno, H. Hayashibe, K. Imai, H. Imagawa, and S. Kiritani. Syntactic structures and prosody in Japanese. *Annual Bulletin of the Research Institute of Logopedics and Phoniatrics, University of Tokyo*, 15:91–108, 1981.

[Ven94] J. J. Venditti. The influence of syntax on prosodic structure in Japanese. In *Working Papers in Linguistics*, Vol. 44, pp. 191–223. The Ohio State University, 1994.

[vWK87] C. van Wijk and G. Kempen. A dual system for producing self-repairs in spontaneous speech: Evidence from experimentally elicited corrections. *Cognitive Psychology*, 19:403–440, 1987.

[VYB94] J. J. Venditti and H. Yamashita-Butler. Prosodic information and processing of temporarily ambiguous constructions in Japanese. In *Proceedings of the International Conference on Spoken Language Processing, Yokohama, Japan*, Vol. 3, pp. 1147–1150, 1994.

[WH85] G. Ward and J. Hirschberg. Implicating uncertainty: The pragmatics of fall-rise intonation. *Language* 61: 747–776, 1985.

[Woo93] A. Woodbury. Against intonational phrases in Central Alaskan Yupik Eskimo. Paper presented at the annual meeting of the Linguistic Society of America, Los Angeles, CA, 1993.

3
Prosody, Models, and Spontaneous Speech

Hiroya Fujisaki

ABSTRACT This paper presents a definition of prosody as the organization of linguistic units within an utterance and a coherent group of utterances, having manifestations both in segmental and suprasegmental features of speech, serving at the same time as a medium for conveying para- and nonlinguistic information. It then discusses the process of spontaneous speech production, emphasizing the role of quantitative generative models in both speech synthesis and speech recognition, examples are illustrated in Japanese. Finally, it discusses the continuum of spontaneity in speech, and briefly touches on the characteristics of speech that become dominant with increased spontaneity.

3.1 What is Prosody? Its Nature and Function

Although there are many views on prosody (e.g., Lehiste [Leh70]), I do not intend to elaborate on their differences here. These views may be broadly classified, after Ladd and Cutler [LC83], into the following two categories:

1. "Concrete"—defining prosody more or less in physical terms, *as those phenomena that involve the acoustic parameters of pitch, duration, and intensity.*
2. "Abstract"—defining prosody more from the point of view of its place in linguistic structure than its phonetic nature, *as phenomena that involve phonological organization at levels above the segment.*

According to Ladd and Cutler, these two approaches can be characterized—or caricatured—by their methodological preferences for either "taking measurements" or "building models."[1]

I consider that the study of human communicative behavior through language belongs to the empirical sciences, where one needs to obtain, first and foremost, clear and objective knowledge of the phenomena through

[1]Here I would use "theories" instead of "models," since the word "model" is being used in a different meaning in this paper.

making measurements. Such knowledge can be more powerful if it is not confined to that obtained only from direct observations but can be generalized in the form of theories to cover instances that are not yet encountered or observed. By *induction*, i.e., going from measurements to theories, we can obtain the underlying, general principles. We can use intuition to *build theories*, i.e., to posit general, underlying principles first, but their validity has to be tested by confirming that their predictions agree with the measurements. The process of obtaining predictions for concrete, individual facts is through *deduction.* Thus the two processes—induction and deduction—are not dichotomous; both can be profitably used in the scientific quest for precise and generalizable knowledge.

Prosody has both measurable manifestations and underlying principles. It is only realized when a message is produced as a coherent string of sounds as speech. I would therefore like to present a third definition:

3. Prosody is the systematic organization of various linguistic units into an utterance or a coherent group of utterances in the process of speech production. Its realization involves both segmental and suprasegmental features of speech, and serves to convey not only linguistic information, but also paralinguistic and non-linguistic information.

Here I define *linguistic information* as the symbolic information that is represented by a set of discrete symbols and rules for their combination. It can be represented either explicitly by the written language, or can be easily and uniquely inferred from context. Linguistic information thus defined is discrete and categorical. For example, the information concerning the accent type of a Japanese word is discrete in the sense that it specifies one of a finite number of possible accent types.

On the other hand, *paralinguistic information* is defined as the information that is not inferable from the written counterpart but is deliberately added by the speaker to modify or supplement the linguistic information.[2] A written sentence can be uttered in various ways to express different intentions, attitudes, and speaking styles which are under the conscious control of the speaker. Paralinguistic information can be both discrete and continuous. For example, the information regarding whether a speaker's intention is an assertion or a question is discrete, but it can also be continuous in the sense that a speaker can express the degree within each category.[3]

---

[2]I am aware that these definitions of linguistic and paralinguistic information are unconventional (e.g., Crystal [Cry69], Laver [Lav94]). However, they are introduced here to deal more systematically with various functions of prosody and prosodic features than conventional definitions do.

[3]It is possible that a speaker's intention can be expressed either as linguistic information or as paralinguistic information, or both. For instance, interrogation can be regarded as linguistic (in the sense defined above) if a speaker uses

Nonlinguistic information concerns such factors as the age, gender, idiosyncracy, physical and emotional states of the speaker, etc. These factors are not directly related to the linguistic and paralinguistic contents of the utterances and cannot generally be controlled by the speaker, although it is possible for a speaker to control the way of speaking to intentionally convey an emotion, or to simulate an emotion, as is done by actors. Like paralinguistic information, nonlinguistic information can be discrete as well as continuous. Gradation within a category is a common feature of para- and nonlinguistic information, as opposed to the essentially discrete and categorical nature of linguistic information (see also Ladd [Lad93]).

3.2 Prosody in the Production of Spontaneous Speech

The relationship between these three types of information and the acoustic-phonetic manifestation of prosody as the organization of various linguistic units (i.e., segments, syllables, words, etc.) is schematically illustrated by the block diagram of Fig. 3.1.

The figure shows the process of spontaneous speech production in four consecutive stages: (1) message planning, (2) utterance planning, (3) motor command generation, and (4) speech sound production. I will later elaborate on the notion of "spontaneous" speech, but here I will use

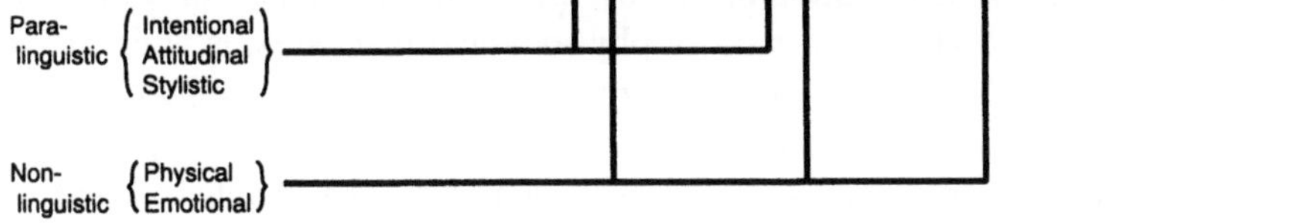

FIGURE 3.1. Processes by which various types of information are manifested in the segmental and suprasegmental features of speech.

---

an interrogative sentence and utters it without a rising intonation, or it can be regarded as paralinguistic if he/she uses a declarative sentence with a rising intonation. Commonly, both types of information are used simultaneously whenever it is feasible.

it somewhat broadly to mean speech produced without referring to a text. It is to be noted that the processes at each stage can overlap in time in spontaneous speech production; for example, the planning of an utterance often starts before the planning of a whole message is complete.

At the stage of message planning, linguistic units and structures are selected on the basis of linguistic information following the rules and constraints of the grammar of spoken language. The organization of the corresponding utterance, however, is determined at the next stage—the stage of utterance planning.

Although my definition of prosody concerns both a single utterance and a string of utterances, let us for the moment restrict ourselves to the prosody of an utterance consisting of one spoken sentence. How does a speaker organize various linguistic units systematically into a meaningful and expressive utterance? Although various languages may differ in the details, it is generally done by three means: accentuation, phrasing, and pausing. "Accentuation" means changing (generally increasing) the relative prominence of a syllable within a word or a group of words. It is usually accomplished by controlling the fundamental frequency, duration, and intensity, although languages differ in the way they use these features. "Phrasing" means grouping together a string of words into a perceptually coherent constituent. It is generally done by controlling the fundamental frequency and the local speech rate. "Pausing" literally means putting a pause after a constituent (which can be one word or more in length), to indicate that the constituents at both sides of the pause should be processed separately. Thus it is often accompanied by phrasing, but not vice versa.

All three ways described above serve to achieve a common basic goal, i.e., to facilitate decoding and understanding the linguistic content of the message on the part of the listener when it is received as an utterance. For instance, accentuation serves to resolve lexical ambiguity as well as to locate focus. Phrasing serves to facilitate parsing of the incoming utterance and to resolve syntactic ambiguity. Pausing serves to provide the listener with the time necessary to process the part of the utterance already received, without being interrupted by the flow of further information. Since, however, all the three require physiological mechanisms which have their own constraints, the actual outcome may not always be ideal from the point of view of message decoding. For example, a pause may happen to be inserted at an inappropriate position because the speaker needs to breathe. Furthermore, in addition to the discrete form of information which these three means can express (accented/unaccented, presence/absence of phrase boundary, presence/absence of pause), they can also express continuous forms of information. For example, the degree of both accentuation and phrasing as well as the duration of pauses can be varied continuously over a wide range, and these provide channels to convey further information which is not specified in the message, namely paralinguistic and nonlinguistic information.

The utterance plan is executed, first as neuromotor commands which drive the physical and acoustic mechanisms of speech production through activities of various phonatory and articulatory muscles. These commands, or equivalently the resulting muscular tensions, can carry both discrete and continuous forms of information, i.e., by their presence or absence as well as by their strength and/or duration.

The consequences of these neuromotor commands (driving forces) are temporal changes in the mechanical structure of the phonatory and articulatory organs. At the final stage of speech sound production, changes in the shape of the phonatory mechanism cause corresponding changes in the tension of the vocal cords, which produce the on/off patterns of the vocal cord vibration as well as temporal changes in the fundamental frequency of its vibration. On the other hand, the changes in the shape of the articulatory mechanism cause corresponding changes in the transfer characteristics of the vocal tract, and the final results are temporal changes in the spectral characteristics of the speech sounds. Because of the smoothing inherent in mechanical systems, the observed characteristics of the speech sounds tend to reveal smooth trajectories, and show little trace of the commands that are the direct indications of utterance organization.

The schematic diagram of Fig. 3.1 shows the complex, multi-stage nature of the process of conversion from information to speech sound characteristics, and explains why it is difficult to find clear and unique correspondence between physically observable characteristics of speech and the underlying prosodic organization of an utterance. In order to infer the underlying organization from the observed characteristics of speech, it is logical to go through the following two steps:

1. Inferring the commands from the speech characteristics,
2. inferring the units and structures of prosody from the commands.

Since step 1 is the inverse operation of the process of speech sound production, it can be most accurately and objectively conducted if we have a quantitative model of the production process. If we succeed in recovering the commands from speech characteristics, step 2 will be greatly facilitated.

3.3 Role of Generative Models

It may be necessary here to clarify what I mean by a model. The word has been used in different meanings in different academic disciplines. In natural sciences, a model means *a formal description of the essential characteristics of the structure of a mechanism or the function of a process.* In humanistic sciences, especially in linguistics, a model means *a specially designed representation of concepts or entities used to explain their structure or function* (Crystal [Cry80]). It is in the former sense that I use the word model in this paper.

Let us next consider the significance of models in their relation to three types of inference used in scientific investigation. Studies on the relationship between the underlying neuromotor commands and the acoustic characteristics of speech provide a good example.

1. Finding a model:
   From a set of observed characteristics of speech and their corresponding underlying commands, one can construct a model by induction.
2. Use of a model in speech syntheses:
   From the underlying commands and a model, one can, by deduction, predict the characteristics of speech to be observed. This is the use of a generative model in speech synthesis.
3. Use of a model in speech recognition:
   From a set of observed speech characteristics and a model, one can infer the underlying commands by abduction. This is the inverse problem of 2. Since it is analytically unsolvable, the solution has to be obtained by the method of Analysis-by-synthesis. This is the use of a generative model in speech recognition.

Figure 3.2 illustrates the differences in the role of a model in these three cases. It is clear that the same generative model can be and should be useful both in speech synthesis and in speech recognition.

3.4 A Generative Model for the F_0 Contour of an Utterance of Japanese

By way of further illustration of the use of quantitative models, I will present a model that has proved to be useful in the study of intonation in the broad sense[4] (e.g., Fujisaki and Nagashima [FN69], Fujisaki and Sudo [FS71b], Fujisaki and Sudo [FS72], Fujisaki [Fuj81], Fujisaki, Hirose and Takahashi [FHT90b]). Since the modelling of the entire process of speech production is beyond the scope of this paper, we will restrict ourselves to the generation of the characteristics that are most closely related to prosody in the case of Japanese, namely the contour of the voice fundamental frequency (henceforth the F_0 contour). Analysis of the F_0 contour allows one to investigate manifestation of prosody both in F_0 and in time.

The F_0 contour of an utterance of Japanese can be regarded as the response of the mechanism of vocal cord vibration to a set of commands which carry all the three types of information mentioned in Section 3.1. The linguistic information that is most closely related to F_0 contours is

[4]Manifestations of various types of information are differently named as tone, pitch accent, or intonation, depending on the size of linguistic units associated, but here I use the word "intonation" in the broad sense to include all of them.

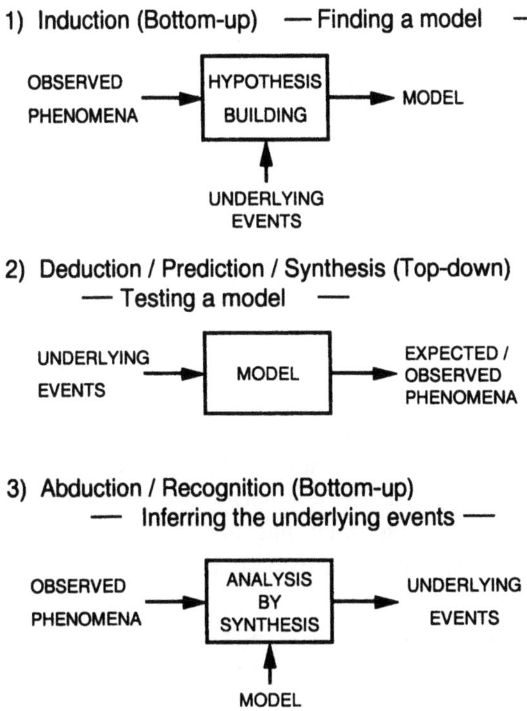

FIGURE 3.2. Roles of generative models in research.

concerned with lexical word accent, syntactic and discourse structures of the utterance. Two different kinds of command have been found to be necessary to account for the formation of an F_0 contour of an utterance of common Japanese; one is an impulse-like command for the onset of a relatively slow rise-fall pattern of the fundamental frequency over a time span roughly corresponding to a syntactic phrase, while the other is a stepwise command for the onset and termination of a relatively rapid rise-fall pattern of the fundamental frequency over a time span corresponding to the accented mora or morae of a word or a string of words.[5] Consequences of these two types of command have been shown to appear as the phrase component and the accent component, each being approximated by the response of a second-order linear system to the respective commands. If we

[5]Strictly speaking, a third kind of command is necessary to account for the rising component of F_0 at the final mora of a phrase, a clause, or a sentence which indicates the speaker's intention of continuation or interrogation. Since, however, the time constant of this third component has been found to be almost equal to that of the accent component, it is assumed that the rising component is also caused by an accent command. This problem has been discussed in more detail elsewhere (Fujisaki, Ohno, and Osame [FOO93]).

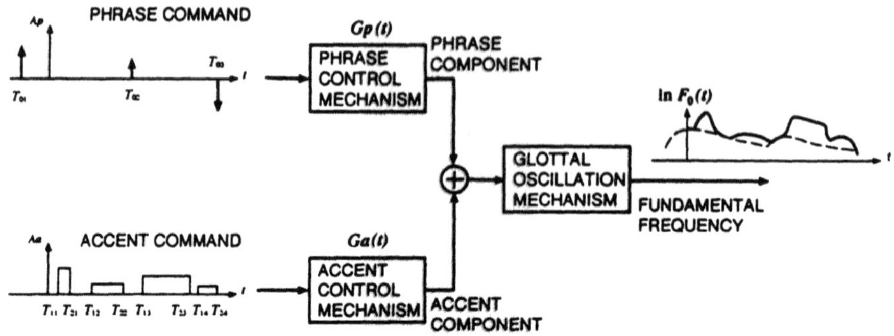

FIGURE 3.3. A functional model for generating F_0 contours of sentences.

represent an F_0 contour as a pattern of the logarithm of the fundamental frequency along the time axis, it can be approximated by the sum of these components. The entire process of generating an F_0 contour of a sentence can thus be modelled by the block diagram of Fig. 3.3.

In this model, the F_0 contour can be expressed by

$$\ln F_0(t) = \ln Fb + \sum_{i=1}^{I} Ap_i Gp(t - T_{0i}) + \sum_{j=1}^{J} Aa_j \{Ga(t - T_{1j}) - Ga(t - T_{2j})\}, \tag{3.1}$$

where

$$Gp(t) \begin{cases} = \alpha^2 t \exp(-\alpha t) & \text{for } t \geq 0 \\ = 0 & \text{for } t < 0 \end{cases}, \tag{3.2}$$

and

$$Ga(t) \begin{cases} = \min[1 - (1 + \beta t)\exp(-\beta t), \gamma] & \text{for } t \geq 0 \\ = 0 & \text{for } t < 0 \end{cases}, \tag{3.3}$$

respectively, indicate the impulse response function of the phrase control mechanism and the step response function of the accent control mechanism. The symbols in Eqs. (3.1), (3.2), and (3.3) indicate

> Fb : asymptotic value of fundamental frequency in the absence of accent components, I : number of phrase commands, J : number of accent commands, Ap_i : magnitude of the ith phrase command, Aa_j : amplitude of the jth accent command, T_{0i} : timing of the ith phrase command, T_{1j} : onset of the jth accent command, T_{2j} : end of the jth accent command, α : natural angular frequency of the phrase control mechanism to the ith phrase command, β : natural angular frequency of the accent control mechanism to the jth accent command, γ : a parameter to indicate the ceiling level of the accent component (generally set equal to 0.9).

A rapid downfall of F_0, often observed at the end of a sentence and occasionally at a clause boundary, can be regarded as the response of the

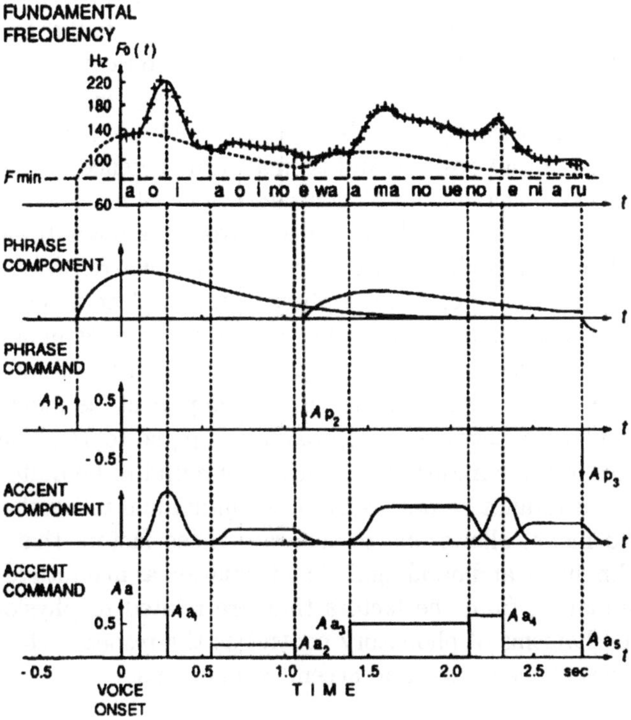

FIGURE 3.4. Analysis-by-synthesis of an F_0 contour of the Japanese declarative sentence: /aoiaoinoewa jamanouenoieniaru/. The figure illustrates the optimum decomposition of a given F_0 contour into the phrase and accent components, and also shows the underlying commands for these components.

phrase control mechanism to a negative impulse for resetting the phrase component.

By the technique of Analysis-by-Synthesis, it is possible to decompose a given F_0 contour into its constituents, i.e., the phrase components and the accent components, and estimate the magnitude and timing of their underlying commands by deconvolution, as shown in Fig. 3.4.

Thus the model can predict and generate from a set of commands, not just a few points on the F_0 contour such as its peaks and valleys subjectively selected, but the entire contour. Moreover, the close agreement of the model's output with the measured F_0 contour, found in this as well as in a number of speech samples analysed, attest the validity of the model. It has been shown that the model has its basis in the actual physiological and physical mechanisms of the larynx (Fujisaki [Fuj88]).

The timings of these commands are found to be closely related to the linguistic contents of the utterance. The accent command is found to start at 40 to 50 ms before the segmental onset of a subjectively high mora and to end also at 40 to 50 ms before the segmental ending of a high mora. The phrase command, on the other hand, is found to be located approximately 200 ms before the onset of an utterance and also before a major syntactic boundary, such as the boundary between the subject phrase and the predicate phrase. In general, the phrase command is largest at the sentence-initial position and is smaller at sentence-medial positions, so that the overall shape of an F_0 contour, disregarding local rises and falls due to accent components, shows a decay from the onset toward the end of the whole utterance. There are cases, however, where pragmatic factors call for the occurrence of a large phrase command at a sentence-medial position. Our analysis also shows that the rate of rise, as indicated by the natural angular frequency β of the accent component, is approximately equal to 20/s, whereas that of the phrase component, as indicated by the natural angular frequency α, is approximately equal to 3/s. The variations in the values of these natural frequencies are found to be quite small from utterance to utterance as well as from one individual to another.

Thus the model allows one to separate those factors that are closely related to linguistic and paralinguistic information as magnitude and timing of the commands, from the factors that are related to physiological and physical mechanisms of phonatory control as the response characteristics, i.e., as shapes of the phrase and accent components.

3.5 Units of Prosody of the Spoken Japanese

Having proved the validity of the model and succeeded in extracting the underlying commands, we are now in a position to go on to the second step of inferring the prosodic units and structures from the extracted commands, as shown in Fig. 3.5.

The fact that a word or a sentence can be uttered with various patterns of intonation clearly indicates that the prosody of an utterance has its own units and structure, which are closely related, but not necessarily identical,

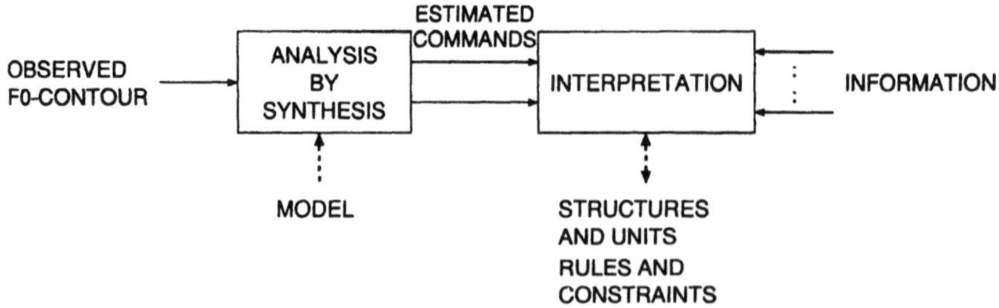

FIGURE 3.5. Processes by which various types of information are manifested in the segmental and suprasegmental features of speech.

to the lexical and syntactic units and structure of the underlying message. Since the phrase components and the accent components are two entities that can be objectively extracted from an F_0 contour, we define various units of prosody of the spoken Japanese on the basis of the observed characteristics of these components, and will then discuss the structure of prosody constructed by these units (Fujisaki, Hirose, and Takahashi [FHT93]).

As far as the word accent and sentence intonation are concerned, the minimal prosodic unit of the spoken Japanese is a "*prosodic word*," which we define as a part or the whole of an utterance that forms an accent type. Thus it has one and only one accent command. Under certain conditions, a string of prosodic words can form a larger prosodic word due to the merger of individual accent commands, a phenomenon defined as "*accent sandhi*" based on a parallelism to "*tone sandhi*" commonly found in tone languages.

The syntactic unit of Japanese which is most closely related to this prosodic word is the "*bunsetsu*," defined as the immediate constituent in the syntax of Japanese, consisting of a content word with or without being followed by a string of function words. However, it is apparent that the syntactic structure of Japanese cannot be accurately described in terms of the relationships among such units as *bunsetsu*. Furthermore, a *bunsetsu* cannot be a unit of prosody since there are cases where a *bunsetsu* is uttered as two or more prosodic words, or where a sequence of several *bunsetsu* is uttered as one prosodic word.

Larger prosodic units are then defined on the basis of phrase components and pauses inserted between two phrase components. The interval between two successive positive phrase commands shall be defined as a "*prosodic phrase*." A prosodic phrase may extend over several prosodic words. Conversely, a prosodic word seldom extends over two prosodic phrases. Furthermore, in longer sentences, several prosodic phrases may form a section delimited by pauses. Such a section shall be defined as a "*prosodic clause*." When a sentence is spoken, it is also terminated by a pause, which is generally much longer than a clause-final pause. The prosodic manifestation of a sentence shall be defined as a "*prosodic sentence*."

The syntactic units that correspond to these three larger prosodic units are the "*ICRLB*," clause, and sentence. The ICRLB is an abbreviation for the "immediate constituent with a recursively left-branching structure," denoting a syntactic phrase which is delimited by right-branching boundaries and contains only left-branching boundaries. Roughly speaking, a parallelism exists between the hierarchy of syntactic units and the hierarchy of prosodic units, as shown in Table 3.1.

It should be emphasized, however, that the correspondence is not exactly one-to-one. As already mentioned, a prosodic word may be much larger than a *bunsetsu* in the case of *accent sandhi*. Likewise, a phrase component is not always resumed at every ICRLB when the latter is comparatively short, so that a prosodic phrase may contain two or more ICRLBs.

TABLE 3.1. Prosodic units of spoken Japanese.

Manifestations	Prosodic units		Syntactic units
An accent component	Prosodic word	sandhi	*"Bunsetsu"* (a content word followed by *n*(≥0) function words)
A phrase component	Prosodic phrase		*ICRLB* (Immediate Constituent with Recursively Left-branching Structure)
Phrase component(s) delimited by utterance-medial pause(s)	Prosodic clause		Clause
Phrase component(s) delimited by utterance-final pauses	Prosodic sentence (utterance)		Sentence

However, prosodic phrases are often used to disambiguate constructional homonymities.

A prosodic clause boundary can occur whenever the speaker takes a breath, regardless of the syntactic structure of the sentence. A speaker will typically take a breath at a clause boundary or at the end of a sentence, but may not always be able to do so depending on the content of the utterance and the speech rate.

3.6 Prosody of Spontaneous Speech

Referring back to Fig. 3.1 in Sect. 3.2, we shall elaborate on the notion of *spontaneous speech.* The word "spontaneous" is understood here to describe the attribute of something that occurs by its own internal force, motivation, etc., rather than by external ones. As already mentioned in Sect. 3.2, spontaneous speech is commonly used to mean speech produced without referring to a text, in contrast to read speech. However, I will go into more detail concerning the information processing that occurs in the speaker in both cases.

The most obvious difference between these two cases concerns the need for message planning. When reading a text aloud, the reader is more concerned with the form of the utterance, since the message is already encoded in the text. On the other hand, in ordinary conversations the speaker has to create the message. Since immediacy is of primary importance, the speaker does not have enough time to complete message planning and utterance planning, and often speaks without a good plan. This also happens, though to a much lesser degree, even in the case of text reading if the reader does not have enough time for text understanding and utterance planning. Rather than simply contrasting spontaneous speech against read speech, I would say that various types of speech form a continuum from the most well-prepared, and therefore highly constrained, speech to the least well-prepared, and therefore highly free and informal speech. We may call it a continuum of the *degree of spontaneity*. Table 3.2 shows five exemplars of speech that can be ordered along this continuum, though each exemplar may have a wide distribution that overlap each other.

The stage of message planning needs further elaboration. It can be subdivided into two consecutive stages: (a) To determine "what to say,"[6] i.e., to select the information content to be expressed by the message, and (b) to determine "how to say," i.e., to select the linguistic units and structures for the message. Both these stages require a varying amount of processing time. For example, when a speaker faces a difficult or unexpected question, stage (a) will require time during which the speaker generally produces a certain kind of filler sound or expresses hesitation. On the other hand, when a speaker has difficulty in finding appropriate words or phrases, it is indicated by another type of filler sound or expression, or by interruptions and re-starts. The occurrence of these phenomena depends very much on the type, complexity, and difficulty of the task which the speakers have to accomplish through the dialogue. However, we can say that the following two trends can be observed as we move on the continuum toward higher spontaneity:

1. Somewhat systematic changes in vocabulary, syntax, as well as speaking styles,
2. increased variations in almost every aspect of the utterance.

TABLE 3.2. A continuum ranging from the most well-prepared to the least well-prepared speech.

Most well-prepared (lowest in spontaneity)
A. Recitation
B. Reading (from text)
C. Simulated dialogue (from text)
D. Controlled dialogue (format/topic/task specified)
E. Free dialogue (format/topic/task unspecified)
Least well-prepared (highest in spontaneity)

[6]The word "say" here refers only to the planning of the message, but not its execution as an utterance.

At the same time, the following changes occur more often in the linguistic aspects of utterances:

1. Specific expressions (clichés) and rhetoric:
   these are often used to keep the conversation going.
2. Elliptic and anaphoric expressions:
   these serve to expedite communication.
3. Abbreviations, especially grouping together several words into one:
   these also serve to expedite communication.
4. Repetitions of important words:
   serves to emphasize and to ensure reliability.
5. Relaxed word order:
   important words tend to be said first.
6. Errors, both corrected and uncorrected:
   indicate that the speaker prefers to keep his/her turn or to be quick in response.
7. Filler sounds:
   indicate that the speaker wants to keep his/her turn and also to show the state of internal information processing.

Almost all these changes listed above are accompanied by some changes in the segmental and/or suprasegmental features of speech. In addition, greater variations are observed in prosody such as the following:

1. Speech rate:
   unimportant words/phrases tend to be uttered at a faster speech rate, with reduced articulation.
2. Accentuation:
   unimportant words tend to be partially or totally deaccented.
3. Paralinguistic modification:
   the speaker seems to rely more heavily on the use of paralinguistic information as a means to convey finer nuances without spending much time to express them linguistically.

A quantitative analysis of these linguistic changes and prosodic variations will be discussed more fully in later sections of this book (see, e.g., the chapters "Speech understanding" by Hirose *et al.*, and "Speaker characterization" by Hirai *et al.*).

References

[Cry69] D. Crystal. *Prosodic Systems and Intonation in English.* Cambridge, UK: Cambridge University Press, 1969.

[Cry80] D. Crystal. *A First Dictionary of Linguistics and Phonetics.* London: André Deutsch, 1980.

[FHT90b] H. Fujisaki, K. Hirose, and N. Takahashi. Manifestation of linguistic and paralinguistic information in the voice fundamental frequency contours of spoken japanese. *Proceedings of the International Conference on Spoken Language Processing, Kobe, Japan*, 1:485–488, 1990.

[FHT93] H. Fujisaki, K. Hirose, and N. Takahashi. Manifestation of linguistic information in the voice fundamental frequency contours of spoken Japanese. *IEICE Transactions on Fundamentals of Electronics, Communications and Computer Sciences*, E76-A:1919–1926, 1993.

[FN69] H. Fujisaki and S. Nagashima. A model for synthesis of pitch contours of connected speech. *Annual Report, Engineering Research Institute, University of Tokyo*, 28:53–60, 1969.

[FOO93] H. Fujisaki, S. Ohno, and M. Osame. Modelling the process of voice fundamental frequency control by paralinguistic information. *Reports of Fall Meeting, Acoustical Society of Japan*, 1:237–238, 1993.

[FS71b] H. Fujisaki and H. Sudo. Synthesis by rule of prosodic features of connected Japanese. *Proceedings of 7th ICA*, 3:133–136, 1971.

[FS72] H. Fujisaki and H. Sudo. A generative model for prosody of connected speech in Japanese. *Proceedings of 1972 Conference on Speech Communication and Processing*, pp. 140–143, 1972.

[Fuj81] H. Fujisaki. Dynamic characteristics of voice fundamental frequency in speech and singing—Acoustical analysis and physiological interpretations —. *Proceedings of the Fourth F.A.S.E. Symposium on Acoustics and Speech*, 2:57–70, 1981.

[Fuj88] H. Fujisaki. A note on the physiological and physical basis for the phrase and accent components in the voice fundamental frequency contour. In O. Fujimura, editor, *Vocal Fold Physiology: Voice Production, Mechanisms and Functions.* New York: Raven, 1988.

[Lad93] D. R. Ladd. Notes on the phonology of prominence. *Working Papers 41, Proceedings of the ESCA Workshop on Prosody, Lund University, Sweden*, pp. 10–15, 1993.

[Lav94] J. Laver. *Principles of Phonetics.* Cambridge, UK: Cambridge University Press, 1994.

[LC83] D. R. Ladd and A. Cutler. Models and measurements in the study of prosody. In A. Cutler and D. R. Ladd, editors, *Prosody: Models and Measurements.* Berlin: Springer-Verlag, 1983.

[Leh70] I. Lehiste. *Suprasegmentals.* Cambridge, MA: M.I.T. Press, 1970.

4
On the Analysis of Prosody in Interaction

G. Bruce
B. Granström
K. Gustafson
M. Horne
D. House
P. Touati

ABSTRACT The research reported here is conducted within the ongoing research project "Prosodic Segmentation and Structuring of Dialogue". The object of study in the project is the prosody of dialogue in a language technology framework. The specific goal of our research is to increase our understanding of how the prosodic aspects of speech are exploited interactively in dialogue—the genuine environment for prosody—and on the basis of this increased knowledge to be able to create a more powerful prosody model. In this paper we give an overview of project design and methods and present some tentative findings.

4.1 Introduction

The vast majority of phonetic studies of prosody have until quite recently been centered upon relatively stereotypic settings in the phonetics laboratory, so-called laboratory speech. In this type of speech material experimental control is high, as relevant parameters can be varied and studied systematically, while the degree of naturalness is often instead quite low. The construction of prosody models currently being used in text-to-speech systems is typically based on the analysis of prosody from such laboratory speech material. Even today there exist few phonetic studies of the prosody of spontaneous speech and dialogue, i.e., the kind of context where prosody has its main function and use. The reason for this bias is to be found in the relative complexity of prosody. Spontaneous speech and dialogue offer such a richness of prosodic variation that its study can be said to presuppose a fundamental understanding of prosody in the more controlled context of laboratory speech.

The object of study in the ongoing project *Prosodic Segmentation and Structuring of Dialogue* (the ProZodiag project) is the prosody of dialogue in a language technology framework [BGG+94b], [BGG+94a]. The project represents cooperation between Phonetics at Lund University and Speech Communication at KTH, Stockholm and is part of the Swedish Language Technology Programme. Related projects within the Language Technology framework are *Intonation in Restrictive Texts: Modelling and Synthesis* [HFJ+93], [Hor94], *Interaction in Speech between Prosody, Syntax, Semantics and Pragmatics* [SEH93], [SH94] and also *Language Technology for Spoken Dialogue Systems* (the Waxholm project) [BCE+93], [CH94]. The focus of our present contribution will be on research methodology.

4.2 Background Work

Research within our project *Prosodic Segmentation and Structuring of Dialogue* is based on earlier work on prosody from different perspectives. The original point of departure for our present research effort is experience from two decades of study of the prosody of prepared speech in a laboratory setting. The intention of using such laboratory speech has not been to study the prosody of reading, but rather to simulate natural, spontaneous speech. Laboratory speech provides us with a high degree of experimental control but presents an artificial situation with often pragmatically strange speech material, where the skill of acting of the informant becomes important.

A second starting point is research conducted within the project *Contrastive Interactive Prosody* (KIPROS) at Lund supported by the Bank of Sweden between 1988–90. The object of study of KIPROS was dialogue prosody in a contrastive perspective in French, Greek, and Swedish. We conducted three types of analysis: analysis of dialogue structure, auditory (prosodic) analysis, and acoustic-phonetic analysis. This project was our first large-scale confrontation with spontaneous speech and dialogue and comprised exploratory testing of the prosody model which was based on experience from extensive work with laboratory speech (see also [Går67]). The focus of the KIPROS project was largely on methodology, which resulted in the development of tools and conventions for prosodic transcription of Swedish and French [BT90], [BT92b]. Experience from the project also made apparent the main difficulties involved in analyzing spontaneous speech where experimental control is low.

A third starting point for the current project is work carried out within the project *Prosodic Phrasing in Swedish* which was also a joint effort between Phonetics in Lund and Speech Communication at KTH, Stockholm, and part of the Language Technology Programme 1990–93. Our cooperation relates to two different research traditions: work in Lund aimed at developing a model for Swedish prosody and work in Stockholm directed towards the development of the prosodic component of a text-to-speech system. The main orientation of this project was directed towards studying how prosody signals phrasing, i.e., grouping of words into phrases. The Prosodic Phrasing project represented a return to the phonetics laboratory and more controlled conditions in the form of analyses of read speech [BGH92], [BGGH93a], [BGGH93b].

4.3 Goal and Methodology

The primary goal of the new project is to increase our understanding of how the prosodic aspects of speech are exploited interactively in dialogue—the genuine environment for prosody—and on the basis of this increased knowledge to be able to create a more powerful prosody model. To be able to achieve this goal the following methodology is being employed:

(1) analysis of discourse/dialogue structure (independent of prosody);

(2) prosodic analysis:

  (a) auditory analysis in the form of prosodic transcription,

  (b) acoustic-phonetic analysis (based on $F0$ and waveform information);

(3) speech synthesis (model-based resynthesis, text-to-speech).

An important methodological starting point of our work is to initially consider prosody and dialogue as potentially independent. This means that we consider it possible to first make separate analyses of prosodic categories and dialogue structuring. Only later are the prosodic analysis and the analysis of discourse/dialogue structure combined in order to find potentially interesting connections. Therefore, we do not a priori anticipate that a particular question intonation is always used by the speaker taking a strong initiative in a conversation, or that the introduction of a new conversation topic is necessarily signalled prosodically by the speaker.

The different types of analysis—analysis of dialogue structure, auditory analysis, acoustic-phonetic analysis—involving both symbol and signal information are combined and synchronized with each other in the same ESPS/waves+ environment. The labelling used (symbol information) consists of an orthographic tier (marking the end of words), a tonal tier (symbols of tonal structure), a boundary tier (symbols of grouping), discourse/dialogue structure tier (e.g., hierarchy of conversational topics), and a miscellaneous tier (with extralinguistic and other information). In our work we are exploiting speech material from the national Swedish prosodic database under development. The dialogues under study cover true spontaneous conversation, spontaneous but more restricted and well controlled dialogues, as well as acted dialogues from scripts. Artificially spliced dialogues, dialogues simulated using text-to-speech synthesis and man-machine dialogues are also exploited in our study of dialogue prosody.

4.4 Prosody in Language Technology

Until fairly recently both speech recognition and speech synthesis have operated with rather narrow contextual windows of analysis. In the case of speech synthesis systems, a window size of one sentence has been typical, and in speech recognition, one utterance consisting of possibly one complete sentence but often limited to just one word in size has been typical. Recent years have, however, seen the development of integrated systems where relatively large contexts can be successfully exploited during the speech recognition part of a man-machine dialogue. An example of this is the project Language Technology for Spoken Dialogue Systems [BCE+93], [CH94].

Up until now, prosody has played only a minor role in affording contextual cues in the speech recognition components of such systems, and in the speech synthesis components, prosodic differentiation based on contextual cues has been very limited. In a man-machine dialogue context it is important that the synthesis is capable of generating relevant context-dependent prosodic distinctions. For instance, in a language like Swedish where accentual defocussing is commonly used as one means of signalling old/given information, a lack of a defocussing strategy in a man-machine dialogue situation may lead to misunderstandings on the part of the human participant [HFJ+93].

As part of the Language Technology Programme, the Prosodic Segmentation project is able to make use of the work being conducted in the other projects within the programme. We are benefiting especially from cooperation with the projects Language Technology for Spoken Dialogue Systems (Waxholm) and Intonation in Restrictive Texts: Modelling and Synthesis. This cooperation entails a two-way effect of benefits: our project is able to draw on the work done in those projects in terms of both theoretical results and material for use in our own analysis work, and in turn the practical results we are producing by way of an enhanced prosody model and rules for the automatic generation of distinctive and contextually relevant prosodic patterns can be tested within the framework of the other projects. The database of the Waxholm project includes samples of spontaneous speech produced in a man-machine dialogue situation. This constitutes important material for the study of the prosody produced spontaneously by humans in such situations.

4.5 Analysis of Discourse and Dialogue Structure

The ultimate goal for prosody research within language technology is to be able to combine phonetic knowledge about prosody with linguistic and other contextual information. It is therefore important that the analysis of this structuring is carried out independently of prosody. The two fundamental aspects are:

1. text/discourse segmentation (topics, focus);
2. dialogue structuring (initiative/response, turns, feedback).

The *text/discourse segmentation* concerns division into conversation topics involving grouping into "speech paragraphs" [BCK80]. This applies to discourse both in the form of dialogue and monologue. It is clear that prosody plays an important role in signalling topic structure, even if different studies show different types of relationships. The textual aspect also involves the distribution of focus in a discourse. It is generally believed that lexical-semantic, syntactic, and discourse/pragmatic factors are all involved in decisions about focus placement and intonational prominence [Hir93b].

Aspects of dialogue structuring include *initiative/response* structure, i.e., traditionally questions and answers, and concern the contribution of the speakers to the development of the dialogue through taking or refraining from taking initiative, responding to initiatives and making reference to what has been said (cf. [LG87]). Prosody plays an important role here, although it is clear that there is a considerable degree of freedom in the way that it is used to signal this aspect of dialogue structure.

The *turn regulating* aspect involves, e.g., taking, keeping, yielding, and competing for the floor in a dialogue (cf. [CP86]). This may partly overlap with the initiative/response structure while still being potentially distinct. It is apparent that this aspect is signalled by different means (verbal, non-verbal, prosodic). The exact contribution of prosody here is not fully understood.

In addition to the above, there is also a *feedback* dimension, indicating the way in which speakers give and seek feedback in a dialogue. Feedback giving (backchannelling) is often noted in dialogue studies while the speaker's feedback seeking (seeking feedback from the listener) has not been given as much attention [Aye94]. It is possible that the feedback dimension can be seen as a subdivision of the initiative/response structure, although we have chosen to regard it as a separate dimension for the present time. We believe that prosody plays an important role in signalling both feedback giving and seeking.

Other dimensions of discourse and dialogue structuring which are expressed prosodically are the signaling of *attitudes/emotions* and *rhetoric activity* [Tou93a], [Tou95].

4.6 Prosodic Analysis

The prosodic analysis consists of an auditory analysis in the form of a prosodic transcription as well as an acoustic-phonetic analysis based mainly on $F0$ and waveform information.

4.6.1 Auditory Analysis

We have witnessed a marked increase in interest in transcription, including prosodic transcription, during recent years. One important reason for this newborn interest arises from new needs for annotation of large speech databases. A starting point was the 1989 IPA Convention in Kiel for the revision of the International Phonetic Alphabet, the first substantial revision in 40 years. The new version of the IPA (cf. [I.P89]) does not, however, contain any specific symbolization of discourse prosody.

Another example of this transcription wave is Tones and Break Indices (ToBI), a system which has recently been developed for the prosodic transcription of American English [SBP+92]. This transcription system provides symbols mainly for intonational prominence and grouping. An innovation in ToBI is the combined auditory and acoustic ($F0$, waveform) analysis, where both types of information are integrated in the prosodic notation.

A vital issue for the construction of a national Swedish prosody database within the Language Technology framework is the choice of prosodic transcription conventions. Discussions of this issue have resulted in an agreement whereby a base module for prosodic transcription based on the IPA comprises a standard for the phonological symbolization of selected prosodic categories. Our point of departure was a prosodic transcription developed originally within the KIPROS project [BT90], [BT92b] and containing a categorization and symbolization of the basic, prosodic functions, prominence, and grouping. The base module for the transcription of Swedish prosody recognizes three levels of prominence (apart from unstressed) and three degrees of grouping [BGG+94a]. This prosodic base transcription has recently been evaluated [SH95]. In addition to this base module different prosody projects within Language Technology are expected to create their own modules according to existing needs.

Our modelling of Swedish intonation within the ProZodiag project takes as its starting point this prosodic core transcription. For the purpose of intonation analysis and synthesis we have developed an additional module for the symbolization of tonal categories. This tonal module recognizes two levels of prominence, for each level of prominence the distinction between the two word accents in Swedish, initial and terminal boundary tones, and two degrees of grouping. For the tonal transcription we are using a similar set of symbols (H, L, *, %) as is used in ToBI [SBP+92]. The intonation model underlying this labelling is elaborated in Sec. 4.6.2. For the symbolization of grouping the following symbols are used: minor group boundary [|], major group boundary [||].

The transcription of the dialogues is made by an expert based on an auditory analysis and tied to the orthographic representation of the dialogue. It results in a sequence of boundary and tonal labels represented on two separate tiers in the ESPS/waves environment. The alignment of the tonal labels is with the *CV* boundary of the stressed syllable, and the alignment of the boundary labels is with the start and end points of the speech or group boundaries.

4.6.2 The Intonation Model

Our intonation model involves categorization with respect to accentuation (prominence) and phrasing (grouping), including boundary signalling and other intonation features (cf. [Bru77], [BG93]). The categories are expressed using tonal turning points (H/L) with association to stressed syllables or boundaries. The main features of the intonation model are given in Table 4.1.

Accent I and accent II are critically timed in relation to foot boundaries, i.e., stressed syllables. In our analysis the two word accents appear to have a distinctively different timing of the same accentual gesture (H(igh), L(ow)) relative to the stressed syllable, accent I being timed earlier than accent II. Thus accent as a higher prominence level than just stress is cued mainly by pitch, although an accented foot is usually also longer than an unaccented foot.

An important grammatical as well as prosodic distinction in Swedish is the one between simplex and compound words. A compound consists of (at least) two feet (stress groups), where only the first foot is accented,

TABLE 4.1. **Representation of prosodic categories in the intonation model**

Prosodic Category	Tonal Turning Points
Unaccented	-
Accent I	HL*
Accent II	H*L
Focal accent I	(H)L*H
Focal accent II	H*LH
Focal accent II compound	H*L... L*H
Initial juncture	%L, %H
Terminal juncture	L%, LH%

while a simplex word consists of only a single foot (stress group). While focal accentuation is primarily determined by semantics and pragmatics (given/new), focal accent is also typically a default choice for a word in a phrase final position. Phonetically, focal accentuation is marked by a more complex accentual gesture, an extra H after the HL for (word) accent. The focal accent H is executed in the same foot (stress group) as the accent HL for a simplex word, while it occurs in the final foot of a compound word. This extra pitch prominence is usually accompanied by increased duration of the word in focus.

Generally, the initial juncture (boundary signal) of a prosodic phrase involves a LH gesture. This LH gesture can be either a separate gesture before the first accent of the phrase or coincide with an accentual gesture, e.g., with the LH of an initial, focal accent. The terminal juncture (boundary signal) of a phrase instead often involves a HL gesture. Correspondingly, this HL gesture can be either a separate gesture, e.g., after a phrase final focal accent, or coincide with the HL of a (non-focal) accent gesture. In longer phrases (with more than two accented words), two post-focal accents within the same phrase will typically occur in a downstep, i.e., the terminal HL gesture can be regarded as being executed in two successive steps, while instead two prefocal accents of a phrase are characterized by the absence of downstep. This tonal signalling of coherence and boundaries for phrasing is also accompanied by temporal signalling as well as by other correlates.

4.6.3 Acoustic-phonetic Analysis

Our acoustic-phonetic analysis comprises standard $F0$ extraction and spectral information in addition to the speech waveform. As indicated above, the analysis is carried out in the ESPS/waves environment which includes transcription and labelling of prosodic and discourse/dialogue categories in multiple tiers [Aye94]. This enables an automatic processing of possible relationships between, for example, prosodic and discourse categories.

A first part of the acoustic-phonetic analysis is the inspection of the $F0$ contours and the qualitative comparison of the signal information with the symbol information in the prosodic transcription. This is done in order to obtain feedback on the tonal transcription as reflected in expected $F0$ patterns. A second part is the quantitative analysis of the $F0$ contours. One example of this is the use of a statistical program which performs calculations of selected $F0$ values for each transcribed phrase. $F0$ data are collected for local, absolute $F0$ minimum and $F0$ maximum values as well as average $F0$ over each transcribed phrase [Tou95].

An important part of the $F0$ analysis is the intonation model which has been developed from extensive studies of laboratory speech. We are using the model in our analysis/synthesis approach which will give us information about deviations between predicted and observed $F0$ contours. The phonological features of the model have been described above, while the phonetic features of our intonation model are outlined in the speech synthesis section below. Both global features in terms of, for example, $F0$ level and $F0$ range and local features in terms of direction and timing of $F0$ events are taken into consideration and interpreted in our current prosody model.

4.7 Speech Synthesis

The methods for speech synthesis used in our project work are the analysis/resynthesis procedure integrated in the ESPS/waves environment as well as the KTH text-to-speech system.

4.7.1 Model-based Resynthesis

The tonal transcription discussed above represents a phonological analysis of prominence and grouping. It also constitutes the input to the resynthesis module. In the implementation of the intonation model in the ESPS/waves environment, the prosodic information contained in the transcription has to be supplemented with phonetic rules which will take care of the more specific timing of prominences, pitch level, and range (including focus realization), $F0$ drift (downdrift, downstep, upstep), as well as the interpolation between turning points.

The use of the analysis/synthesis method in the present framework has a double purpose. The first, more direct goal is to verify/falsify the phonological analysis as reflected in the prosodic transcription. This will give us feedback on the correctness of the transcription and will reveal incorrect auditory judgments about focus placement, prominence levels, phrasing, and the like. The second, more long term goal is to use the analysis/synthesis tool for developing our intonation model in a dialogue prosody framework. Recurring deviations between predicted and observed $F0$ values will give us an indication in what direction the model can be modified. As our prosodic transcription covers only prominence and grouping, other aspects of intonation such as those related to topic strucutre are thus not explicitly modelled in our current resynthesis.

To be able to generate an $F0$ contour, the transcription labels have to be split into a sequence of turning points (L and H). This is performed following a set of simple rules [BGF+95]. It is important to note that the only input to the system consists of the transcription labels and their time alignment. This gives us a segmentation of the speech signal into approximately stress groups or feet. We have no other segmental information available, nor any voiced/voiceless distinctions. This, of course, limits the amount of information which can be included in the rules for placing the turning points, as we cannot refer to, e.g., vowels or syllables as points of reference in the speech signal. Preceding H's are placed a fixed number of milliseconds before the location of the label. Succeeding turning points are spaced equally between the locations of the current transcription label and the next label. This solution may seem to be ad hoc but is not without motivation in production data. In a study by Bruce [Bru86] variability in the timing of the pitch gesture for focal accent relative to segmental references was demonstrated. Instead, disregarding segmental references and using the beginning of the stress group as the line-up point, there appeared to be a high degree of constancy in the timing of the whole focal accent gesture. It should be noted that the actual numbers in milliseconds of the implementation are at this point chosen as test values partly based on earlier work on tonal stylization [Hou90], [HB90].

For resynthesis, we have been exploiting two different methods: the LPC synthesis included in the ESPS/waves+ software package [MG76] and an implementation of the PSOLA synthesis system in the same environment [MC90], [MD95]. The PSOLA synthesis seems to be well-suited for our present purpose.

4.7.2 Text-to-speech

The other approach is to exploit our existing KTH synthesis system, which is based on the RULSYS development system [CGK90], [GN92], [CGH91]. Using an experimental version of this system which includes an extended set of prosodic markers, we have a flexible tool for manipulating prosodic parameters [BG89]. During the initial phases we are studying the prosody of humans in a man-machine dialogue situation, using the KTH database of labelled speech material collected as part of the Waxholm project [BCE+93], [CH94].

The model is a parametric one. This means that we have defined a set of prosodic parameters corresponding to those we observe in our speech material. By manipulating the parameter values we are able to generate $F0$ and durational patterns closely resembling those of our speech material. The advantages of such a parameter-based model are several: it allows us to test perceptual properties of the different parameters, it is easy to specify and model new patterns when such appear in the speech material, and we are able to model such differences which are due to factors other than strictly phonological ones, for instance such as are due to speaker attitudes and emotions. Above all, in the context of dialogue modelling, it is possible to specify prosodic variation that can be attributed to the dialogue situation and to model this variation.

The parametric model basically specifies the phonetic shape of utterances. Linked with this is a mapping procedure whereby relevant phonological categories can be mapped. The parametric categories are, on the whole, the same as the ones we use in our analysis/resynthesis approach.

4.8 Tentative Findings

The point of departure for our work on a more powerful prosody model is our present model which goes back to Bruce [Bru77] and which is in essence implemented in our model-based resynthesis and text-to-speech systems. To achieve our goals within this project we are studying a wider range of speech situations than those on which the present model was based, in particular speech produced in dialogue situations of different kinds, both man-man and man-machine dialogues [BGG+95].

In one of our studies of man-man dialogues we have, among other things, investigated the differences that exist between spontaneous and read speech [Bru95]. This has been done by having the participants of a spontaneous dialogue re-enact the same dialogue a few weeks later from a (slightly edited) written transcript. This method, which has been successfully used in other studies, e.g., [Går67], [Aye94], allows us to more easily identify such factors that can be presumed to be characteristic of spontaneous dialogue, as opposed to read speech. The two versions of the dialogue are, not surprisingly, auditively clearly distinct. One of the observations of this study has been that the acted version exhibits a more coherent speaking style as compared to the spontaneous version which, on the other hand, exhibits a more lively, interactive style. A broad prosodic transcription of the two versions reveals some interesting differences in accentuation and focus locations as well as in phrasing between the versions, some of which appear to reflect more stable differences between the two speaking styles. The most stable of these striking differences seem to be related to those of phrasing. There is a tendency for a phrase, as signalled by pitch and other cues, to accommodate more words in the read version than in the spontaneous one. This may be thought of as due to the difference in planning between the speaking styles. The chunking into smaller units characteristic of the spontaneous speech is likely to be a reflection of the on-line planning.

It is clear, from our study, as has been demonstrated often before, that characteristically different pitch patterns are powerful tools in signalling both the textual aspects (in particular topic structure) and the feedback dimension, of dialogue structure. The complexities of the use of prosody in spontaneous speech can be illustrated by the following example: in an area of transition between two conversational topics, one of the speakers signals the topic shift by using a marked increase of $F0$ level in the last utterance of the first topic. The effect is that, although this utterance belongs textually to the first topic, and thus points backwards, prosodically it is grouped with the next utterance, and thus points forwards. The raised $F0$ level may also be thought of as part of a turn-keeping signalling by the speaker. In the read version, on the other hand, the topic shift is a typical, marked increase in $F0$ at the discourse boundary.

A more quantitative analysis of $F0$ patterns has not yet been completed within the ongoing project. As an illustration of what kind of findings can be expected, we present some data from a monologue section of the dialogue rererred to above. Figure 4.1 shows $F0$ values ($F0$ min, $F0$ max and $F0$ mean) of successive phrases within the dialogue section for one of the two speakers (female speaker) as well as the corresponding $F0$ averages of the section. The most apparent finding is the greater variability in $F0$ maxima and the more extreme values for the sucesssive phrases in the spontaneous version as compared to the read, acted version. The difference in $F0$ max values may be thought of as effects of differences in on-line planning and interactivity between the two versions.

Compared to the prosody of real-life human dialogues, that of the human participants in the man-machine dialogue situations that we are studying exhibits on the whole a rather limited range of prosodic variation. This pertains to both $F0$ range, focus assignment, and tempo. It seems likely that this is, at least in part, related to the limitations in prosodic variation on the part of the machine partner. The main notable exception is when the machine does not understand or misunderstands what the human says. Characteristically, the human then often exhibits a wider $F0$ range, and commonly other features typical of agitated or angry speech, such as increased intensity. One variant involves a widened range in the focus domain, combined with a narrowed range in a higher than average register. This kind of signalling could be used as a criterion to switch to a human operator in a practical machine-man system.

4.9 Final Remarks

One important motivation for our research approach is the need for extra descriptive parameters related to the description of dialogue. Our prosody model was based originally on laboratory speech data. In order to adequately describe the prosodic properties of spontaneous speech and dialogue, it is clear that it is necessary to introduce additional descriptive parameters. The purpose of the parametric model is precisely this, to establish a set of parameters by which all human-produced speech, be it spontaneous or read, in the form of monologue or dialogue, can be described, and such that using the same parametrical approach it is possible to generate essentially the same prosodic characteristics in a speech synthesis system. Furthermore, looking more into the future, by coupling this model to dialogue handling systems, it should be possible to assist in the process of both automatic speech recognition and in the automatic generation of adequate prosodic contours in the speech synthesis part of such systems.

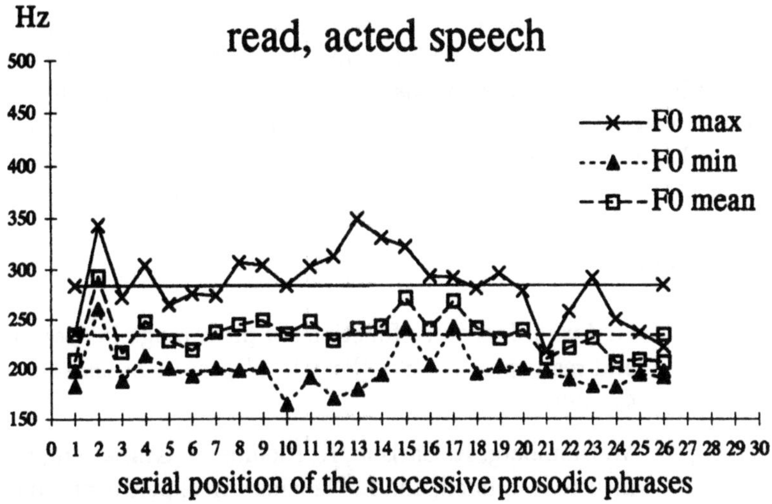

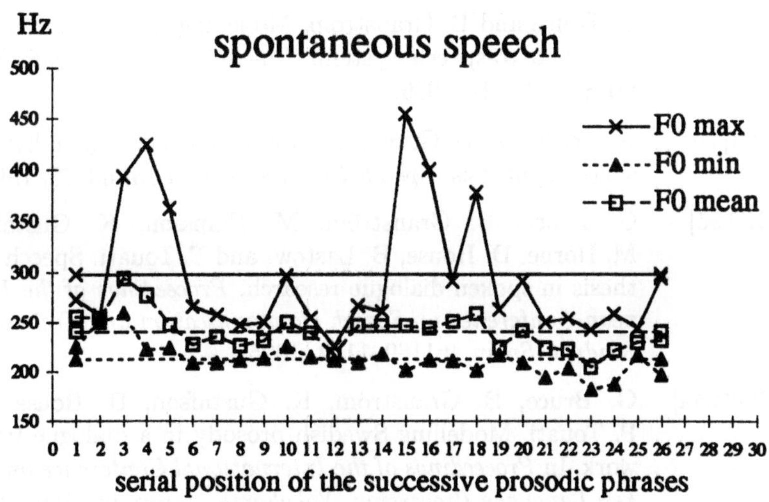

FIGURE 4.1. Comparison of read, acted speech (upper part) and spontaneous speech (lower part). $F0$ values in Hz ($F0$ max, $F0$ min, and $F0$ mean) for successive prosodic phrases of a monologue section of a dialogue as produced by a female, Swedish speaker. The horizontal lines represent the corresponding $F0$ average values for the section.

Acknowledgments

This work was carried out under a contract from the Swedish Language Technology Programme (HSFR-NUTEK). Gayle Ayers, Entropic Research

Laboratory, Inc. Washington D.C., was a guest researcher in Lund during several months during 1993 and 1994 and has contributed to the project. Marcus Filipsson and Birgitta Lastow (Lund) have been developing the analysis / resynthesis system. Johan Frid recently joined the project and is involved in the analysis of dialogues.

References

[Aye94] G. Ayers. Discourse functions of pitch range in spontaneous and read speech. *OSU Working Papers in Linguistics*, 44:1–49, 1994.

[BCE+93] M. Blomberg, R. Carlson, K. Elenius, B. Granström, S. Hunnicutt, R. Lindell, and L. Neovius. An experimental dialogue system: Waxholm. *RUUL 23, Fonetik -93*, pp. 49–52, 1993.

[BCK80] G. Brown, K. L. Currie, and J. Kenworthy. *Questions of Intonation.* London: Croom Helm, 1980.

[BG89] G. Bruce and B. Granström. Modelling Swedish intonation in a text-to-speech system. *Proceedings Fonetik-89*, STL-QPSR 1:17–21, 1989.

[BG93] G. Bruce and B. Granström. Prosodic modelling in Swedish speech synthesis. *Speech Communication* 13, 63–73, 1993.

[BGF+95] G. Bruce, B. Granström, M. Filipsson, K. Gustafson, M. Horne, D. House, B. Lastow, and P. Touati. Speech synthesis in spoken dialogue research. *Proceedings of the European Conference on Speech Communication and Technology, Madrid, Spain*, 2:1169–1172, 1995.

[BGG+94a] G. Bruce, B. Granström, K. Gustafson, D. House, and P. Touati. Modelling Swedish prosody in a dialogue framework. In *Proceedings of the International Conference on Spoken Language Processing, Yokohama, Japan*, pp. 1099–1102, 1994.

[BGG+94b] G. Bruce, B. Granström, K. Gustafson, D. House, and P. Touati. Preliminary report from the project "Prosodic segmentation and structuring of dialogue." *Working Papers 43, Fonetik -94*, pp. 34–37, 1994.

[BGG+95] G. Bruce, B. Granström, K. Gustafson, M. Horne, D. House, and P. Touati. Towards an enhanced prosodic model adapted to dialogue applications. *Proceedings ESCA Workshop on Dialogue Management*, Aalborg, Denmark, pp. 201–204, 1995.

[BGGH93a] G. Bruce, B. Granström, K. Gustafson, and D. House. Interaction of F_0 and duration in the perception of prosodic phrasing in Swedish. In B. Granström and L. Nord, editors, *Nordic Prosody VI*, pp. 7–22. Stockholm: Almquist & Wiksell, 1993.

[BGGH93b] G. Bruce, B. Granström, K. Gustafson, and D. House. Prosodic modelling of phrasing in Swedish. *Working papers 41, Proceedings of the ESCA Workshop on Prosody, Lund University, Sweden*, pp. 180–183, 1993.

[BGH92] G. Bruce, B. Granström, and D. House. Prosodic phrasing in Swedish speech synthesis. In G. Bailly, C. Benoît, and T. R. Sawallis, editors, *Talking Machines: Theories, Models, and Designs*, pp. 113–125. Amsterdam: Elsevier Science, 1992.

[Bru77] G. Bruce. *Swedish Word Accents in Sentence Perspective.* Lund: Gleerup, 1977.

[Bru86] G. Bruce. How floating is focal accent? In Gregersen and Basbøll, editors, *Nordic Prosody IV*, pp. 41–49. Odense: Odense University Press, 1986.

[Bru95] G. Bruce. Modelling Swedish intonation for read and spontaneous speech. In *Proceedings of the 13th International Congress of Phonetic Sciences, Stockholm, Sweden*, Vol. 2, pp. 28–35, 1995.

[BT90] G. Bruce and P. Touati. On the analysis of prosody in spontaneous dialogue. In *Working Papers 36*, pp. 37–55. Department of Linguistics, Lund University, 1990.

[BT92b] G. Bruce and P. Touati. On the analysis of prosody in spontaneous speech with exemplifications from Swedish and French. *Speech Communication*, 11:453–458, 1992.

[CGH91] R. Carlson, B. Granström, and S. Hunnicutt. Multilingual text-to-speech development and applications. In W. Ainsworth, editor, *Advances in Speech, Hearing and Language Processing*, pp. 269–296. London: JAI Press, 1991.

[CGK90] R. Carlson, B. Granström, and I. Karlsson. Experiments with voice modelling in speech synthesis. *Speech Communication*, 10:481–490, 1990.

[CH94] R. Carlson and S. Hunnicutt. Dialog management in the Waxholm system. *Working Papers 43, Fonetik -94*, pp. 46–49, 1994.

[CP86] A. Cutler and M. Pearson. On the analysis of prosodic turn-taking cues. In C. Johns-Lewis, editor, *Intonation in Discourse*, pp. 139–155. London: Croom Helm, 1986.

[Går67] E. Gårding. Prosodiska drag i spontant och uppläst tal. in G. Holm, editor, *Svenskt talspråk*. pp. 40–85, Uppsala: Almqvist & Wiksell, pp. 40–85, 1967.

[GN92] B. Granström and L. Nord. Neglected dimensions in speech synthesis. *Speech Communication*, 11:459–462, 1992.

[HB90] D. House and G. Bruce. Word and focal accents in Swedish from a recognition perspective. In Wiik and Raimo, editors, *Nordic Prosody V*, pp. 156–173. Turku: Turku University, 1990.

[HFJ+93] M. Horne, M. Filipsson, C. Johansson, M. Ljungqvist, and A. Lindström. Improving the prosody in TTS systems: Morphological and lexical-semantic methods for tracking 'new' vs. 'given' information. *Working Papers 41, Proceedings of the ESCA Workshop on Prosody, Lund University, Sweden*, pp. 208–211, 1993.

[Hir93b] J. Hirschberg. Studies of intonation and discourse. *Working Papers 41, Proceedings of the ESCA Workshop on Prosody, Lund University, Sweden*, pp. 90–95, 1993.

[Hor94] M. Horne. Generating prosodic structure for synthesis of Swedish intonation. *Working Papers 43, Fonetik 94*, pp. 72–75, 1994.

[Hou90] D. House. *Tonal Perception in Speech*. Lund: Lund University Press, 1990.

[I.P89] I.P.A. Report on the 1989 Kiel convention. *Journal of the International Phonetic Association*, 19 (2):67–80, 1989.

[LG87] P. Linell and L. Gustavsson. Initiativ och respons. Om dialogens dynamik, dominans och koherens. *Studies in Communication*, 15, 1987.

[MC90] E. Moulines and F. Charpentier. Pitch-synchronous waveform processing techniques for text-to-speech synthesis using diphone. *Speech Communication*, 9:453–467, 1990.

[MD95] G. Möhler and G. Dogil. Test environment for the two level model of Germanic prominence. *Proceedings of the European Conference on Speech Communication and Technology, Madrid, Spain*, 2:1019–1022, 1995.

[MG76] J. Markel and A. Gray, Jr. *Linear Prediction of Speech.* Berlin: Springer, 1976.

[SBP+92] K. Silverman, M. Beckman, J. Pitrelli, M. Ostendorf, C. Wightman, P. Price, J. Pierrehumbert, and J. Hirschberg. ToBI: a standard for labelling English prosody. In *Proceedings of the International Conference on Spoken Language Processing, Banff, Canada*, Vol. 2, pp. 867–870, 1992.

[SEH93] E. Strangert, E. Ejerhed, and D. Huber. Clause structure and prosodic segmentation. *RUUL 23, Fonetik -93*, pp. 81–84, 1993.

[SH94] E. Strangert and M. Heldner. Prosodic labelling and acoustic data. *Working Papers 43, Fonetik -94 Department of Linguistics, Lund University*, pp. 120–123, 1994.

[SH95] E. Strangert and M. Heldner. Labelling of boundaries and prominences by phonetically experienced and non-experienced transcribers. *PHONUM*, 3:85–109, 1995.

[Tou93a] P. Touati. Overall pitch and direct quote-comment structure in French political rhetoric. *RUUL 23, Fonetik -93*, pp. 98–101, 1993.

[Tou95] P. Touati. Pitch range and register in French political speech. In *Proceedings of the 13th International Congress of Phonetic Sciences, Stockholm, Sweden*, Vol. 4, pp. 244–247, 1995.

Part II

Prosody and the Structure of the Message

5 Introduction to Part II

Anne Cutler

5.1 Prosody and the Structure of the Message

Most of the speech that speakers produce and listeners hear is spontaneous, and intended for the purpose of communicating. Apart from the occasional monologue, or the mutterings of a deranged passerby, the greater part of the speech we experience assumes the availability of a listener. Speakers use prosodic means (among others) to communicate to listeners the structure of the message that they wish to impart.

Intonational prominence, in particular, is a prosodic device which does principal service as an indicator of message structure. Moreover, it is a phenomenon of considerable cross-linguistic generality. The analyses reported in the contributions in this section are based on several different languages—American English, Dutch, Japanese. One finding which appears across languages (e.g., in Nakatani's analysis of English, and Nakajima and Tsukada's analysis of Japanese—is that a shift in discourse topic is accompanied by raised F0. The same result has been reported for Scottish English, by Brown, Currie, and Kenworthy [BCK80] and for speech of American parents to children, by Menn and Boyce [MB82]. Indeed, Bolinger [Bol78] listed obtrusions for prominence as (along with the expression of closure) a truly language-universal use of prosody.

The relationship between intonational prominence and message is exploited by listeners during the processing of spoken language. Thus listeners accord a high priority to the task of detecting where sentence accent falls in a speaker's utterance; preceding prosodic cues enable listeners to direct attention to accents [Cut76]. If part of the normally available prosodic information is absent, listeners will make use of what remains [CD81]; but it seems that no one prosodic dimension is paramount in signalling accent location, because conflict between different sources of prosodic information (e.g., F0 and rhythm) leaves listeners unable to predict where accent will fall [Cut87].

The importance of seeking accent location is explained as a search for focussed, or semantically central, aspects of the speaker's message [CF79]. In fact listeners are extremely efficient at processing the mapping of discourse structure onto accentuation patterns, and extremely sensitive to mismatch in the mapping [BM83, TN87, FH87, TH94, vDL94].

Since the target of most listening is spontaneously uttered speech, it is reasonable to assume that the processing abilities of listeners as evidenced in these laboratory studies have been developed in the spontaneous situation. Yet the experimental studies cited above have been carried out almost without exception using speech materials which have been carefully constructed for the purpose and read from text in a laboratory situation. It is thus reasonable to ask whether we are as yet in possession of the full story regarding the processing which listeners apply to the speech they hear in most everyday situations.

Spontaneous speech and read speech differ with regard to prosodic structure: the former has, in particular, shorter prosodic units and more frequent pauses and hesitations (see, e.g., Crystal & Davy [CD69]). Thus listening procedures which involve the tracking of prosodic contours of some extended duration may be less well served by the average spontaneously spoken utterance. Indeed, Mehta and Cutler [MC88]) found that the pattern of listeners' responses in a phoneme detection task performed on spontaneously uttered materials differed from the response pattern obtained with exactly the same materials produced as read speech. In particular, an effect of response facilitation for phoneme targets occurring later in the sentence, which appears consistently with read-speech materials including those of Mehta and Cutler's own study, disappeared with the spontaneous utterances. This effect has been variously interpreted as reflecting prediction of target location from syntactic, semantic, or prosodic structure; since the materials in Mehta and Cutler's read versus spontaneous conditions did not differ syntactically or semantically, the failure to find the position effect in one of two conditions which differed prosodically certainly supports a prosodic interpretation, and suggests that prosodic prediction may be of limited applicability with spontaneous input.

On the other hand, Mehta and Cutler did find response facilitation for targets on accented as opposed to unaccented words in spontaneous utterances. Note that intonational prominence—obtrusion of an intonational peak from median F0 across an utterance—tends in fact to be greater in spontaneous than in read speech [vB90, vB91]. Mehta and Cutler argued that the prosodic characteristics of spontaneous speech (such as shorter prosodic units and hence more frequent occurrence of relative accent) allow rich opportunities for the exercise of some processing strategies in the listener's repertoire, but poorer opportunities for the exercise of others. Moreover, there may of course exist processing strategies of particular and exclusive usefulness for the processing of spontaneous speech which have as yet not been revealed by experimental investigation. Certainly the modelling of the prosodic structure of spontaneous speech is, given listener sensitivity to effects at this level, an enterprise likely to pay off in the construction of user-friendly synthesis and recognition systems.

References

[BCK80] G. Brown, K. L. Currie, and J. Kenworthy. *Questions of Intonation.* London: Croom Helm, 1980.

[BM83] J. K. Bock and J. R. Mazzella. Intonational marking of "given" and "new" information: Some consequences for comprehension. *Memory & Cognition*, 11:64–76, 1983.

[Bol78] D. L. Bolinger. Intonation across languages. In J. H. Greenberg, editor, *Universals of Human Language.* Stanford: Stanford University Press, 1978.

[CD69] D. Crystal and D. Davy. *Investigating English Style.* London: Longman, 1969.

[CD81] A. Cutler and C. J Darwin. Phoneme-monitoring reaction time and preceding prosody: Effects of stop closure duration and of fundamental frequency. *Perception & Psychophysics*, 29:217–224, 1981.

[CF79] A. Cutler and J. A Fodor. Semantic focus and sentence comprehension. *Cognition*, 7:49–59, 1979.

[Cut76] A. Cutler. Phoneme-monitoring reaction time as a function of preceding intonation contour. *Perception & Psychophysics*, 20:55–60, 1976.

[Cut87] A. Cutler. Components of prosodic effects in speech recognition. In *Proceedings of the 11th International Congress of Phonetic Sciences, Tallin, Estonia*, Vol. 1, pp. 84–87, 1987.

[FH87] C. A. Fowler and J. Housum. Talkers' signalling of 'new' and 'old' words in speech, and listeners' perception and use of the distinction. *Journal of Memory & Language*, 26:489–504, 1987.

[MB82] L. Menn and S. Boyce. Fundamental frequency and discourse structure. *Language & Speech*, 25:381–383, 1982.

[MC88] G. Mehta and A. Cutler. Detection of target phonemes in spontaneous and read speech. *Language and Speech*, 31:135–156, 1988.

[TH94] J. Terken and J. Hirschberg. Deaccentuation and persistence of grammatical function and surface position. *Language & Speech*, 37:125–145, 1994.

[TN87] J. Terken and S. G. Nooteboom. Opposite effects of accentuation and deaccentuation on verification latencies for given and new information. *Language & Cognitive Processes*, 2:145–163, 1987.

[vB90] F. J. Koopmans van Beinum. Spectro-temporal reduction and expansion in spontaneous speech and read text: The role of focus words. In *Proceedings of the International Conference on Spoken Language Processing, Kobe, Japan*, Vol. 1, pp. 21–24, 1990.

[vB91] F. J. Koopmans van Beinum. Spectro-temporal reduction and expansion in spontaneous speech and read text: Focus words versus non-focus words. *Proceedings of the ESCA Workshop on Phonetics and Phonology of Speaking Styles*, Vol. 5, pp. 1–36, 1991.

[vDL94] W. van Donselaar and J. Lentz. The function of sentence accents and given/new information in speech processing: Different strategies for normal-hearing and hearing-impaired listeners? *Language and Speech*, 37:375–391, 1994.

6
Integrating Prosodic and Discourse Modelling

Christine H. Nakatani

ABSTRACT The integration of prosodic information into discourse processing algorithms requires new principles concerning its interaction with other sources of linguistic structure, such as grammatical function and lexical form. This paper presents high level algorithms describing these interactions for reference processing and the modelling of attentional state. The algorithms show how intonational prominence information may guide search and inference procedures during the discourse processing of referring expressions.

6.1 Introduction

A better understanding of the role of prosody in natural language understanding can aid in the assessment of the gains to be had from computing the prosodic characteristics of speech. This paper argues that the process of prosodic interpretation is not essentially separate from that of other nonprosodic linguistic factors such as grammatical function or lexical form. All serve to cue inferences in discourse processing, such as marking changes in attentional state and establishing relations among referents. We focus on intonational prominence as one source of linguistic information contributing to discourse processing decisions. High level discourse processing algorithms sketched in this paper provide a partial specification for the attentional state processing component of a natural language understanding system. These algorithms illustrate the potential contribution of prosody in such a system, thereby motivating work on prominence recognition and heuristic approaches to the modelling of discourse required by the algorithms. The algorithms may also be implemented in message-to-speech systems, in which discourse structure and meaning are directly encoded, for the purpose of capturing meaningful contrasts in prominence.

The algorithms treat in detail the interactions of intonational prominence and other linguistic factors such as lexical form, grammatical function, and discourse structure. Previous studies have shown that the accentuation of referring expressions can be correlated with discourse structural properties

[Ter84], and that taking discourse structure into account can improve the performance of pitch accent assignment algorithms [Hir93a]. On the other hand, these same studies as well as others show that accent is determined partly by lexical form and partly by grammatical function, among other linguistic factors (cf. [Bro83, Fuc84, Alt87, TH94]). These findings are related in the discourse processing algorithms by two fundamental principles: first, the meanings conveyed by choices of accentuation, lexical form, and syntactic structure are separate but interacting; and second, these choices must be interpreted against the background of a dynamic model of attentional state.

In Sec. 6.2, we present the framework of attentional state modelling that we assume in our analyses. In Sec. 6.3, the algorithms and principles embodied in them are presented. We discuss related work in Sec. 6.4 and conclude with an overview of some open issues in Sec. 6.4.

6.2 Modelling Attentional State

In modelling attentional state, we utilize the discourse structure theory of Grosz and Sidner [GS86]. Unlike most existing theories of information status or focus of attention, this theory assumes a fundamental qualitative distinction between global and local salience. A second important aspect of Grosz and Sidner's theory is that it characterizes notions of discourse salience relative to a dynamically unfolding record of mutual beliefs established during the discourse.

Grosz and Sidner propose three interrelated components of discourse structure: intentional structure, linguistic structure, and attentional state. Briefly, a discourse is comprised of discourse segments whose hierarchical relationships are determined by intentional structure and realized by linguistic structure. Discourse processing proceeds at two levels, global and local. The global level concerns relationships among discourse segments, while the local level concerns relationships among utterances within a single segment. In Grosz and Sidner's framework, the notion of salience is formalized by attentional focussing mechanisms that are claimed to underlie discourse processing in general. The attentional state component dynamically records the entities and relations that are salient at any point in the discourse. The present analyses require rudimentary definitions of four attentional statuses provided by the Grosz and Sidner model: primary local focus or Cb, secondary local focus or Cf members (exclusive of the Cb), immediate global focus, and non-immediate global focus.

The global level of attentional state is modelled as a last-in first-out focus stack of focus spaces, each containing representations of the entities and relations salient within the discourse segment corresponding to the focus space [Gro77]. The focus space at the top of the focus stack is termed the immediate focus space. Pushes and pops of focus spaces obey the hierarchical segmental structure of the discourse. An empty focus space is pushed onto the stack when a segment begins; entities are recorded in the immediate focus space as the discourse advances until the discourse segment closes and its focus space is popped from the focus stack. Those entities represented in the top focus space on the stack are in immediate global focus. Entities represented elsewhere on the stack are in non-immediate global focus and are said to be less accessible as referents than entities in the immediate focus space.

It is important to note that there are three discourse segment boundary types that each entail specific manipulations of the focus stack. First, there is the push-only move corresponding to the initiation of an embedded segment, or subsegment. When an embedded segment opens, a new focus space is pushed onto the stack, on top of the focus space of its embedding segment. Second, there is the pop-only move corresponding to the completion of an embedded segment. Upon the close of an embedded segment, the focus space of the embedded segment is popped from the top of the focus stack. The focus space of the *embedding* segment becomes the immediate focus space, and entities within it are said to be immediately accessible. Finally, there is the pop-push move corresponding to the transition between sister segments. In this case, one segment ends and its focus space is popped; immediately, a new focus space is pushed for the next sister segment.

A *. . . so Freud had a few affairs with Fliese*
so big deal you know what I'm saying
he knocked up Minnie Bernais
he was married to Martha and knocked up his sister-in-law

B *and they gave her hey-*
she had an abortion in one of these [clap]

alright he was human too alright . . .

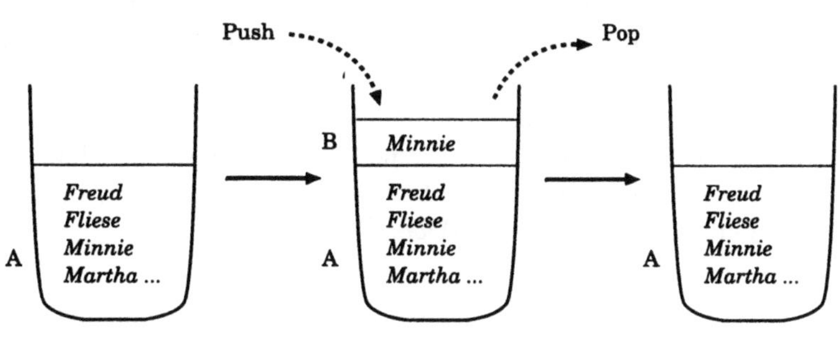

FIGURE 6.1. Illustration of global focussing mechanisms.

Figure 6.1 illustrates the push-only and pop-only manipulations of global focus on an excerpt from a narrative that is described in the following section. At the end of the fourth line, "... and knocked up his sister-in-law", the global focus stack contains one focus space with contents as shown in the left diagram. When Segment B begins, a new focus space is pushed as represented in the center diagram. At the end of Segment B, its focus space is popped and the stack holds only the previous focus space for Segment A as shown in the right diagram. When the pronoun *he* is encountered after the pop, the entities in the preceding embedded segment are no longer on the focus stack and are therefore not available for pronoun reference. The pronoun instead refers to an entity in the focus space of the outer segment that is resumed, namely *Freud.*

The local level of attentional state is modelled by centering mechanisms [Sid79, JW81, GJW83, GJW95]. All of the salient discourse entities realized in an utterance are recorded in the partially ordered set of forward-looking centers (Cf list) for that utterance. The Cf list specifies a set of possible links forward to the next utterance. The ranking of the Cf list members reflects their relative salience in the local discourse context at the time of the uttering of that particular utterance. Each utterance also has a single backward-looking center (Cb), which is the most central and salient discourse entity that links the current utterance to the previous utterance. The Cb of an utterance U_n, Cb(U_n), is defined as the highest-ranking member of the Cf list of the prior utterance, Cf(U_{n-1}), that is realized in U_n. Centering theory currently stipulates that the Cf list members are ordered based on grammatical function and surface position [GGG93, GJW95]. The highest ranking member of the Cf list is called the preferred center (Cp), and is the most likely candidate to become the Cb of the following utterance.

Figure 6.2 illustrates how centering operates on a sequence of utterances. The narrative extract is centrally about Freud, who remains the Cb throughout. Centering constructs are computed within each discourse segment.

To summarize, Grosz and Sidner's global and local attentional state mechanisms distinguish two levels of global salience and two levels of local salience. A discourse entity may be globally salient by virtue of its being represented on the focus stack in either the immediate focus space or a non-immediate focus space. Entities in the immediate focus space are claimed to be relatively more accessible than those in focus spaces deeper in the stack. At the local level, the Cb is claimed to be more salient than non-Cb members of the Cf list. Psychological evidence for this claim is provided by anaphor resolution experiments [HD88, GGG93].

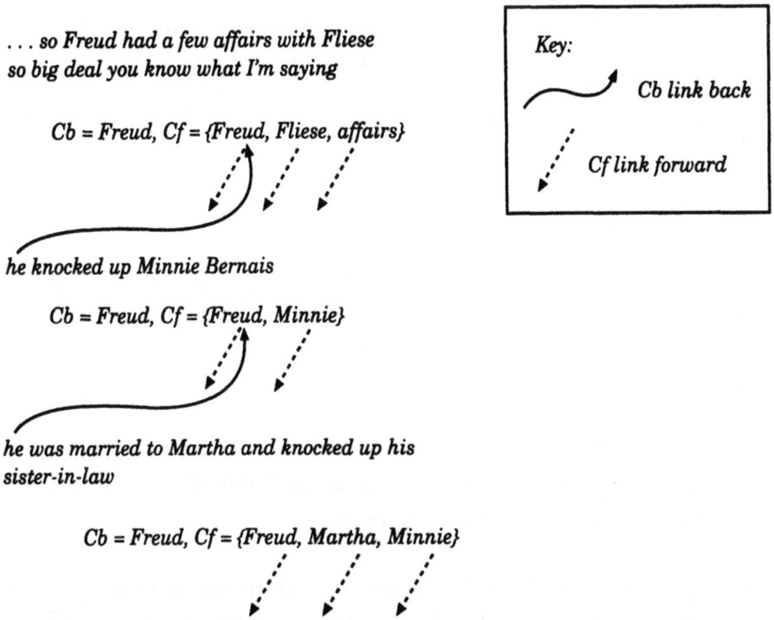

FIGURE 6.2. Illustration of local focussing (centering) mechanisms.

6.3 Accent and Attentional Modelling

The meaning of intonational prominence in discourse is a well-studied problem. However, the specific association between accent placement and Grosz and Sidner's attentional state model was first theorized by Hirschberg and Pierrehumbert [HP86]. They claimed that accent placement must be determined based on the local discourse context or segment in which the accentable item occurs. In later work, they proposed that intonational prominence marks information that the speaker intends to be predicated, or added to the mutual beliefs held between speaker and hearer [PH90b]. Since their work, Cahn [Cah90] and Kameyama [Kam94] examined the role of pitch accent in centering algorithms. These later studies concentrated on the problem of accented pronouns and proposed principles to account for their processing from a theoretical perspective.

This paper builds on Hirschberg and Pierrehumbert's original proposal and provides a more general treatment of the role of intonational prominence in discourse processing than do the studies on accented pronouns. It defines detailed interactions among intonational prominence and other linguistic factors known to contribute to the determination of discourse salience. Further, the principles and algorithms developed in this paper derive from corpus analysis. The corpus consisted of 20 min of unrestricted spontaneous narrative monologue, from an American male speaker.[1] While the distributional and linguistic analyses of intonational prominence in that corpus are reported in detail elsewhere [Nak96], the major findings are summarized in the following section.

6.3.1 *Principles*

Results of the narrative study motivated the adoption of several principles governing the discourse processing of spoken referring expressions.[2] The first concerns the role of lexical form in the attentional processing of referring expressions.

(1) The lexical form of a referring expression indicates the level of attentional processing, i.e., pronouns involve local focussing while full lexical forms involve global focussing.

This principle was formulated in earlier work on centering theory [GJW83] and was strongly supported by data in the narrative study. It reflects the primacy of lexical form of referring expression in determining attentional status (cf. [Bro83, GHZ89, Pri88]).

However, lexical form is not the sole determinant of attentional status. Research in linguistic pragmatics has indicated that subject position tends to hold given information, while non-subject positions tend to hold new information [Pri88]. Early work on centering theory [GJW83, Kam85] has shown that grammatical function contributes to the ranking of the Cf list as well as to the identification of the Cb. This leads to the formulation of the second principle.

(2) The grammatical function of a referring expression reflects the local attentional status of the referent, i.e., subject position generally holds the highest ranking member of the Cf list, while direct object hold the next highest rank in the Cf list.

The third and final principle treats intonational prominence and represents an original contribution of the narrative study.

---

[1]The spontaneous narrative monologue was collected by Virginia Merlini for the purpose of studying American gay male speech and was made available by Mark Liberman at the University of Pennsylvania Phonetics Laboratory. The analysis of 200 referring expressions serves as the basis for this paper.

[2]For the purposes of this paper, the term full lexical form refers to the class of proper names, definite noun phrases, and indefinite noun phrases. The term pronominal refers to third person pronouns. Items considered intonationally prominent bear H* or complex pitch accents in Pierrehumbert's system of American English intonation [Pie80].

(3) The intonational prominence of a referring expression serves as an inference cue to shift attention to a new Cb, or to mark the global (re)introduction of a referent; non-prominence serves as an inference cue to maintain attentional focus on the Cb, Cf members, or global referents.

In some cases, a shift in attention is associated with a push and/or pop of a focus space. While previous research has shown a relationship between intonational prominence and discourse salience, the narrative study used attentional state modelling to distinguish between salience at the global and local levels of discourse structure. The conclusions of this study were that accent serves as an inference cue, marking the preservation of and changes in attentional status, and that accent function must be interpreted against a dynamic background of the attentional state model as well as additional linguistic factors, following the aforementioned principles for lexical form, grammatical function, and intonational prominence.

6.3.2 Algorithms

The results of the narrative study suggest new and more specific ways in which accent information contributes to language understanding. These results are incorporated in two algorithms, one for processing full lexical forms shown in Figure 6.3 and the other for processing pronominal expressions shown in Figure 6.4. The algorithms, which can be viewed as two parts of a single whole, reflect the primacy of lexical form in determining the discourse processing of referring expressions. The different cases of referring expressions (e.g., pronominal intonationally prominent subject) receive different treatments with respect to referent search and update of the attentional state. The input that is assumed for the algorithm consists of a referring expression, marked for lexical form, intonational prominence, and grammatical function; and the current attentional state, represented by the focus stack at the global level and the Cb and the Cf list at the local level. The output of the algorithm is an updated attentional state, with referential indices capturing the referential act of the processed expression. Recent focus spaces include the linearly preceding segment as well as the neighboring focus space on the focus stack.

To show how to apply the algorithms, we consider a few examples of accented subject pronouns that occur in the narrative. The first example appears in Figure 6.1. The pronoun *he* in the last line of text ("alright HE was human too alright") bears a H* prominence. To interpret this pronoun, we follow the processing steps for prominent subject pronouns given in Figure 6.4. The Cb of the previous utterance is *Minnie* and the Cf list is {*Minnie abortion*}. The first test thus fails due to the gender conflict between the pronoun *he* and the Cb of the previous utterance. The second test also fails because the Cf list of the previous utterance does not

If expression is full lexical form, then

(1) If expression is subject, then

If expression is intonationally prominent, then

(a) create new discourse entity e_i;
(b) push new focus space;
(c) add entity e_i and information predicated of it to new focus space;
(d) add entity e_i to Cf list as Cp.

If expression is intonationally non-prominent, then

(a) search for referent e_i in immediate focus space or recent focus spaces, checking the parallel subject referent of the previous utterance first;
(b) add entity e_i and information predicated of it to immediate focus space (if not already there);
(c) add entity e_i to Cf list as Cp.

(2) If expression is object, then

If expression is intonationally prominent, then

(a) create new discourse entity e_i;
(b) add entity e_i and information predicated of it to immediate focus space;
(c) add entity e_i to Cf list after entity realized by subject.

If expression is intonationally non-prominent, then

(a) search for referent e_i in immediate focus space or recent focus spaces;
(b) add entity e_i and information predicated of it to immediate focus space (if not already there);
(c) add entity to Cf list after entity realized by subject.

FIGURE 6.3. Attentional state processing for full lexical forms.

contain a male referent. So the third clause is executed. The focus space for Segment B is popped and the pronoun refers to the entity that is the Cb of the immediately prior utterance in Segment A, namely *Freud*. Finally, the local focusing structures are updated, with *Freud* at the head of the Cf list.

Two more examples of H* accented subject pronouns occur in the following context:

> *They all put FREUD on a pedestal*
> *HE is an icon okay*
> *HE can do no wrong*

The first accented pronoun, in "HE is an icon," exemplifies the second case in the processing of prominent subject pronouns. The Cf list of the prior utterance contains *Freud*, but not as the Cb. So, the pronoun refers to an entity already in local focus but not primarily so. Prominence on the first *HE* shifts attention to *Freud* as the central character in a subsegment that elaborates on the first utterance in this sequence. A new focus space is pushed for this subsegment and *Freud* is entered as the new Cp. The next pronoun, in "HE can do no wrong", refers to the Cb of the previous utterance. As suggested by the first clause in the processing of prominent subject pronouns, the context can be viewed as emphatic (corroborated by an increase in acoustic energy in this case). Indeed, analysis of the intentional structure shows that the asserted proposition, that Freud is considered infallible, is central to the speaker's argumentation in this story and is expressed at several different points in the narrative.

6.4 Related Work

Kameyama, in a study focussing on the phenomenon of accented pronouns, developed the complementarity hypothesis [Kam94]. The major claims of this work are that there are underlying preferences for the hypothetical unaccented version of a pronoun, and that these preferences can be used to derive the interpretation preferences for a given accented pronoun. In particular, attentional constraints apply during the computation of the interpretation preferences for the hypothetical unaccented pronoun. Once these preferences are determined, the preferences for the accented pronoun are simply computed by reversing the original preference list of the hypothetical unaccented version. This processing, it is important to note, takes place in Kameyama's framework of total pragmatics. This means that there are additional semantic, pragmatic, and world knowledge constraints contributing to the processing. While this richness is lacking in the proposals in this paper, it is clear that accent information plays a fundamentally different role in Kameyama's account. In the relevant component of her algorithm, accenting information is considered subsequent to attentional constraints. In contrast, the current algorithms incorporate accent information directly into attentional state modelling at the same stage of processing as syntactic and lexical information.

Terken and Nooteboom conducted psycholinguistic experiments to test the role of intonational prominence in discourse processing by humans

If expression is pronominal form, then

(1) If expression is subject, then

If expression is intonationally prominent, then

(a) if expression refers to Cb of previous utterance, then check for emphatic or contrastive context and add entity e_i to current Cf list as Cp;

(b) else, if Cf list of previous utterance contains referent e_i, then push new focus space and add entity e_i to current Cf list as Cp;

(c) else, pop focus space and check coreference of entity e_i and Cb of new immediate focus space; add entity e_i to current Cf list as Cp.

If expression is intonational non-prominent, then

(a) if expression refers to Cb, or other member of Cf list of previous utterance, then add referent e_i to current Cf list as Cp;

(b) else, if referent e_i is not in Cf list, then pop focus space and check coreference of entity e_i and Cb of new immediate focus space; add entity e_i to current Cf list as Cp.

(2) If expression is object, then

If expression is intonationally prominent, then

(a) search Cf list of previous utterance for referent e_i;

(b) add entity e_i to current Cf list after entity realized by subject;

(c) check for contrast relationship with parallel object referent, or inferable entity, in previous utterance.

If expression is intonationally non-prominent, then

(a) search Cf list of previous utterance for referent e_i;

(b) add entity to current Cf list after entity realized by subject.

FIGURE 6.4. Attentional state processing for pronominal forms.

[TN87]. They concluded that reference resolution proceeds differently for accented and unaccented expressions, namely that listeners assume that an unaccented expression refers to a member of a "restricted set of activated entities" in the discourse context, while the interpretation of an accented expression is not constrained in this manner [TN87, p. 148]. They further hypothesized that mental representations of entities were constructed differently for accented and unaccented referring expressions. They proposed that representations were built bottom-up for accented items, allowing the hearer to use the content of the expression to resolve the reference. In contrast, unaccented expressions were said to be resolved top-down, by taking as candidates for reference resolution the restricted set of activated entities. These two hypotheses contradict each other in the cases of accented pronominal forms and unaccented full forms (Jacques Terken, personal communication). That is, for an accented pronoun, the limited semantic content of the pronoun makes reference resolution by strictly bottom-up processing implausible. Similarly, for unaccented full forms, the use of semantically contentful lexical items is unnecessary (and unexplained) if reference resolution proceeds strictly top-down.

The spirit of Terken and Nooteboom's proposals is maintained in the proposed algorithms, while the remaining contradictions are removed by considering two levels of attention, each providing a distinct set of activated entities for reference resolution. For full forms, the relevant set of activated entities is formalized in terms of the structured contents of the global focus stack. For pronominal expressions, the relevant set of restricted entities is formally cast in terms of the centering constructs computed at the local level of discourse processing. That is, the lexical content of a referring expression signals its level of salience, while accentual prominence conveys further shades of given/newness within the appropriate focussing structures.

Conclusion

The algorithms presented in this paper integrate prominence information into attentional state processing at the global and local levels, following principles that nevertheless generalize across both levels of discourse structure. Intonational prominence serves to mark new entities in either local and global focus, while non-prominence signals that the associated entity is to be maintained in either local or global focus.

To build on the findings of this study, one could investigate additional linguistic factors in relation to prominence, as well as additional aspects of prominence itself. We are focussing on the following research problems. First, sparse data in the narrative corpus did not allow for thorough analyses of the cases of prominent object pronouns and non-prominent subject full forms. Thus, the treatment of these cases is based as much on observations in the literature as on the corpus analysis. It is hypothesized that the phenomena of contrastive and emphatic accent, as well as special effects deriving from parallel syntactic conditions, occur in these configurations. Similarly, we hypothesize that emphatic or contrastive contexts license prominence marking for subject pronouns that refer to the Cb of the previous utterance instead of shifting attention to a new Cb. Second, whether pitch accent type further specifies the discourse functions of prominence is a long-standing yet intriguing question that should be empirically examined. The narrative monologue contained too few tokens with complex pitch accents to draw general conclusions. Third, further study is needed to determine when hierarchical and not simply linear discourse segmental structure makes a difference to prominence interpretation. Finally, there are more factors to be studied and more fine grained distinctions to be made among classes of referring expressions. For example, surface order is one possible factor influencing accentuation. Preliminary investigation on a new multi-speaker corpus of spontaneous task-oriented direction-giving monologues suggests that both intonational phrase position and surface order position affect accent placement less strongly than do grammatical function and lexical form. In the new corpus as well, differences in the distributions of prominence arise between proper names and full noun phrases. As the number of factors integrated into discourse processing algorithms increases, a better understanding of the relationship of prosody to various linguistic structures will emerge.

Acknowledgments

I am indebted to Barbara Grosz and Julia Hirschberg for numerous stimulating and helpful discussions on this research, as well as to Jacques Terken and participants in the April 1995 ATR International Workshop on Computing Prosody for useful feedback at an earlier stage of this work. Thanks also to Barbara Grosz for suggestions for improving this manuscript and to Alan Capil for editorial expertise. This research was partially funded by a National Science Foundation (NSF) Graduate Research Fellowship and NSF Grant Nos. IRI 94-04756, CDA 94-01024, and IRI 93-08173 at Harvard University.

References

[Alt87] B. Altenberg. *Prosodic Patterns in Spoken English: Studies in the Correlation between Prosody and Grammar for Text-to-Speech Conversion.* Lund: Lund University Press, 1987.

[Bro83] G. Brown. Prosodic structure and the given/new distinction. In A. Cutler and D. R. Ladd, editors, *Prosody: Models and Measurements*, pp. 67–78. Berlin: Springer, 1983.

[Cah90] J. Cahn. *The effect of intonation on pronoun referent resolution.* MIT Media Lab, Cambridge MA, 1990. (unpublished manuscript).

[Fuc84] A. Fuchs. "Deaccenting" and "default accent." In D. Gibbon and H. Richter, editors, *Intonation, Accent and Rhythm*, pp. 134–164. Berlin: Walter deGruyter, 1984.

[GGG93] P. C. Gordon, B. J. Grosz, and L. A. Gilliom. Pronouns, names, and the centering of attention in discourse. *Cognitive Science*, 17:311–347, 1993.

[GHZ89] J. Gundel, N. Hedberg, and R. Zacharski. Givenness, implicature and demonstrative expressions in English discourse. *Chicago Linguistics Society 25. Parasession on Language in Context*, pp. 89–103, 1989.

[GJW83] B. J. Grosz, A. K. Joshi, and S. Weinstein. Providing a unified account of definite noun phrases in discourse. *Proceedings of the 21st Annual Meeting of the Association for Computational Linguistics*, 1983.

[GJW95] B. J. Grosz, A. K. Joshi, and S. Weinstein. Centering: a framework for modelling the local coherence of discourse. *Computational Linguistics*, 21(2), 1995.

[Gro77] B. J. Grosz. The representation and use of focus in dialogue understanding. *Technical Report 151, SRI International, Menlo Park, CA*, 1977.

[GS86] B. Grosz and C. Sidner. Attention, intentions, and the structure of discourse. *Computational Linguistics*, 12:175–204, 1986.

[HD88] S. B. Hudson-D'Zmura. *The structure of discourse and anaphor resolution: the discourse center and the roles of nouns and pronouns.* Ph.D. thesis, University of Rochester, Rochester, NY, 1988.

[Hir93a] J. Hirschberg. Pitch accent in context: Predicting prominence from text. *Artificial Intelligence*, 63:305–340, 1993.

[HP86] J. Hirschberg and J. Pierrehumbert. The intonational structuring of discourse. *Proceedings of the 24th Annual Meeting of the Association for Computational Linguistics*, pp. 136–144, 1986.

[JW81] A. K. Joshi and S. Weinstein. Control of inference: Role of some aspects of discourse structure centering. *Proceedings of International Joint Conference on Artificial Intelligence*, pp. 385–387, 1981.

[Kam85] M. Kameyama. *Zero anaphora: the case of Japanese.* Ph.D. thesis, Stanford University, Palo Alto, CA, 1985.

[Kam94] M. Kameyama. Stressed and unstressed pronouns: complementary preferences. *Proceedings of the FOCUS and NLP Conference*, 1994.

[Nak96] C. Nakatani. Discourse structural constraints on accent in narrative. In In J. P. H. van Santen, R. Sproat, J. Olive, and J. Hirschberg, editors, *Progress in Speech Synthesis.* New York: Springer-Verlag, 1997.

[PH90b] J. Pierrehumbert and J. Hirschberg. The meaning of intonational contours in the interpretation of discourse. In P. Cohen, J. Morgan, and M. Pollack, editors, *Intentions in Communication.* Cambridge, MA: MIT Press, 1990.

[Pie80] J. B. Pierrehumbert. *The phonology and phonetics of english intonation.* Ph.D. thesis, Massachusetts Institute of Technology, Distributed by the Indiana University Linguistics Club, 1980.

[Pri88] E. Prince. The ZPG letter: Subjects, definiteness, and information-status. In S. Thompson and W. Mann, editors, *Discourse Description: Diverse Analyses of a Fund Raising Text.* Amsterdam: Elsevier Science, 1988.

[Sid79] C. Sidner. Toward a computational theory of definite anaphora comprehension in English. *MIT Technical Report AI-TR-537*, 1979.

[Ter84] J. M. B. Terken. The distribution of pitch accents in instructions as a function of discourse structure. *Language & Speech*, 27:269–289, 1984.

[TH94] J. Terken and J. Hirschberg. Deaccentuation and persistence of grammatical function and surface position. *Language & Speech*, 37:125–145, 1994.

[TN87] J. Terken and S. G. Nooteboom. Opposite effects of accentuation and deaccentuation on verification latencies for given and new information. *Language & Cognitive Processes*, 2:145–163, 1987.

7
Prosodic Features of Utterances in Task-Oriented Dialogues

Shin'ya Nakajima
Hajime Tsukada

ABSTRACT This paper describes some characteristic patterns and prosodic features of utterances in task-oriented cooperative dialogues. To analyse utterances in natural conversation systematically, we first define some basic analysis units in conversation and taxonomies for dialogue structure. We then analyse how well prosodic features can be correlated with dialogue structures. The results show that by controling the pitch range, speakers indicate the topic structure and utterance perspective. We then investigate two major patterns of utterances: supportive utterances preceding the main utterance (pre-supportive pattern), and subordinative utterances following the main utterance (post-supportive pattern). The results show that in the topic-shifting context, the speaker tends to use the pre-supportive pattern, while in the topic-continuation context, post-supportive patterns are mainly used. Moreover, in topic-shifting dialogues, the supportive utterances are shorter, and acknowledgment/confirmation exchange is made more frequently than the case of topic-continuation dialogues. Finally, we discuss how these results can be utilized for prosodic parameter generation.

7.1 Introduction

The speech generation process plays an important role in a speech dialogue system. Specifically when the system needs to convey a complicated message to the user, it is important to formulate the utterances so that the user can understand them easily, and to monitor how well the user understands each message. The final goal of this work is to develop a speech dialogue system which achieves these points. As a first step towards this goal, we investigate some characteristics of the utterance patterns in task-oriented dialogues.

7.2 Speech Data Collection

As the task for speech dialogue collection, we used a *divided maze problem*; a maze is divided into two pieces and each is given to a participant. The two participants have to find, in cooperation, the path from start to goal only by speech communication. The maze is selected from [Phi76] and includes three-dimensional roads, stairs, and castles. Trolls are placed on some roads to indicate that the road cannot be traversed in the direction against the troll.

We collected a Japanese speech dialogue of 45 min duration as uttered by two participants (a male and a female players). The both speakers speak in Tokyo dialect.

7.3 Framework for Analysis

Since grammatical units such as *sentences* are absent in spontaneous conversations, we must first determine what is the basic unit of conversation to analyse the discourse structure systematically. We refer to this unit as the **utterance unit (UU)**, and it was defined in Nakajima and Allen [NA93b]. It can be described briefly as follows:

Utterance unit (UU): an utterance that is bounded by pauses or intonational phrase resettings.

In the task-oriended dialogues like the one used here, where acknowledgments and interruptions are frequently made by the participants, the speaker normally makes a number of utterances to achieve a single communication goal. We refer to the sequence of utterances to achieve a single communication goal as the **dialogue unit (DU)**. (The notion of dialogue unit was proposed in Allen *et al.* [AS91]. However, DU proposed here is a more global unit than the one defined in [AS91].) The dialogue unit can be defined as follows:

Dialogue unit (DU): a sequence of UUs which is made by one or both participants to achieve a single communication goal.

In other words, a DU begins with an informing or requesting utterance and finishes when the interlocuter perfectly understands that information or request. Each DU must include one or more informing/requesting utterances. It may also include confirmation/acknowledgment utterances.

An example of speech dialogue analysis based on UU and DU is shown in Figure 7.1. The communication goal of this example is "inform the other player that the right most stairs can be reached (in his piece of maze)." (The utterances enclosed by "[]" are UUs. The dialogue examples shown in the rest of this paper are all extracted from our speech database.)

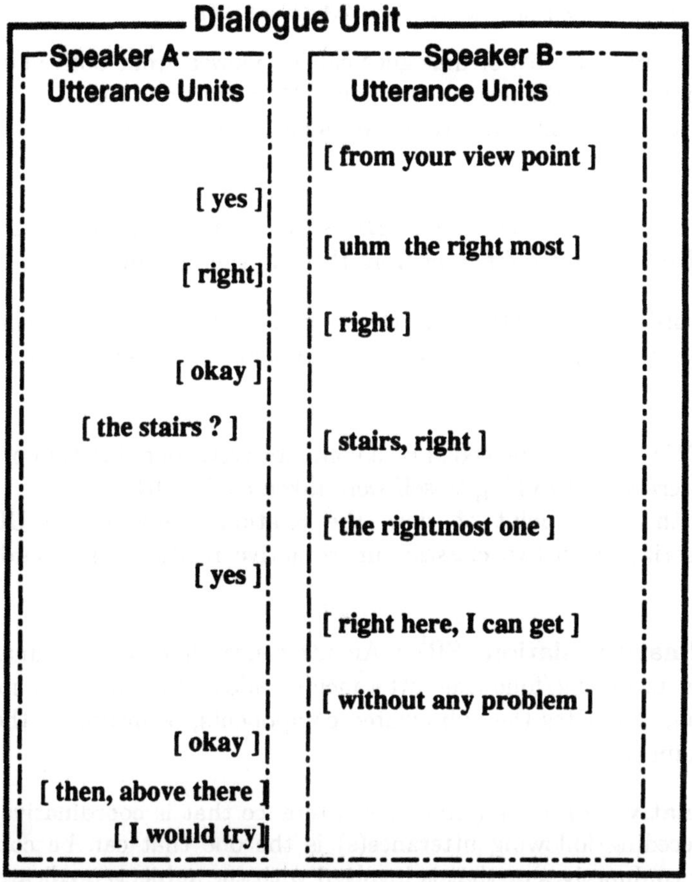

FIGURE 7.1. An example of speech dialogue analysis

Our speech data included some utterances whose communication goals were ambiguous or unclear, and they were excluded from the following analysis.

7.4 Topic Structure and Utterance Pattern

This section investigates the correlation between the structural aspects of dialogue such as topic shifting, and prosodic features. In the following subsection, first we define the terms for topic shifting and the relationship between utterances. We then analyse how prosodic features can be correlated with these structural aspects. Finally, we discuss the relation between topic shifting and the utterance patterns/durations.

7.4.1 Topic Shifting and Utterance Relation

The structural features of dialogues can be viewed from both global and local points. From the former viewpoint, DUs can be categorized into two classes: topic shift DU and topic continuation DU. These are defined as follows:

Topic shift DU (TS DU) : a TS DU is the DU in which the objects being talked about are different from those in the previous DU.

Topic continuation DU (TC DU) : a TC DU is the DU where the objects being talked about are the same as those in the previous DU.

The DUs that were preceded by a comparatively long duration of silence or by utterances of talking-to-self were taken as TS DUs.

From the local point of view, the relations between the UUs can be categorized into two classes: subordinative relation and coordinative relation.

Subordinative relation (SR) : An utterance that is subordinative to the preceding/following utterance(s) adds some relevant information, completes the abbreviated components, or introduce the backgrounds.

Coordinative relation (CR) : An utterance that is coordinative to the preceding/following utterance(s) is the one that can be connected by conjunctives or that is involved with the subject/predicate of the whole sentence.

These relations are depicted in Fig. 7.1.

7.4.2 Dialogue Structure and Pitch Contour

This section discusses the correlation between the structural aspects of dialogue and the prosodic features by using the above notions. As the prosodic features, we analysed the onset and maximal peak of pitch contour (F0). Hereafter, the former is refered to as F_o, and the latter is called F_p.

The speech database consists of 1405 UUs and of these, 712 UUs were uttered by the male participant. The following analysis used 302 of his UUs.

Figure 7.2 shows F_p and F_o average values of topic shift DU, topic continuation DU, coordinative UU, and subordinative UU. (The first UU in each DU was used to determine F_p / F_o values.)

As can be seen in the figure, the highest F_p and F_o values occur in topic shift DU, while subordinative UU yields the lowest values. The values of coordinative UU and topic continuation DU are almost identical and

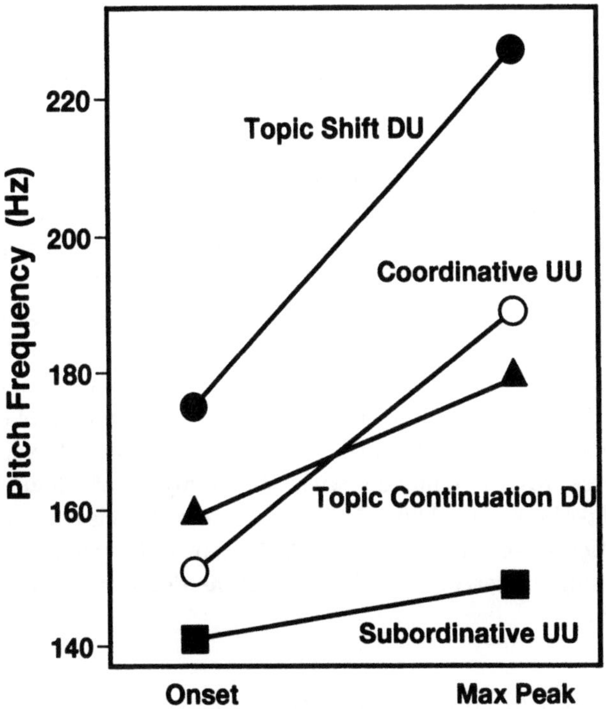

FIGURE 7.2.

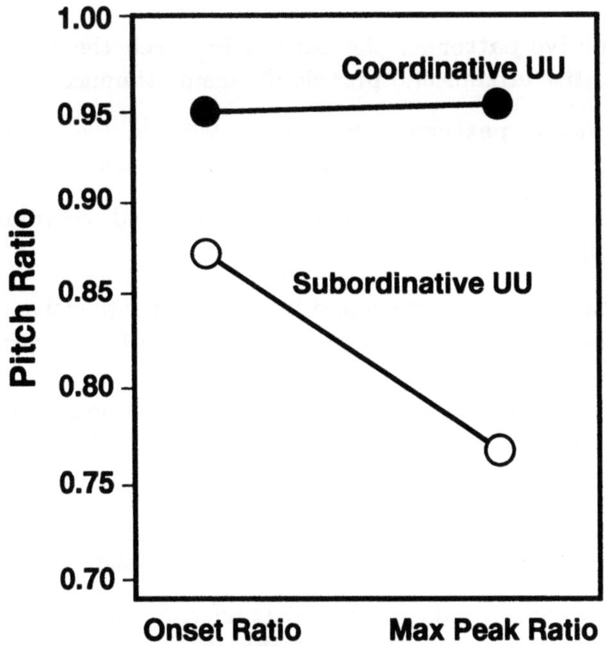

FIGURE 7.3.

roughly speaking, they split the values of topic shift DU and subordinative UU.

These results suggest that; speakers use higher F0 values to indicate to the hearer that the current topic differs from the the previous utterances. Comparatively lower F0 values indicate that the current utterance is subordinative to the previous/following utterance(s).

To make the second characteristic clearer, we analysed the ratio of the current UU's F_p (or F_o) to that of the previous one. Hereafter, the ratio is refered to F_p (or F_o) declination ratio. Figure 7.3 shows the average declination ratios of F_p and F_o; hereafter called R_p and R_o, respectively, of coordinative/subordinative UUs. As shown in the figure, R_p and R_o of suboordinative UU are both smaller than those of coordinative UU. In particular, R_p of subordinative UU is much lower than that of coordinative UU. This result coincides with the interpretation given above.

Since these results coincide with the analysis of English dialogues made by Nakajima and Allen [NA92], they can be viewed as the general characteristics of conversational speech. Moreover, they also are compatible with the results of read sentences prosody as determined by Hakoda and Sato [HS80].

7.4.3 Topic Shifting and Utterance Pattern

In terms of the flow of information conveyed by the UUs, the utterance sequence in the DUs can be classified into the following three patterns:

Pre-supportive pattern : the pattern in which the subordinative or supportive utterance(s) precede the main utterance.

Post-supportive pattern : the pattern wherein the subordinative or supportive utterance(s) follow the main utterance.

Continuous pattern : the pattern of continued coordinative utterance(s).

This section focuses on pre-supportive and post-supportive patterns and discusses their functions and the relation between the patterns and topic shifting.

The characteristics of supportive patterns in questioning utterances were discussed by Ishikawa *et al.* [IT92]. From our speech dialogues, the functions of the supportive utterances in pre-supportive patterns can be summarized as

(a) to introduce background or motivation of the main utterance;

(b) to more precisely indicate the object refered to in the main utterance, or guide the hearer's focus to the object refered to in the main utterance;

(c) to presummarize the contents of the main utterance.

Examples containing these three types of functions are shown in examples 1-(a), (b), and (c), respectively.

From the self-repairing point of view, Ohtsuka *et al.* [OO93] studied the post-supporive patterns, viewing them as the incremental elaboration processes. The major functions of the supportive utterances in post-supportive patterns can be summarized as follows:

(a) Add some relevant information fragment, or complete the abbreviated objects in the main utterance.

(b) Elaborate or clarify ambiguous phrases or descriptions in the main utterance.

(c) State the same semantic content in other words.

Examples of these three types of utterances are shown in examples 2-(a), (b), and (c), respectively.

Example 7.1: Supportive utterance examples in the pre-supportive patterns; The original Japanese utterances are shown in parentheses.

(a)

H: [one more thing I'd like to ask you] (moo hitotsu kikitainogaa) [is. . .] (desunee) {Main question follows}

A: [yes] (hai) [yes] (hai)

(b)

H: [you can see the point where the map is cut vertically, can't you?] (Mannakade kou tateni kireterutokoro arimasuyone)

A: [yes] (hai)

H: [beneath that point, you can find . . .] (sono shitano houni) {Leading hearer's focus continues}

(c)

H: [The paths I can go through, in my map,] (kochirakara ikeru michiga zenbude) [are 4 in total] (4-hon arundakedomo)

A: [hnn-hnn] (hai) [yes] (hai)

H: [the first one is. . .] (soregaa. . .) {The speaker explains each path}

Example 7.2: Supportive utterance examples in the post-supportive patterns.

(a) A: [the third one] [the third one from the right edge]

(b) A: [from that point, a little bit,] [about 1.5 cm,] [downward, can you find a staircase?]

(c) A: [at the lowest hanging point] [in your map, at the most dented point]

TABLE 7.1. Topic shifting and utterance pattern.

	PreP	PosP	ContP
Topic Shift DU	14	5 (-2)	4
Topic Cont DU	1	4	7

PreP: pre-supportive pattern, PosP: post-supportive pattern, ContP: continuation

In the rest of this section, we discuss the relation between the utterance patterns and topic shifting. Table 7.1 shows the number of pre-/post-supportive pattern occurences in the topic-shift and topic-continuation DUs. In this analysis only the outmost patterns (when represented in a tree structure) are counted; i.e., the inner or embedded patterns are excluded.

As can be seen in the table, the pre-supportive patterns occur most frequently in the topic shift DUs, while the topic continuation DUs generally use post-supportive or continuation patterns. This result can be interpreted as follows.

In topic shifting contexts, the speaker introduces a new object which may be located at a complicated point on the map, or, sometimes changes the mode of speech (for instance, from "informing" to "questioning"). Thus, to assist the hearer's understanding, he first utters some supportive utterances that introduce part or all of the backgrounds of the main utterance, or more precisely identify the location of the object that he wants to talk about in the main utterance.

In fact, the speaker used five post-supportive patterns in the topic shifting contexts. Two of them led to the hearer misunderstanding the speaker. That is, these two cases resulted in *communication failure.*

7.4.4 Topic Shifting and Utterance Duration

This subsection discusses topic shifting and utterance duration. Figure 7.4 shows the UU duration histogram of topic shift and topic continuation DUs. (The frequency of each column is normalized by the total number and is plotted as percentages.)

As can be seen in the figure, the frequency distribution of the UU duration in topic shift DUs has two major peaks; the one at the short range

(0.6–0.8 s) and the other at the longer range (1.6–2.0 s). Topic shift DUs, on the other hand, have one major peak around 1.4 s and the frequency declines abruptly for durations shorter than 0.8 s.

This result can be interpreted as follows. In the topic shift contexts, the speaker (or informer) tends to produce shorter utterances with short pauses in the pre-supportive phase, to prompt the hearer (or receiver) to make acknowledgment or confirmation. In other words, in the earlier phase of topic shift context, the dialogue participants have to build up a shared world view, block by block. As this view develops, the informer can produce longer utterances and no longer has to prompt for acknowledgment at short intervals.

To clarify this point, we analysed the average duration of the supportive UUs in topic shift DUs and that of the UUs in topic continuation DUs. We also investigated the duration of the UU sequences at the end of which the interlocuter actually made acknowledgment/confirmation utterances (hereafter refered to as *Acknowledged Segment* (AS)). The results are shown in Table 7.2. The corresponding measurements in terms of syntactic phrases (*bun-setsu* in Japanese) are shown in Table 7.3.

As can be seen in these tables, in the pre-supportive phase of topic shift DUs, both UU and AS are shorter, measured by duration or number of syntactic phrases. These results support the conclusions given above.

Another point to note here is that, the difference between the topic shift and the topic continuation contexts are greater when measured by syntactic phrase number than by duration; the syntactic phrase ratio increases by 27% for topic continuation context, while that in terms of duration, is just 11%.

This fact means that in the topic continuation context, the speaker tends to have a higher speech rate (1.64 phrases per UU) than in the topic shift context (1.45 phrases per UU) and this suggests the existence of *fixed-length conversation rhythm*.

TABLE 7.2. UU/AS duration average in topic shift and topic continuation DUs [s].

	TS DU (ratio)	TC DU (ratio)
UU duration	1.41 (1.00)	1.57 (1.11)
AS duration	2.06 (1.00)	2.30 (1.11)

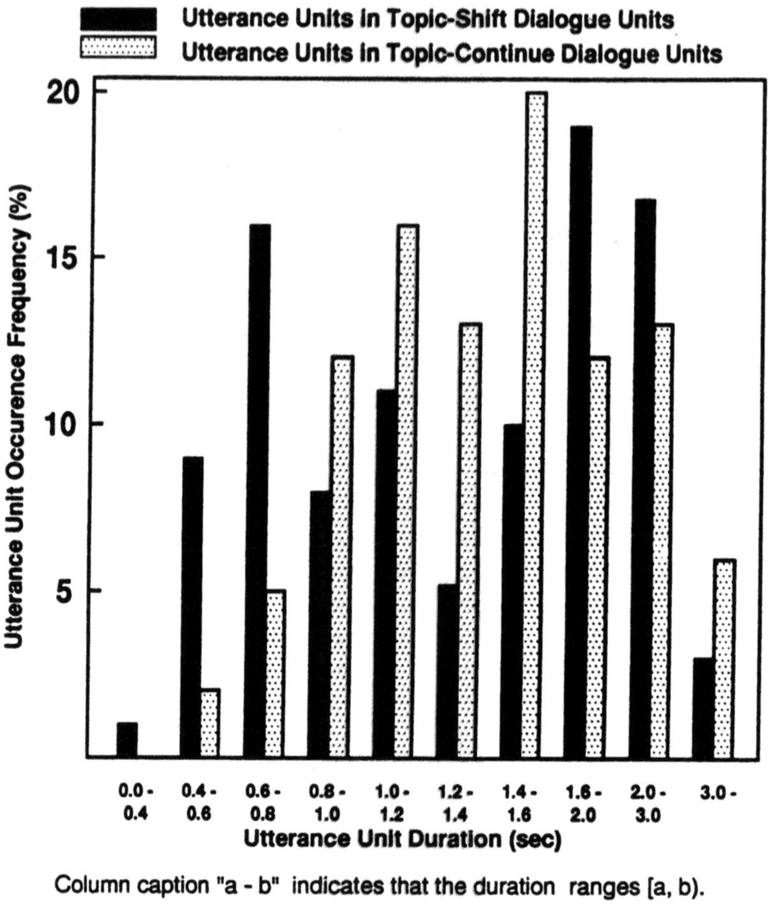

FIGURE 7.4. UU duration histogram: topic shift DU and topic continuation DU.

TABLE 7.3. Number of syntactic phrases in UU/AS of topic shift and topic continuation DU.

	TS DU (ratio)	TC DU (ratio)
Phrases / UU	2.04 (1.00)	2.58 (1.26)
Phrases / AS	2.98 (1.00)	3.78 (1.27)

7.5 Summary and Application

In this section, we first summarize the results given in the previous section. We then discusses the application of prosodic parameter generation based on the results given in Sec. 4.2.

7.5.1 Summary of Results

The results described above can be summarized as follows.

(1) Dialogue structure and F0 contour:

(a) Onset and maximal peak F0 values are higher at topic shift DUs, lower at suboordinative UUs, and medial at topic continuation DUs and coordinative UUs.

(b) Maximal peak declination ratio is much lower at sub-oordinative UUs (0.75) than at coordinative UUs (0.95).

(2) Utterance pattern and topic structure:

(a) Utterance patterns can be classified into two major types: the pre-supportive pattern and the post-supportive pattern.

(b) In the topic shifting contexts, speakers prefer to produce pre-supportive patterns, while in the topic continuation contexts, post-supportive patterns are mainly used.

(c) UU durations of the topic shift DUs are shorter than that of topic continuation DUs. This prompts for the hearers' acknowledgment / confirmation.

In the following subsection, we propose prosodic parameter generation rules based on result (1).

7.5.2 Prosodic Parameter Generation

As for prosodic parameter generation, we focus on the peak F0 assignment rules for a set of UUs which are produced to establish a single communication goal. The set of UUs is here refered to as *the utterance block* (UB). In the peak assignment algorithm, four constant values should be pre-determined: the maximal peak F0 value in topic shift DU (FP TS), that in topic continuation DU (FP TC), the peak declination ratio in subordinative utterance (R SUB), and that in coordinative utterance (R COR). All these constants can be determined via the prosodic parameter analysis given in Sec. 4.

(1) Determine the representative FP value of UB (FP_{UB}): If the context is topic shift, set FP_{UB} to FP TS. Otherwise, set FP_{UB} to FP TC.

(2) Determine FP of each UU (FP_{UU}) in the UB:

(a) If the UB consists of a single UU, set FP_{UU} to FP_{UB}, and terminate the procedure.

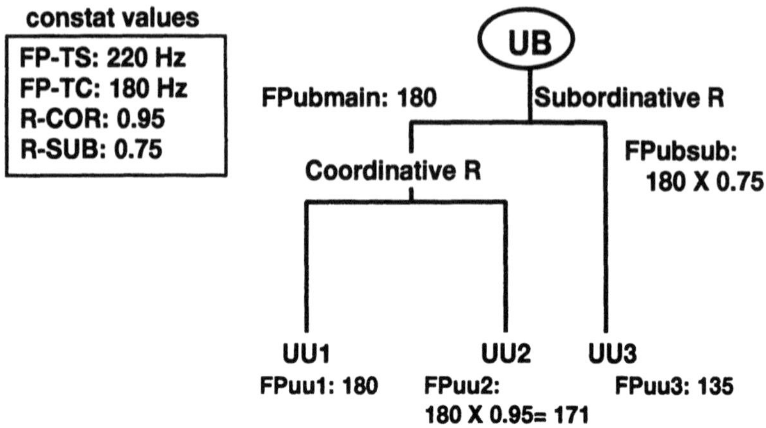

FIGURE 7.5. An Example of Peak F0 Assignment

(b) If the UB includes two coordinative sub-UBs (UB1 and UB2), set FP_{UB1} to FP_{UB}, and set FP_{UB2} to $FP_{UB} \times$ R COR. Then apply this algorithm to UB1 and UB2 recursively.

(c) If the UB includes a main sub-UB (UBmain) and a sub-ordinative sub-UB (UBsub), assign FP_{UB} to FP_{UBmain} and set FP_{UBsub} to $FP_{UB} \times$ R-SUB. Then apply this algorithm to UBsub and UBmain recursively.

In the above algorithm, we assume that a UB includes two sub-UBs at most. However, it is straightforward to generalize two sub-UBs to N. An application example of this algorithm is given in Figure 7.5.

Conclusion

In task oriented natural conversation, the speaker uses prosodic features to convey the topical/relational structure of utterances. Speech with higher F0 range indicates topic shifting and lower F0 range suggests sub-ordination of the utterances. From the results of prosodic analysis, we developed a maximal peak F0 assignment algorithm.

In topic shift contexts, the speaker tends to use the pre-supportive patterns, in which utterances are comparatively shorter, while in the topic continuation contexts, post-supportive patterns are mainly used.

References

[AS91] J. F. Allen and L. K. Schubert. An overview of the TRAINS project. *Proceedings of the Third International Forum on the Frontier of Telecommunications Technology*, 1991.

[HS80] K. Hakota and H. Sato. Prosodic rules in connected speech synthesis. *Trans. IECE Japan*, J63-D(9):715–722, 1980. (in Japanese).

[IT92] Y. Ishikawa and T.Kato. Generation rule for dialogue inquiry. *Proceedings of the 6th Meeting of JSAI*, 1992. (in Japanese).

[NA92] S. Nakajima and J. F. Allen. Prosody as a cue for discourse structure. *Proceedings of the International Conference on Spoken Language Processing, Banff, Canada*, pp. 425–428, 1992.

[NA93b] S. Nakajima and J. F. Allen. A study on prosody and discourse structure in cooperative dialogues. *Phonetica*, 50:197–210, 1993.

[OO93] H. Ohtsuka and M. Okada. Self-repairing phenomena in spontaneous speech. *Proc Spring Meeting Acoustical Society of Japan*, pp. 2–4–14, 1993.

[Phi76] D. Phillips. *Graphic and Op-art Mazes*. New York: Dover, 1976.

8

Variation of Accent Prominence within the Phrase: Models and Spontaneous Speech Data

Jacques Terken

ABSTRACT
Various models have been proposed to account for judgments of the relative prominence of pitch accents in relation to $F0$ variation. Two topics are addressed in this paper. The first topic is how pitch accents need to be realized in order to obtain the appropriate prominence patterns. In order to answer this question, relevant data and models for prominence perception are summarized. It is tentatively concluded that the prominence associated with $F0$ peaks is judged relative to the local $F0$ range, as signalled by the pitch at utterance onset. No model for prominence perception proposed so far can account for the available data, and more insight is needed into the issue of pitch range estimation before real progress can be made. The second topic concerns the assumption of free gradient variability underlying models of prominence perception: it is assumed that the prominence associated with pitch accents may vary freely and in a gradient way from accent to accent within the phrase. Prominence ratings collected for fragments of spontaneous speech provide no evidence of a constraint prohibiting such variation. Some implications of these findings are considered.

8.1 Introduction

The notion of "prosodic prominence" concerns the acoustic properties by which certain elements in the speech stream are perceived to stand out from their neighboring elements. The concerned properties are duration, amplitude, $F0$[1], and spectral characteristics, e.g., vowel quality.

[1] Throughout this paper, I will use the term $F0$ as a shorthand for the acoustic property of which *pitch* is the perceptual correlate. If T is the length of the interval between two successive excitation pulses in seconds, then we define $F0$ as $1/T$; i.e., $F0$ is the property shown graphically as the output of pitch or $F0$ extraction algorithms. In cases where the distinction between the acoustic and perceptual perspective is irrelevant, the term "pitch" will be used.

Phonological treatments have agreed on at least two types of prominence for Germanic languages such as English and Dutch. The notion of *stress* is needed to account for the difference between noun-verb pairs such as *pérmit-permít.* The acoustic correlates of stress are duration, amplitude, and vowel quality. The notion of *accent* is needed to account for differences between expressions such as *i have plans to LEAVE* and *i have PLANS to leave* (the location of accents is indicated by capitalization). In order to simplify matters, it may be said that the primary phonetic correlate of accent as distinguished from stress is a clearly perceptible pitch change in or near the accented syllable. (And, needless to say, in isolation the distinction between *pérmit* and *permít* will also involve a difference in accentuation.)

In addition, there is evidence that speakers may produce finer gradations in prominence by varying the realization of accents. For instance, the realization of the accents on *Anna* and *Manny* in the utterance *Anna came with Manny* shows systematic differences depending on the context [LP84]. However, attempts to establish phonological categories for such finer shades of prominence have failed so far. Some of the intuitions involved have been accounted for as the outcome of syntagmatic prominence relations (strong-weak relations), but in general the adequate treatment of prominence at the phrase level is the subject of ongoing debate [Lad94]. Thus, as observed above, only two categories of prominence are agreed upon: stress and accent; the use of the term "prominence" is usually understood in relation to these categorical distinctions, and finer gradations are treated as intra-categorical variation. As a result, these finer gradations, especially those involving accentuation, have been neglected, or explained away as reflecting paralinguistic influences and therefore not belonging to the subject domain of phonology. However, precisely these finer gradations within the "accent" category will be the focus of this paper, and I will use the expression "variation of accent prominence" to make this explicit. So far, it remains a puzzle how prominence patterns involving variation in accent prominence should be characterized, what are acceptable prominence patterns, and which function they serve, i.e., how they relate to the linguistic and paralinguistic properties of messages.

The issue has become more urgent due to developments in speech synthesis, where it has become obvious that the generation of adequate prominence patterns is important for the acceptability of synthetic speech. Synthetic speech sounds dull or even unacceptable without adequate modelling of prominence, and listeners may be distracted by the inappropriate prosody. In addition, the different kinds of prominence in human speech help the listener to process the information in the speech signal both at the lexical, semantic, and pragmatic level. Therefore, inadequate modelling of prosodic prominence in synthetic speech will slow down the listener's comprehension of the incoming message.

In this paper, two main questions are addressed. In the first place it is asked how perceived variations in accent prominence relate to $F0$ variation.

In the second place, it is asked whether there exist restrictions on acceptable prominence patters; more specifically, we ask whether variation in accent prominence within the phrase is in any way constrained (the question concerning the factors governing the assignment of accents has been dealt with elsewhere, e.g., [Hir92, HJ91, Ter94], and will not be addressed here). The structure of the paper will be as follows. In the first section, I will discuss evidence from experiments on the perception of accent prominence, and summarize existing models. We shall see that most experiments and models assume that prominence may vary freely and in a gradient way from accent to accent. In the second section, I will present an experiment which tests the validity of this assumption with respect to spontaneous speech.

8.2 *F*0 and Variation of Accent Prominence

The line of research to be summarized here has started with the experiments reported in [Pie79], which focussed on the relation between *F*0 variation and pitch judgments for accent peaks in utterances containing two accented syllables. However, it will be more convenient to look first at a number of experiments investigating the relation between *F*0 and prominence in single-accent utterances.

8.2.1 Intrinsic Prominence of Single Accents

The notion of prominence in single-accent utterances can be intuitively understood in relation to emphasis: speakers can easily speak a word at different degrees of emphasis when asked to do so [LP84]. Likewise, listeners can easily express their judgment about the degree of prominence for such words on a rating scale [RG93]. Other experiments have used pairwise comparison [RG85] or adjustment techniques [HR94, HvG91]. In the latter case, listeners adjusted the *F*0 peak of the accented syllable in a test stimulus in a number of steps, until it was judged to carry the same prominence as the accented syllable in a reference stimulus.

The outcomes of the experiments conducted so far agree in showing that the prominence associated with an accented syllable is proportional to the size of the *F*0 change: greater *F*0 changes tend to elicit higher prominence ratings.

This outcome raises several further questions. First, there is the question of which points in the *F*0 contour are used to compute the magnitude of perceived changes in pitch.

A simple answer is that listeners compute the size of pitch changes on the basis of the peaks and valleys in the utterance. However, the low points in the contour may vary considerably due to extraneous factors such as the temporal distance between pitch peaks.

In order to overcome this problem, several models of intonation have adopted the notion of a baseline, i.e., a line connecting the local $F0$ minima, to provide a more stable basis for estimating the low points, e.g., [FH84, tHCC90]. The magnitude of a perceived change in pitch can then be expressed by taking the distance between the $F0$ maximum and the value of the virtual baseline at that time. But one may question whether this procedure adequately models the listener's behavior. For instance, most experiments employ stimuli with synthesized $F0$ contours in which the baseline is clearly defined. However, in utterances taken from natural speech, the local $F0$ minima do not neatly fall on a straight line, so that the baseline is not directly observable from $F0$ measurements. Therefore, it is unclear whether listeners can reliably compute the course of the baseline in natural speech.

So far, only a few experiments have been devoted to the question of how accurately the baseline can be perceived, or to the question of what is the relevant information for estimating the course of the baseline. Unpublished data suggest that frequency variations for local $F0$ minima are less easily detected than variations for local maxima [Slu95a]. Furthermore, it was found that the position of the utterance-final low (the end frequency) did not affect prominence judgments [RG93]. On the other hand, it was found that the prominence associated with a given rise-fall accent decreased as the onset frequency increased [HR94, RG93], although there was an effect of onset length: raising the onset affected the perceived prominence associated with peaks only if the onset was sufficiently long (approximately 400-500 ms); with short onsets (below 200 ms) there was no effect of onset height [RG93].

Alternative models have suggested that perceived prominence is a function of the distance between the tonal targets associated with an $F0$ change (Ladd and Morton, personal communication) or the distance between the pitch levels associated with successive vowel nuclei [HR94]. However, these accounts do not explain why no effect of onset $F0$ was found in [RG93] for early peaks, as one might expect that the initial syllable contains sufficient information for the listener to determine the pitch of the tonal target or the vowel nucleus.

At present, the best guess seems that the listener uses the pitch at utterance onset to determine the relative position of the contour in the speaker's overall $F0$ range, and that this information is used in some, as yet unidentified way, to evaluate the position of the peak. That is, it appears that the position of the peak is evaluated relative to the position of the current contour in the overall range of the speaker, as estimated from the onset pitch: a high onset suggests that the current contour is high in the speaker's overall range, and a low onset suggests that the current contour is low in the speaker's overall range. This seems plausible since the position of the contour in the overall range of the speaker may vary due to factors such as the emotional state of the speaker or to discourse structure.

In this way, within-phrase prominence variation can be separated from register shifts across phrases. It appears evident that, as the listener has more onset information available, his estimate of the position of the current contour in the overall range of the speaker will become more accurate. It may be assumed that as the estimate of this relative position becomes more reliable, it provides a better basis for judging the prominence associated with $F0$ peaks.

A further question concerns the frequency scale to be used for expressing the magnitude of $F0$ changes. Since female speakers on average speak at higher frequencies than male speakers, comparisons of typical sizes of $F0$ changes in female and male speakers are strongly affected by the scale one uses: a linear (Hz), logarithmic (semi-tone) or yet another scale. In this respect, the outcomes of different experiments are less consistent: whereas [RG85] provides evidence for the adequacy of the linear (Hz) scale, the results of [HvG91] seem to support an ERB scale representation (which was not included in the computations in [RG85]).

8.2.2 Relative Prominence of Successive Accents

The outcomes of the experiments with single-peak utterances cannot simply be generalized to utterances containing several accents: phonetic and phonological factors such as declination and downstep preclude a direct mapping of the size of $F0$ changes onto perceived prominence [GR88, Ter91, Ter94]. Also, the relation between $F0$ and the perceived prominence for the second peak (P2) is influenced by the frequency of the first peak (P1) [GR88, LVJ94]. It is still unclear how these influences can be integrated into a single model. Below, I will summarize some relevant findings.

Further evidence about the role of local range variation (i.e., the distance between the peaks and valleys in a contour) and the role of the baseline comes from a number of experiments which have used adjustment or judgment methods. In these experiments, listeners were required to adjust the frequency of one peak relative to the other, or to judge whether the two peaks are associated with equal prominence [Ler84, Pie79, Ter91]. In general, the following findings seem to be fairly well established. In the first place, an increase in P1 frequency should be accompanied by an increase in P2 frequency in order to maintain equal prominence. In the second place, the second peak (P2) should have a lower frequency than the first peak (P1) in order to be judged as giving equal prominence. In the third place, this difference between P1 and P2 appears to increase as a function of P1 frequency. From these findings it may be concluded that perceived prominence increases as the peaks get higher in the speaker's range, and that the relevant range is wider near utterance onset than near the end.

In most of these experiments the slope and position of the baseline were kept constant, and raising the peak implied increasing the size of the $F0$

change. Subsequent studies [RRT94], in which P1 frequency and the slope of the baseline were varied independently, showed that the baseline slope by itself influenced the relative prominence of P1 and P2 only to a small extent: even with considerable changes in the slope of the baseline only small adjustments in P2 frequency were needed with a given P1 frequency to maintain equal prominence. Since changes in the slope of the baseline affected the size of the $F0$ change for P1 much more than for P2[2], it appears that distance to the objective baseline (i.e., defined in terms of actual $F0$ minima) is not a relevant factor in determining perceived prominence, or at least is not very important.

Hence, an alternative interpretation may be given, along the lines proposed by [RG93]. It appears that relative prominence is judged on the basis of the $F0$ information contained in the contour, but that the position of the contour in the overall range of the speaker, as estimated on the basis of the onset pitch is taken into consideration. In cases where the onset is very short, listeners may have difficulty to estimate the actual size of a pitch change, and they may assume some default pitch level as a reference to compute a size for P1. Even though this interpretation is very speculative, it may be noticed that it supposes that the actual size of an $F0$ change may not be the only factor affecting prominence, but that $F0$ peaks are interpreted relative to the position of the current contour in the overall range of the speaker. This raises a new set of questions regarding the adequate representation of $F0$ range, and about the way in which listeners perceive range variation within and across speakers.

Evidence about the existence of peak-to-peak effects was obtained in a set of experiments using rating techniques [GR88, LVJ94]. In this paradigm, listeners rate the prominence of individual accents on a scale, usually a ten-point scale. The ratings may then be used to identify what kind of prominence differences can be perceived, and how adjacent accents affect each others' perceived prominence. There is some evidence from these experiments that variation in the height of P1 affects the perceived prominence associated with a given P2. Furthermore, it has been suggested that the direction of the effect may depend on the height of P2. In an experiment reported in [LVJ94] it was found that an increase in the height of P1 resulted in an increase in the prominence associated with P2, but only for low P2; for high P2, an increase in P1 height resulted in a decrease of the prominence associated with P2. However, the reliability of these findings needs to be further established: the sizes of the effects vary considerably across experiments; in addition, the findings for high P2 in [LVJ94] were inconsistent with those in [GR88], where the effect for high P2 was in the

[2]In the experiments, the end frequency of the baseline was held constant while varying the slope, to do justice to the observation that utterance end frequency is a very stable speaker characteristic.

same direction as for low P2. Evidently, as long as the inconsistency has not been cleared up, it makes no sense to look for an explanation. Both sets of results agree, however, in that they provide a further complication of the relation between $F0$ variation and perceived prominence.

8.2.3 Discussion

Before drawing conclusions from these findings, a methodological point must be made. Most experiments conducted so far have been done with rather simple utterances, containing one or two peaks. With respect to double-peak utterances, it remains unclear whether the findings will generalize to all cases of adjacent accents, or whether they reflect a special effect of the first peak which would affect all further accents in a multi-accent utterance. Also, in most experimental stimuli the peak in single-peak utterances and the first peak in double-peak utterances are usually close to the utterance onset, which may impede the listener's ability to accurately estimate the pitch of the fragment preceding the first peak, relative to which the peak frequency might be interpreted [RG93]. Thus, more experiments are needed to obtain a more complete view. As long as the relevant data are lacking, a comprehensive model remains beyond reach. With these restrictions in mind, the following conclusions may be drawn from the findings obtained so far.

In the first place, the height of $F0$ peaks is an important determinant of perceived prominence, more so than the distance between the $F0$ peak and some base level. However, as mentioned before, this interpretation may be restricted to situations where the first accent is close to the utterance onset. Still, a general model should apply also to these situations.

In the second place, these peaks seem to be evaluated in relation to the position of the current contour in the overall range of the speaker. That is, listeners appear to able to make a fairly good estimate of the $F0$ range available to or exploited by a speaker, and to derive quite specific phonetic expectations about where $F0$ peaks and valleys should be in the overall $F0$ range in different situations. Several production studies have shown that the $F0$ characteristics of utterances are quite consistent and replicable within and across speakers [LP84, GR88, dBGR92], and it seems quite plausible that listeners have learned to exploit such regularities.

Of course, this view raises many questions. Most important, there are questions as to the adequate description of $F0$ range in speech, and to the sorts of information that are relevant to listeners for estimating the $F0$ range. Several models have been proposed recently, which agree in describing $F0$ range variation in terms of a two-component model, e.g., [Lad90, dBGR92, Ter93]: one component captures local $F0$ range variation, i.e., the distance between $F0$ maxima and minima; the other component captures the relation between the local $F0$ range and the overall range available to the speaker, which is represented by a lower reference level.

Once an adequate descriptive model is available, we may collect data on $F0$ range variation in different speaking modes and investigate how listeners may exploit such knowledge in interpreting the $F0$ characteristics of actual utterances.

In the second place, a more comprehensive account is needed of peak-to-peak effects on $F0$ range variation, and of their effects on the prominence judgments by the listeners. For instance, as has been mentioned before, we do not know whether the influence of the pitch of one peak on the judged prominence of a subsequent peak is really a peak-to-peak effect or is an effect associated with the initial peak.

8.3 Variation of Accent Prominence in Spontaneous Speech

8.3.1 Introduction

The findings reviewed so far support a model of prominence perception in which the prominence associated with accented syllables may vary for each accent independently in a gradient way (this assumption has been named the free gradient hypothesis [Lad94]). However, it appears that such a model makes quite strong assumptions about the discriminative and interpretative abilities of the listeners. Numerous factors can be thought of (and have indeed been suggested) that may potentially influence relative prominence of accents, including syntactic factors such as grammatical class and grammatical function, semantic factors, and pragmatic factors. If accent prominence may vary freely, it would mean that all factors can at the same time influence the prominence of individual accents. It can be easily seen that this poses considerable problems to the listener. Thus, the assumption of free gradient variability may not be very realistic.

For illustration, let us look at a phrase contour containing two accents, where the pitch of the second accent peak is lower than that of the first peak. In a model which allows prominence to vary within the phrase, this contour is compatible with two different interpretations. In one interpretation, the two accents signal equal emphasis but the pitch of the second peak is lower than that of the first peak because downstep has applied. In the other interpretation, no downstep has applied and the lower pitch of the second peak signals less emphasis. In other words, there would be no unique solution for this $F0$ contour. Although it cannot be excluded that such ambiguities indeed occur and that listeners solve them by taking into consideration information from other sources such as syntax and semantics, the situation seems unattractive in view of all the potential influences on $F0$ variation. So, it needs to be determined whether the assumption of free gradient variability is realistic at all. Further investigations are needed to identify potential constraints on variation in relative prominence. The existence of such constraints would help the listener to make sense of the acoustic information.

In the remainder, I will test a strong version of the hypothesis that prominence variation within the phrase is constrained: the null hypothesis to be tested is that the accented words within a phrase are not allowed to vary with respect to prominence. (Note that this would imply that if the speaker wants to pronounce an accented word with more or less prominence than the preceding accented word, he can do so only by inserting a phrase boundary in between.) That is, here we address the issue of free variability. The issue of gradient variability is discussed elsewhere [Lad94].

In order to test the hypothesis that accented words within a phrase do not vary with respect to prominence, I analysed materials from a set of elicited spontaneous monologues. The materials were taken from a set of task-oriented monologues, which had been collected for a different purpose [Ter84]. The task involved subjects giving instructions to listeners as to how to put together a two-dimensional front view of a house from a set of ready-made cardboard pieces. Each instruction consisted of a number of phrases instructing a listener what to do with one of the pieces in the set. For illustration, a monologue segment is shown below. For orientation: the speaker has already instructed the listener to lay down a large black square serving as the front. The segment starts at the point where the speaker tells the listener to take a piece with "living room window" written on its back. (The experimental situation was set up such that the speaker could neither see nor hear the listener.)

> Then we take the living room window
> We turn this colored side up
> and lay it bottom left
> leaving some space below it
> so that the long side runs parallel to the bottom of the house.

When listening to the instruction monologues, I got the impression that prominence did not vary strongly within phrases; in line with the null hypothesis, the variation in prominence seemed to be much larger across phrases than within. Thus, it appeared that variation in prominence to express degrees of emphasis is primarily realized by varying the $F0$ range for a whole phrase; impressionistically, one might conclude that, once the $F0$ range for a phrase has been fixed, variation in $F0$ range for individual accents within a phrase no longer relates to prominence variation, but instead reflects the effect of phonological factors such as downstep. In order to get more reliable prominence judgments, a perceptual rating task was conducted to determine whether indeed prominence varies little or not within phrases containing multiple accents.

8.3.2 Method

The instruction monologues of two speakers (one male, one female) were segmented into utterance-like units on the basis of content and melodic and temporal criteria (boundary tones, pause, and final lengthening). For each speaker, 12 fragments were selected to be presented in the rating task (a full list is given in the Appendix). The location of accented words was determined on the basis of a formal intonational analysis, following the description of Dutch intonation in [tHCC90]. In all, the phrases for speaker A contained 40 accented words, those for speaker B 38 accented words.

Sixteen listeners with backgrounds in speech, hearing, and linguistics (native Dutch speakers) were asked to rate the prominence of the accented syllables on a ten-point scale, with "1" for "no prominence" and "10" for "strong prominence." Judges listened to each fragment and wrote their prominence ratings for the accented words on answer forms containing the written versions of the fragments. The words to be rated were marked by underlining in the written texts. Fragments were presented in scrambled order per speaker, with different orders for different listeners. Stimuli were presented through headphones at a comfortable loudness level. Listeners were allowed to listen to each fragment as often as desired. The task took between 10 and 15 min.

The choice for this method was based on two considerations. (1) Other methods such as ranking prominence are difficult for the listener if the fragments contain larger number of accents, and do not provide information about the sizes of the differences in relative prominence, if any. (2) By obtaining prominence ratings from a panel of listeners and taking the mean ratings more reliable estimates are obtained than if just a single judge is used. In fact, this method has been used successfully in different investigations, both with respect to prominence rating and rating the strength of prosodic boundaries [RG85, LVJ94, PS94].

The rating task was restricted to accented words only, since the question under investigation was about the difference in prominence between words containing pitch accents.

8.3.3 Data Analysis

Twelve of the 24 fragments contained an intonation phrase boundary (as determined by melodic criteria). Since the main question under investigation concerns the presence or absence of prominence variation *within* the phrase, including these fragments as wholes in the analyses might bias the results towards an affirmative outcome (i.e., that prominence may indeed vary from accent to accent). Therefore, the 24 fragments were segmented into intonation phrases (in the Appendix, the location of major phrase boundaries within fragments is marked by %), and the intonation phrase instead of the fragment was taken as the domain of analysis. In this way, 26 intonation phrases containing two or more accents were obtained to answer the question whether there is prominence variation within the phrase. Thirteen phrases contained two accents, nine contained three accents, three contained four accents, and one contained five accents.

8.3.4 Results and Discussion

Mean prominence judgments for the individual target words (k=78) are given in the Appendix (means are taken over the ratings of 16 judges). The agreement between judges was assessed by computing the mean of the pairwise correlations between judges, and was found to be 0.50. All pairwise correlations between judges were significant at the 0.05 level, 90% were significant at the 0.0001 level. Accordingly, it may be concluded that judges behaved uniformly, and that the mean ratings provide reliable estimates of the accent prominence of individual words. Of course, this should not conceal the fact that listeners may disagree about which word has highest prominence in a particular phrase, especially if the difference in prominence is relatively small. This observation fits in with conclusions of earlier work [Cur80, Cur81, tH81], and appears to support the method of working with a panel of judges rather than with individual judgments. However, it also implies that the communicative function of gradient variability is constrained, in that only fairly large differences will be perceived consistently in the same way by all listeners.

The main question under investigation concerns the presence or absence of prominence variation for accented words within a phrase. Since there were phrases containing two, three, four, and five accents, respectively, the data for the four types of phrases (2-accent, 3-accent, 4-accent, and 5-accent phrases) were entered into separate analyses of variance, one for each type, to deal with the unequal number of accented words per phrase. In each analysis phrases and accents were taken as fixed factors and listeners as the replication factor, in a repeated-measurements design (since the 5-accent class contained only one phrase, the factor "phrases" was dropped in the corresponding analysis).

In order to obtain an indication of the size of the difference in accent prominence that may be considered significant (i.e., where we may conclude that there is indeed a difference in accent prominence between two accented words within a phrase), Tukey's honestly significant difference (HSD) was calculated, which is computed from the studentized range statistic Q and the error mean square [FT89]. The error mean square was obtained from a one-way analysis of variance on the complete set of prominence ratings, with words ($k = 78$) as a fixed factor and judges as the replication factor, and amounted to 1.51 (the effect of "words" was significant at the .00005 level, $F_{77,1155} = 15.1$). The estimated Q statistic for $k = 78$ for $\alpha = .01$ equalled 8.7 (it could not be determined exactly since table values are given only for values of k up to 15, but it was estimated by extrapolating from the table values). On the basis of these values, a critical difference was obtained of 0.7, so that a difference in rated accented prominence of more than 0.7 between accented words within a phrase will be taken to be significant.

The effects of primary interest in the separate analyses of variance are the main effect of "words", and the interaction between words and phrases. A significant "word" effect would provide clear evidence that the prominence of accented words in a phrase may vary. However, in itself it would not provide sufficient evidence for unconstrained variation, as it might reflect systematic effects related to word position or some other factor. Therefore, the interactions are of greater importance. If the interactions are significant, it means that the differences in prominence between accented words within phrases are not uniform across phrases (in the extreme case, the interaction may be significant in the absence of a significant effect of the words factor, if different phrases show opposite prominence patterns).

The effect of words and the interaction between words and phrases were both significant at the .0005 level in all cases but one. For the 2-accent phrases, $F_{1,15} = 6.61$ ($p = .02$) and $F_{12,180} = 43.12$, for the words factor and the phrases x words interaction, respectively. For the 3-accent phrases, $F_{2,30} = 34.49$ and $F_{16,240} = 12.46$ for words and the interaction, respectively. For 4-accent phrases $F_{3,45} = 14.81$ and $F_{6,90} = 7.99$ for words and phrases x words, respectively. For the 5-accent phrase, $F_{4,60} = 10.6$ for the words effect (as explained above, there was only one 5-accent phrase, so there was no interaction term in this case).

With these results we conclude in the first place that the null hypothesis (i.e., that all accented words within a phrase should have the same accent prominence) is rejected: accented words within a phrase may differ with respect to their prominence. Indeed, in 22 out of 26 major phrases containing two or more accents, the difference between the largest and smallest prominence is larger than Tukey's HSD of 0.7.

Furthermore, the differences are not constant for different phrases. Not only is the size of the difference in prominence variable, but also its sign: in some phrases the most prominent accent is the first one in the phrase, in other phrases it is the last one. Thus, accent prominence is not constrained in such a way that the second accent should always be a certain amount less prominent than the first accent, and the third a constant amount less prominent than the second, and so forth. Of course, this does not necessarily imply that the variation is fully unpredictable. For instance, certain syntactic or semantic properties might be typically associated with particular degrees of prominence, so that the prominence patterns would be predictable from the syntactic or semantic properties of the phrase. However, this issue is beyond the scope of the present paper, and a much larger corpus would be needed to address the question concerned. In general, however, it appears that there is no phonological constraint which would prohibit or constrain variation of accent prominence within the phrase, and which would facilitate the listener's task of interpreting the prominence pattern posed by the accents in the phrase.

In a first attempt to establish an association between judged prominence variation and acoustic variation, the $F0$ maxima in the accented syllables to be judged were measured. The Appendix lists for each target word the mean prominence and the associated $F0$ peak. For some words, occurring in phrase-final position and containing a pre-boundary rise, no clear accent-related $F0$ maximum could be determined. For these words the Appendix gives two $F0$ points, the $F0$ at the amplitude maximum and the $F0$ max. Product-moment correlation coefficients were computed between the $F0$ and prominence data, for the male and female speaker separately, excluding the cases containing the pre-boundary rises. Correlation coefficients were 0.51 and 0.71 for the male and female speaker, respectively. Thus, there is a clear trend for higher $F0$ peaks to be associated with higher prominence, as shown in Figure 8.1. However, as might be expected the relation is far from perfect. As mentioned in earlier sections, both from production and perception studies with read-aloud speech and isolated utterances it is well-known that there are other factors in addition to $F0$ peak height which affect prominence judgments: position in the utterance, vowel or syllable duration, vowel identity, phrasal pitch range, and so on. In addition, non-phonological, e.g., semantic, factors may also play a role in prominence judgments. Clearly, further investigations are required. But at least the current exploratory study has shown that the outcomes of such investigations are relevant to modelling the perception of prosodic variation not only for experimental stimuli presented in the laboratory but also for spontaneous speech.

8.3.5 Limitations

A word needs to be said here about the limitations of the current study. As outlined in Sec. 8.3.1, there are many potential influences on the height of pitch peaks, and the listener's task to interpret a particular sequence of pitch peaks is facilitated to the extent that different influences cannot co-occur. In particular, it was outlined that paralinguistic factors such as information value and the phonological property of downstep may interfere.

The rationale of the current study was that prominence judgments might be used to establish the existence of such cooccurrence restrictions. In particular, it was assumed that prominence judgments might be used to determine whether there exists a constraint that prohibits successive accents within the phrase to differ in prominence. The reasoning was that, in that case, variation in the height of pitch peaks within the phrase could not reflect prominence variation, and therefore might be interpreted unambiguously in terms of phonological and phonetic properties such as downstep and declination. However, this reasoning is valid only if non-downstepped and downstepped accents are judged to be equally prominent.

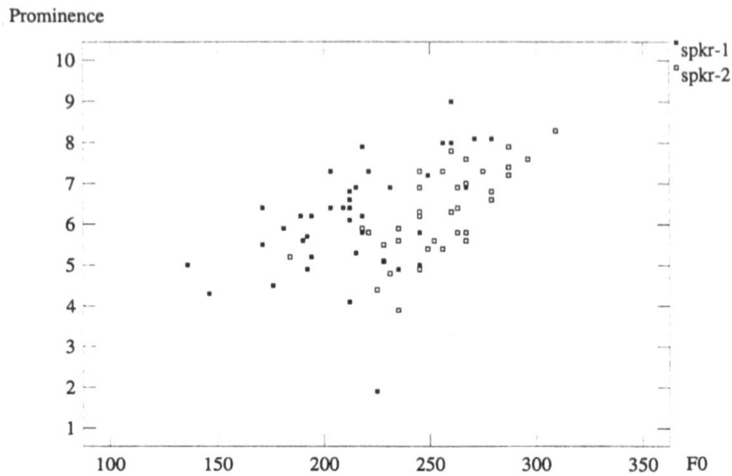

FIGURE 8.1. Mean prominence ratings for accented target words on a 10-point scale, as a function of $F0$ peak height, for speaker 1 (male) and speaker 2 (female).

That is, it is based on the assumption that the judges assigned prominence ratings to accented words after phonological interpretation rather than before. This assumption may not be valid, however, and listeners may indeed have assigned different prominence ratings for downstepped accents than for non-downstepped ones, *ceteris paribus*. The validity of this assumption therefore needs to be assessed in further investigations.

Nevertheless, the finding that both the size and the direction of a difference in accent prominence within the phrase may vary, shows that the pattern of results cannot be explained in terms of downstep only: the application of downstep would reduce the prominence associated with an accent by a fixed amount and always in the same direction, since downstep is assumed to be constant and to operate from left to right. Thus, the current results are compatible with the interpretation that the variation in the height of pitch peaks reflects the influence of many factors operating simultaneously. This in turn implies that the interpretation of the pattern of pitch peaks by the listener is an integral part of the activities making up the speech understanding process. Otherwise, it is hard to see how potential ambiguities might be solved in an efficient way.

Conclusion

The first part of this paper started from the observation that the term "prominence" is used in two different ways. In the first place there is the phonological hierarchy of discrete prominence categories such as reduced, un-reduced, stressed, and accented. In the second place, there is more gradient variation in the prominence of accented syllables, for instance in relation to the magnitude of $F0$ changes. The main part of the paper focused on the second kind of prominence, and addressed the question of how the perceived prominence for a given pitch accent might be predicted from information about phonetic characteristics, in particular variation in $F0$. The results of perception experiments that were summarized led to the conclusion that we do not yet completely understand how perceived prominence varies with $F0$ variation. Also, it became clear that the models implied by perception studies make strong assumptions about the discriminative and interpretative powers of the listeners.

One of these assumptions is that prominence can vary freely and in a gradient way from accent to accent in a phrase containing multiple accents. This assumption was addressed in a small study described in the second part, involving a prominence rating task for fragments taken from spontaneous speech. The outcomes supported the assumption that the speaker may indeed assign different degrees of accent prominence within a phrase. This finding rules out a constraint which prohibits variation of accent prominence within the phrase, which would urge the speaker to keep accent prominence within the phrase at a constant level and to start a new phrase each time he wanted to bring about a change in prominence. The potential ambiguities which arise at the level of prosodic structure due to the absence of such a constraint, make it likely that prosodic information is already at an early stage of the comprehension process integrated with other sources information, such as lexical and syntactic, in order to allow the listener to arrive at the interpretation of the message in an efficient way.

References

[Cur80] K. L. Currie. An initial search for tonics. *Language and Speech*, 23:329–350, 1980.

[Cur81] K. L. Currie. Further experiments in the search for tonics. *Language and Speech*, 24:1–28, 1981.

[dBGR92] R. Van den Berg, C. Gussenhoven, and A. C. M. Rietveld. Downstep in Dutch: Implications for a model. In G. J. Docherty and D. R. Ladd, editors, *Papers in Laboratory Phonology*

II: Gesture, Segment, Prosody, pp. 335–359. Cambridge, UK: Cambridge University Press, 1992.

[FH84] H. Fujisaki and K. Hirose. Analysis of voice fundamental frequency contours for declarative sentences of Japanese. *J. Acoust. Soc. Japan (E)*, 5:233–242, 1984.

[FT89] G. A. Ferguson and Y. Takane. *Statistical Analysis in Psychology and Education*. New York: McGraw-Hill, 1989.

[GR88] C. Gussenhoven and A. C. M. Rietveld. Fundamental frequency declination in Dutch: testing three hypotheses. *Journal of Phonetics*, 16:355–369, 1988.

[Hir92] J. Hirschberg. Using discourse context to guide pitch accent decisions in synthetic speech. In G. Bailly, C. Benoît, and T. R. Sawallis, editors, *Talking Machines: Theories, Models, and Designs*, pp. 367–376. Amsterdam: Elsevier Science, B. V, 1992.

[HJ91] M. Horne and C. Johanson. Lexical structure and accenting in English and Swedish restricted texts. *Working Papers*, 1991.

[HR94] D. J. Hermes and H. H. Rump. Perception of prominence in speech intonation induced by rising and falling pitch movements. *J. Acoust. Soc. Am.*, 96:83–92, 1994.

[HvG91] D. J. Hermes and J. C. van Gestel. The frequency scale of speech intonation. *J. Acoust. Soc. Am.*, 90:97–102, 1991.

[Lad90] D. R. Ladd. Metrical representation of pitch register. In J. Kingston and M. E. Beckman, editors, *Papers in Laboratory Phonology I: Between the Grammar and Physics of Speech*, pp. 35–57. Cambridge, UK: Cambridge University Press, 1990.

[Lad94] D. R. Ladd. Constraints on the gradient variablity of pitch range (or Pitch Level 4 Lives!). In P. A. Keating, editor, *Phonological structure and phonetic form. Papers in Laboratory Phonology III*, pp. 43–63. Cambridge, UK: Cambridge University Press, 1994.

[Ler84] L. Leroy. The psychological reality of fundamental frequency declination. *Antwerp Papers in Linguistics*, 40, 1984.

[LP84] M. Liberman and J. Pierrehumbert. Intonational invariance under changes in pitch range and length. In M. Aronoff and R. Oehrle, editors, *Language Sound Structure*. Cambridge: MIT Press, 1984.

[LVJ94] D. R. Ladd, J. Verhoeven, and K. Jacobs. Influence of adjacent pitch accents on each other's perceived prominence, two contradictory effects. *Journal of Phonetics*, 22:87–99, 1994.

[Pie79] J. Pierrehumbert. The perception of fundamental frequency declination. *J. Acoust. Soc. Am.*, 66:363–369, 1979.

[PS94] J. R. De Pijper and A. A. Sanderman. On the perceptual strength of prosodic boundaries and its relation to suprasegmental cues. *J. Acoust. Soc. Am.*, 96:2037–2047, 1994.

[RG85] A. C. M. Rietveld and C. Gussenhoven. On the relation between pitch excursion size and prominence. *Journal of Phonetics*, 13:299–308, 1985.

[RG93] A. C. M. Rietveld and C. Gussenhoven. Scaling prominence. *Proceedings Dept. of Language and Speech*, 16/17:86–90, 1992/1993.

[RRT94] B. H. Repp, H. H. Rump, and J. Terken. Relative perceptual prominence of fundamental frequency peaks in the presence of declination. *Technical Report, Instituut voor Perceptie Onderzoek*, 1994.

[Slu95a] A. C. M. Sluijter. Een perceptieve evaluatie van een model voor alinea-intonatie met synthetische spraak (a perceptual evaluation of a model for paragraph intonation with synthetic speech). *Internal Report 801, Instituut voor Perceptie Onderzoek*, 1995.

[Ter84] J. M. B. Terken. The distribution of pitch accents in instructions as a function of discourse structure. *Language & Speech*, 27:269–289, 1984.

[Ter91] J. Terken. Fundamental frequency and perceived prominence of accented syllables. *J. Acoust. Soc. Am.*, 89:1768–1776, 1991.

[Ter93] J. Terken. Baselines revisited: Reply to Ladd. *Language and Speech* 36:453–459, 1993.

[Ter94] J. Terken. Fundamental frequency and perceived prominence of accented syllables. II. Non-final accents. *J. Acoust. Soc. Am.*, 95:3662–3665, 1994.

[tH81] J. 't Hart. Differential sensitivity to pitch distance, particularly in speech. *J. Acoust. Soc. Am.*, 69:811–821, 1981.

[tHCC90] J. 't Hart, R. Collier, and A. Cohen. *A Perceptual Study of Intonation.* Cambridge, UK: Cambridge University Press, 1990.

Appendix: Prominence Ratings in Spontaneous Speech. Materials and Detailed Results

The following is a list of speech fragments used in the prominence rating test. For each speech fragment, identified by (spkr...) the first line gives the approximate English translation, staying close to the Dutch text, with words to be rated for prominence printed in capitals; (-dim) in the English gloss stands for the Dutch diminutive suffix; the second line gives the Dutch text, with information about the presence of hesitations (h), and the duration of filled pauses (fp) and silent pauses (p) in milliseconds interspersed; the third line gives the mean prominence rating, aligned with the target words; the fourth line gives information about the height of $F0$ peaks and phrasing, with boundaries of Intonation phrases marked by %. For fragments spanning several lines this structure is repeated.

```
(spkr1-019)
the DOOR is                     a BIT
de deur staat <fp:400> <p:720> een stukje <p:1280> <fp: 250>
    4.9                            5.0
    235                            245

to the LEFT of the MIDDLE
       links van het midden
       8.0           5.9
       256           181

(spkr1-020)
so let's say first you divide the paper
dus laten we zeggen je deelt het papier eerst <fp: 160> <p: 340>

in TWO    and then you set        THERE          in the LEFT HALF
in tweeen en dan zet je <p: 330> daar <p: 350> in het linker vakje
    5.7                          6.4                   6.2    4.5
    192  %                       171                   189    176

in the MIDDLE the FRONTDOOR
 in het midden de voordeur
        5.5       5.0
        171       136

(spkr1-022)
oh i HAVE a                      CHIMNEY(-dim) or something here
oh ik heb hier zelfs nog een schoorsteentje of zo
      1.9                    6.9
      225                    267

(spkr1-024)
```

```
uhm                     the LIVING-ROOM is in MOST houses DOWNSTAIRS
uhm <fp: 650> <p: 1075> de woonkamer  is in de meeste huizen beneden
                           8.1                 4.1             6.1
                           279       %         212             212
```

```
(spkr1-025)
and that we set in               NEXT to the FRONTDOOR
en die zetten we dus in <p: 900> naast de voordeur
                                 7.3     5.2
                                 203     194
```

```
(spkr1-031)
so it hangs at the UPPER side of the window    of the LIVING-ROOM
dus die hangt bovenaan het raam <p: 270> van de woonkamer
              7.9                                4.3
              218                                146
```

```
(spkr1-032)
 uhm then we may place the two BEDROOM-WINDOWS you've got
uhm dan kunnen we de twee slaapkamerramen die je hebt <p: 150>
                           9.0
                           260                          %
```

```
more or less SYMMETRICALLY over the DOOR and the LIVING-ROOM-WINDOW
zo'n beetje symmetrisch boven de <h> deur en het woonkamerraam plaatsen
               6.4                   5.6         6.2
               209                   190         194
```

```
(spkr1-035)
yes the BEDROOM-WINDOWS are the SAME SIZE
ja de slaapkamerramen zijn even groot
       6.2                 6.4   6.8
       218                 212   186-271
```

```
(spkr1-036)
and the CURTAIN with the TEETH fits EXACTLY
en het gordijn met de tandjes past er precies op
          5.3         8.1                5.1
          215         271                228
```

```
(spkr1-037)
and ..                      that curtain that i appended
en van die <fp: 400> <p: 590> dat gordijn dat ik
```

```
in the LIVING-ROOM-WINDOW that has also TEETH(-dim)
bij de woonkamerraam had gehangen dat heeft dus ook tandjes <p: 200>
       7.2                                          7.3
       249                                          221      %
```

```
in contrast with the other which is somewhat more ROUNDED
 in tegenstelling voor die ander die een beetje ronder is
                                                6.9
                                                231

(spkr1-041)
and that        i place NEATLY in the MIDDLE of
en die <p: 1720> leg ik mooi in het midden van <p: 150> <fp: 325>
                        6.9         6.6
                        215         212

    the GREEN top which constituted the roof of the HOUSE
 uh de groene   top die <h> het dak van het huis vormde
        6.8                                 6.4
        212                                 203

(spkr1-042)
the GREEN FLOWER-BOX(-dim)
het groene bloembakje <p: 320>
     5.8    8.0
     245    260

we'll put in the LIVING-ROOM-WINDOW on the WINDOW-SILL
zullen we in het woonkamerraam zetten  op de vensterbank
                  5.8                           4.9
                  218                  %        192

(spkr2-005)
and             ONE side against the UPPER-SIDE of the black SQUARE
en <p: 490> met een kant tegen de bovenkant van het zwarte vierkant aan
                6.3               5.4                        6.4
                260               256                        263         %

so that one gets an OBLIQUE point UPWARD
zodat je een schuine punt omhoog krijgt
             7.3            5.2
             256            184

(spkr2-007)
and that we place at the RIGHT side in the BLACK square BOTTOM-RIGHT
en die zetten we rechts in het zwarte vierkant  rechtsonder
                 8.3           6.6                     7.8
                 309           279            %        260

(spkr2-008)
so that the SHORT side of the front-door
zodat de smalle kant van de voordeur <p: 610>
         7.4
         287
```

```
lies against the BOTTOM of the black SQUARE
tegen de onderkant van het zwarte vierkant aan zit
         7.9                       5.8
         287                       267

(spkr2-014)
so that the UPPER-SIDE of the LIVING-ROOM-WINDOW is the same height
zodat de bovenkant van het woonkamerraam  gelijk zit
         7.6               5.6
         296               267             %

as the UPPER-SIDE of the FRONT-DOOR
met de bovenkant van de voordeur
       5.5              6.9
       228              263

(spkr2-017)
and that we PUT in the LEFT UPPER-CORNER
en dat plaatsen we in de linker bovenhoek <p: 300>
       3.9               7.2    5.6
       235               287    252         %

about STRAIGHT above the LIVING-ROOM-WINDOW
ongeveer recht boven het woonkamerraam
         5.9             6.2
         235             245

(spkr2-018)
so that the LEFT side of the LIVING-ROOM-WINDOW
zodat de linker <h> kant van het woonkamerraam
         6.8                    5.1
         279                    228

and the left side of the BEDROOM-WINDOW
en de linkerkant van het slaapkamerraam op
                         6.9
                         245

are in one LINE           vertically seen
een lijn liggen <p: 1160> verticaal gezien
    6.5
    184-221     %

(spkr2-028)
so that it's WIDER than it's HIGH
zodat ie breder is dan dat ie hoog is
         7.6                  5.9
         267                  218
```

```
(spkr2-031)
and the DOOR-WINDOW we place in the FRONT-DOOR
en het deurraam  plaatsen we in de voordeur
        5.8                        7.3
        263   %                    275

(spkr2-039)
or the          living-room-window we place un-
of dat <p: 225> woonkamerraam plaatsen we on- <p: 270>

the curtain sorry we place DOWN-UNDER in the LIVING-ROOM-WINDOW
dat gordijn sorry  plaatsen we onder in het woonkamerraam
                               7.0          4.9
                 %             267          245

(spkr2-040)
so that the CLOUDS(-dim)  are at the same height as the LOWER    line
zodat de wolkjes <p: 100> gelijk liggen met de onderste <p: 1285> lijn
         5.4                                   5.6
         249                                   235

of the LIVING-ROOM-WINDOW
van het woonkamerraam
        6.3
        245

(spkr2-046)
and we do that the SAME with bedroom-window TWO
en dat doen we hetzelfde met slaapkamergordijn twee
               4.4                             7.1
               225                             200-342

(spkr2-056)
and we put the TREE to the LEFT         of the FRONT
en de boom die zetten we links <p: 245> naast de voorgevel
      4.8                7.3                     5.8
      231                245                     221
```

9 Predicting the Intonation of Discourse Segments from Examples in Dialogue Speech

Alan W. Black

ABSTRACT In the area of speech synthesis it is already possible to generate understandable speech with discourse neutral prosody for simple written texts. However, at ATR-ITL we are researching speech synthesis techniques for use in a speech translation environment. Dialogues, in such conversations, involve much richer forms of prosodic variation than are required for the reading of texts. For our translations to sound natural it is necessary for our synthesis system to offer a wide range of prosodic variability, which can be described at an appropriate level of abstraction. This paper describes a multi-level intonation system which generates a fundamental frequency (F_0) contour based on input labelled with high level discourse information, including speech act type and focussing information, as well as part of speech and syntactic constituent structure. The system is rule driven but rules (and parameters) are derived from naturally spoken dialogues. Two experiments using this model are described, testing its accuracy. First results are given for a system to predict ToBI intonation labels from discourse information use a CART decision tree. Second a detailed investigation of the intonational variation of the word "okay" in different discourse contexts is presented.

9.1 Introduction

This paper presents a framework for generating intonation parameters based on existing natural speech dialogues labelled with that intonation system, and marked with high level discourse features.

The goal of this study is to predict the intonation of discourse segments in spoken dialogue for synthesis in a speech-translation system. Spontaneous spoken dialogue involves more use of intonational variety than does reading of written prose, so the intonation specification component of our speech synthesizer has to take into account the prosody of different speech act types, and must allow for the generation of utterances with the same variability as found in natural dialogue.

For example, the simple English word *"okay"* is heard often in conversation but can perform different functions. Sometimes it has the meaning "*I understand*", sometimes "*do you understand?*", and other times it is used as a discourse marker indicating a change of topic, or as an end-of-turn marker signalling for the other partner to speak. Different uses of a word may have different intonational patterns.

Predicting F_0 directly from speech act and discourse level labels is too grand, especially as there are already a number of existing intonation systems that offer a suitable level of abstraction over an F_0 contour (e.g., ToBI ([SBSP92]), RFC and Tilt ([Tay94], [TB94]), or the Fujisaki model ([Fuj83]). Instead of creating yet another intonation system, we can predict parameters for some existing system (which in turn will be used to render the actual F_0 contour). Because of reasons discussed in more detail later we cannot afford to choose any one particular intonation system, as finding labelled data for training is not easy. Thus this overall system does not commit itself to one particular intonation parameter system but does commit itself to some abstract intonation system. Even though these existing intonation systems may represent conceptually different aspects of intonation, they all offer a level of abstraction from which varied F_0 patterns may be generated.

Therefore, in this paper we are primarily concerned with a level of discourse intonation "above" these intonation parameter systems. That is a system that will predict intonation parameters (for whatever intonation system being used) from higher level discourse information such as speech act, discourse function, as well as syntactic structure and part of speech information. In ToBI, terms that is *which* pitch accents and boundary tones have to be predicted based on discourse level labelling. Figure 9.1 positions this work in the process of generating an F_0 contour in our speech synthesis system.

Intonation systems in general allow parameters to be specified to represent the variation required for intonation in speech dialogues. However, although these varied parameters may be specified by hand, being able to predict such variation automatically is much harder. It is that task that we are addressing here.

Different intonation systems offer different parameters which can be modified, the following is a non-exhaustive list of the sort of parameters we wish to predict.

(1) pitch accent type;

(2) boundary tone type (both start and end);

(3) start and end points for phrases (speaker normalized);

(4) pause duration (at least in simple cases);

(5) reset and declination rates;

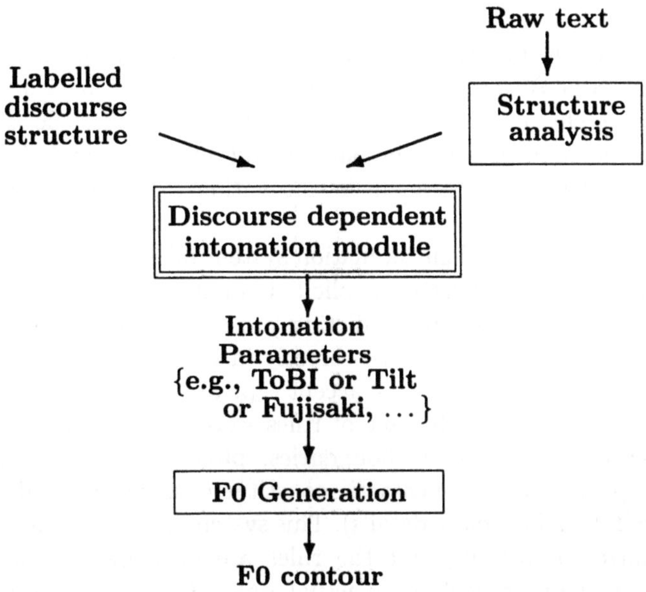

FIGURE 9.1. Overall Intonation prediction system.

(6) pitch ranges.

Although the above parameters are to some extent speaker dependent they may be represented without specifying absolute Hertz values for F_0. We can specify various pitch ranges in normalized terms, such as starting and end values with respect to a speaker's range and thus only need specific speaker information at a later stage in F_0 generation.

Predicting the *position* and *type* of appropriate pitch accents and boundaries tones that convey the desired intentions is a non-trivial problem, and determining what information is necessary in order to predict such parameters is still an on-going research topic. Many researchers are actively working in this area, for example, see [Bec96b, Sec. 2] for a discussion of such work.

In order to be able to predict appropriate dialogue intonation we need the input utterance to be labelled so that distinctions which are not implicit in the orthographic words alone may be realized appropriately. With more appropriate information included, a greater diversity in the realization of the intonation is possible.

The sort of information suggested as affecting intonation is

(1) focusing information (global and local);

(2) new and old information;

(3) speech act (including discourse function);

(4) contrastive and emphatic markings.

In addition, specific words such as "*only*" are known to have specific effects on prosodic patterns. Also varying intonation can be used to mark discourse function, such as change of topic and end of turn. For example [Hir93a] discuss the relationship between cue phrases and intonation including how the use of the word "*now*" varies in its intonation realization in varying discourse contexts.

Thus our discourse dependent intonation system takes explicit discourse features as input and generates explicit intonation parameters. This involves the more basic tasks of predicting prosodic phrasing and accent positioning which we will not discuss directly in this paper where we will concentrate on the issues of choice of accent type and boundary tone type.

An initial simple hand-crafted set of rules were written which predict intonation parameters (prosodic boundaries, pitch accents, and phrase accents) from part of speech, syntactic constituent structure, and speech act labels (see [BT94a] for more details). This system is adequate for simple high level control of prosody but the rules are developed by personal intuition rather than derived from actual data. Hence they are prone to whims of the writer and require skill to amend. A more data-driven approach is required to make this system more general.

To determine the degree of relationship between different uses of a word or phrase, and different intonational contours, we analysed a number of spontaneous conversations between clients and an agent discussing queries about travel to a conference site. Two analyses of this data are presented using different intonation systems.

9.2 Modelling Discourse Intonation

In order to build models predicting intonation parameters from discourse features, our data must be labelled with both the parameters we wish to predict and the discourse features we wish to predict from. Finding large quantities of prosodically labelled data is non-trivial and the further constraint that it is labelled with discourse features makes it harder. We have performed some initial work on predicting intonation accent type using the Boston University FM Radio database [OPSH95a] but it is not certain we can generalize from read news announcer speech to the type of spontaneous dialogue speech we wish to model.

The database used in both the following experiments, was collected using the ATR Environment for Multi-Modal Interactions (EMMIs) and consists of a total of 17 English dialogues ranging from 2 to 8 min between a single agent and different clients asking directions and information about a conference. The dialogues were transcribed and labelled with phonemes using an automatic aligner. They were then manually labelled with speech act classes based on those described in [Ste94], and with prosodic labels using the ToBI system.

When the EMMI database was collected, two types of interaction were recorded: (a) multi-modal, where the agent and client could see each other via video, speak through an audio channel, and a display allowed maps to be mutually seen; and (b) by telephone alone. For this analysis only the agent side of the nine multi-modal dialogues were used—as the same agent was used in all dialogues, but the clients changed.

Two different systems were used to investigate the relationship between the discourse labels and the observed intonation patterns: one using the hand labelled ToBI system; and another using a purely automatic intonation labelling system.

9.2.1 Analysis with ToBI Labels

In addition to phonemes, words and ToBI intonation labels (pitch accents, phrase accents, boundaries tones and break indices), the dialogues were further labelled with IFT (broad class speech acts) and discourse acts (fine detailed discourse acts). There are 22 IFT classes and 58 discourse acts [SFT94]. This speech act labelling was done for research in discourse structure but we will show that they are also relevant in predicting prosody, even though they were not specifically designed for that purpose. The agent side of the dialogue was **chunked** into discourse act sized sections giving a total of 630 chunks, consisting of a total of 5101 words.

Initially the distribution of pitch accents was investigated. By pitch accents in ToBI we include any label containing a `*`. 1770 (35%) words were labelled with one or more pitch accents. Of these 1676 (95% of accented words) were labelled with `H*` alone. The next most frequent accent type was `L+H*` which appeared only 39 (2%) times. The next was `L*+H` at nine times.

Using a CART technique [BFOS84], decision trees were built to predict pitch accent type for each word. It was assumed pitch accented position was known but type was not. Various trees were built using features such as accentedness of current and adjacent words, function/content type of a window of words, ToBI break index, position phrase, IFT etc. But no tree could predict more than `H*` for any accented word. Better results were hoped for but there does not seem to be enough differentiation in the input to reliably predict accents other than `H*` as there are so few examples of non-`H*` accents. Accents such as `L+H*` and `L*+H` have been suggested as being used for emphasis along some scale [WH85], but without such marking in our input no learning system will detect that. This result shows that without appropriately labelled data, in sufficient quantities, prediction will not be possible.

A second investigation was to predict boundary tones at the end of discourse act sized chunks. 389 (62%) chunks were terminated with a boundary tone, the other chunks were not terminated with a prosodic phrase break or only a phrase accent. ToBI labelling has four sequences of phrase accent and boundary tones found at the end of chunks: L-L%, H-L%, L-H%, and H-H%. The distribution of these four tones is

Tone	Occur	Percentage
L-L%	173	44%
H-L%	110	28%
L-H%	76	20%
H-H%	30	8%

This distribution changes for different discourse acts. For example, the **instruct** discourse act and **do-you-understand** discourse act have the following distributions;

Instruct

Tone	Occur	Percentage
L-L%	13	46%
H-L%	11	39%
L-H%	2	7%
H-H%	1	3%

Do-you-understand

Tone	Occur	Percentage
L-L%	3	23%
H-L%	2	15%
L-H%	6	46%
H-H%	2	15%

A CART decision tree was then built to predict ending tone. Various features were used but the best results were achieved from the following factors:

(1) most frequent ending tone for current discourse act;

(2) break index preceding final word;

(3) break preceding the word preceding the final word;

(4) preceding IFT;

(5) current IFT;

(6) current discourse act.

This produces a decision tree of depth 16 that can correctly predict the ending tone of discourse act sized prosodic phrase given the above features 60% of the time. If we simply select the most frequent ending tone the accuracy drops to 49%, or if we always predict L-L% then accuracy is 44%.

This decision tree was used in the ATR CHATR speech synthesis system [BT94b], to predict suitable ending tones for different discourse acts. Although no formal listening tests were done, this end tone prediction system plus a simple pitch accent prediction of only H* produces more varied dialogue speech than is possible when no speech act information is available.

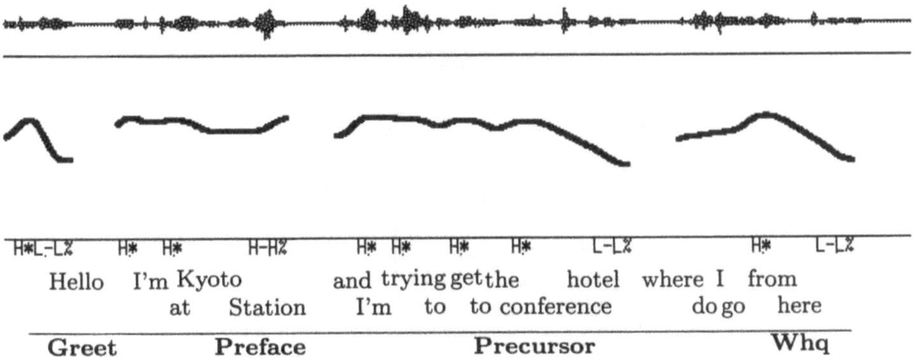

FIGURE 9.2. Intonation parameters predicted from discourse acts.

For example given the dialogue sentence, "*Hello, I'm at Kyoto station and I'm trying to get to the conference hotel, where do I go from here*", when labelled with discourse act information it contains four chunks, **Greeting**, **Preface**, **Precursor**, and **Whq.** The ending tones predicted using the above described decision tree (and the resulting F_0 generated from those predicted ToBI labels) are show in Figure 9.2.

Particularly note the prediction of the H-H% tone after *station.* The contributing factors used in predicting this include, that it is within a **Preface** discourse act but also that it is preceded by a **Greet** and that the phrase it is in, is more than one word long.

The above method is only a start at building high level intonation prediction systems based on labelled natural dialogue data. More work is required but that will require more detailed labelled data from which we can learn mappings.

9.2.2 Analysis with Tilt Labels

In this second test we used the RFC and Tilt intonation system ([Tay94], [TB94]). RFC and Tilt encode the pitch patterns found in the speech without explicitly identifying linguistic intonation events as ToBI does. Tilt makes no distinction between boundary tones, phrase accents and pitch accents. Its main advantage is that it can automatically label data. The process of Tilt labelling is achieved by the following process. The F_0 is extracted from the speech waveform using a pitch tracker, and then median smoothed. The smoothed contour is RFC labelled ([Tay94]) segmenting the contour into a sequence of rise, fall, and connection elements, each with a duration and amplitude specification. The phonetic labels are used for syllabification, and aligned with the RFC elements. The elements are then converted to a series of *tilt* events separated by *connections.* The canonical form of a tilt event is a simple "hat" shape, with equal degrees of rise

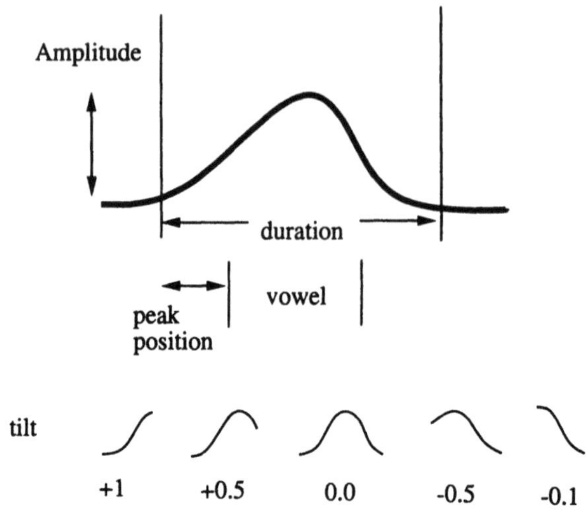

FIGURE 9.3. Tilt accent parameters.

and fall, which can be modified by four continuous parameters: amplitude, duration, accent peak position with respect to the vowel, and tilt, which describes the relative height of the rise and fall of the event. -1 denotes a fall with no rise while 1 denotes a rise (with no fall). 0 denotes equal rise and fall while other values state that the rise and fall are of different heights (cf. upstep and downstep). Figure 9.3 show a canonical tilt even showing the different dimension of the four parameters.

Although no formal tests were done on the accuracy of labelling this data, measurements on other data have been carried out ([Tay94] [TB94]).

To show how tilt labelling can be used in the prediction of events from the same form of input as specified in the previous example, we again looked at the EMMI dialogue data. Specifically we looked at how the word "*okay*" is realized. In all the dialogues (both multi-modal and telephone only modes) there are 140 occurrences of the word *"okay"* spoken by the agent. 112 of which appear alone in a prosodic phrase. These examples fall into 12 discourse act classes, only four of which occured more than twice. These four are: **frame** (37 occurrences), **ack** (31), **d-yu-q** (22), and **accept** (10). **Frame** marks the end of a discourse segment, **ack** is a general acknowledgment, **d-yu-q** is a do-you-understand question, and **accept** as in an immediate reply to a question. It should be noted that these discourse act types were not defined for differentiating intonational classes, they were independently motivated to represent discourse function, so it is not necessarily the case that all classes are distinguished by different intonational patterns.

The following table shows the mean start and end F_0 values (standard deviations are shown in parentheses) for these examples for each discourse act type. The values are normalized and given in number of standard deviations from the mean. (Note that the means for the start and end values are calculated separately, and thus cannot be directly compared.)

Discourse act	Accept	d-yu-q	Ack	Frame
No. of occurs	10	22	31	37
Start	-0.10 (0.64)	-0.23 (1.2)	-0.73 (1.3)	-0.13 (1.38)
End	0.10 (0.85)	0.92 (0.96)	-0.11 (0.79)	-0.47 (0.86)

All the start values are below mean start value, this is probably because longer phrases in general start higher and all these phrases are short. Student t-tests confirm that end values for **frame** examples are significantly lower than end values of other examples ($t = 3.9$, $df = 98$, $p < 0.001$). Also the end values of **d-yu-q** discourse acts are significantly higher that the other discourse acts ($t = 5.55$, $df = 98$, $p < 0.001$), as would be expected for a question.

The second set of results concerns the tilt event description. In most cases there is one tilt event (i.e., one accent) in the prosodic phrase. The following table shows the mean tilt parameter (and standard deviation) for each discourse act class:

Discourse act	Accept	d-yu-q	Ack	Frame
Tilt	0.45 (0.89)	0.74 (0.55)	0.19 (0.93)	-0.28 (0.79)

The tilt parameter indicates the amount of rise and fall at that point in the F_0 contour. Values near zero represent events with equal rise and fall, values closer to 1.0 represent rise only while values closer to -1.0 represent a fall with no preceding rise. Thus we can see **frame** examples have significantly more downward tilt than the other discourse acts ($t = 4.13$, $df = 98$, $p < 0.001$), while **d-yu-q** examples are predominantly rising events ($t = 3.68$, $df = 98$, $p < 0.001$)

These three results show a significant difference between different renderings of "*okay*". **Frame** examples tilt more downward ending lower than other acts. **Ack** examples tend to start lower and not tilt as much ending higher. **D-yu-q** start relatively neutral but rise up to significantly higher values than other examples.

These parameters (start, end and tilt) can be used directly in the intonation specification of our synthesis system. For example, a **d-yu-q** labelled "*okay*" can be assigned a start value -0.23 standard deviations from the mean F_0 and event's tilt parameter a value of 0.74.

9.3 Discussion

It is important to realize that although it may be possible to predict so-called "default intonation" for plain text any variation from default emphasis, focus, discourse function, etc. would have to be derived from the text. The additional discourse features are not intonational features but describe discourse function and are necessary to predict more appropriately varying intonation. At ATR within a framework of telephone translation a much richer input is available as output from the translation process, so IFT, focus, etc. are available directly as input information, with no special processing required to predict them.

Because the number of non-H* pitch accents in the EMMI database is so small, it seems unlikely that a more complex pitch accent prediction model than simply predicting H* (or the Tilt equivalent) for accented words, can be found based on this current data and its labelling. Even with a larger database with more variation in pitch accents, in order to differentiate between pitch accent types we would most probably need richer labelling of the input data identifying focus, new and old information, contrastive marking, emphasis, etc.

We do not yet wish to choose between the two methods of labelling intonation system presented here, in fact we are likely to add to them. Lack of prosodically labelled data is probably our greatest hurdle. Any database labelled with any reasonable form of prosodic labelling cannot be ignored. Tilt labelling has the advantage of being automatically derivable from waveforms, though does not explicitly distinguish between pitch accents, phrase accents, and boundary tones. Automatic ToBI labelling is under consideration ([WC94]) and would aid us greatly in labelling of more databases. Although hand labelling is resource intensive it is becoming easier with appropriate tools. Also as it is becoming a standard it is likely that more suitable data will soon become widely available.

The framework presented here has been designed to be language independent and to some extent intonation theory independent. A Japanese version of the same speech dialogue database has been recorded and is currently being ToBI labelled, and we will apply similar analysis techniques to that data.

9.4 Summary

This paper discusses the synthesis of intonation for dialogue speech. It presents a framework which allows prediction of intonation parameters (for various intonation theories) from input labelled with factors describing discourse function. If factors such as speech act, syntactic constituent structure, focus, emphasis, part of speech, etc. are labelled in the input then more varied intonation patterns can be predicted.

Rather than writing translation rules directly, techniques for building such rules from prosodically labelled natural dialogue speech are presented. Two analyses of aspects of the ATR EMMI dialogue database are presented showing how speech act information can be used to distinguish different intonational tunes. The main conclusion we can draw from these analyses is that discourse act can play a significant role in predicting intonational pattern.

The initial results look promising and we will continue to expand the system for English and also for Japanese. There are still questions as to which modelling techniques to use but at present the greatest problems lie in labelling. Both in the task of actually labelling data, and in deciding on what level of detail to label the data.

References

[Bec96b] M. Beckman. A typology of spontaneous speech. In *Computing Prosody: Approaches to a Computational Analysis of the Prosody of Spontaneous Speech.* New York: Springer-Verlag, 1997. This volume.

[BFOS84] L. Breiman, J. Friedman, R. Olshen, and C. Stone. *Classification and Regression Trees.* Pacific Grove, CA: Wadsworth & Brooks, 1984.

[BT94a] A. W. Black and P. Taylor. Assigning intonation elements and prosodic phrasing for English speech synthesis from high level linguistic input. In *Proceedings of the International Conference on Spoken Language Processing, Yokohama, Japan,* Vol. 2, pp. 715–718, 1994.

[BT94b] A. W. Black and P. Taylor. CHATR: A generic speech synthesis system. *Proceedings of COLING-94,* II:983–986, 1994.

[Fuj83] H. Fujisaki. Dynamic characteristics of voice fundamental frequency in speech and singing. In P. MacNeilage, editor, *The Production of Speech,* pp. 39–55. Berlin: Springer-Verlag, 1983.

[Hir93a] J. Hirschberg. Pitch accent in context: Predicting prominence from text. *Artificial Intelligence,* 63:305–340, 1993.

[OPSH95a] M. Ostendorf, P. Price, and S. Shattuck-Hufnagel. The Boston University Radio News Corpus. Technical Report ECS-95-001, Electrical, Computer and Systems Engineering Department, Boston University, Boston, MA, 1995.

[SBSP92] K. E. A. Silverman, E. Blaauw, J. Spitz, and J. Pitrelli. Towards using prosody in speech recognition/understanding systems: Differences between read and spontaneous speech. *Proceedings DARPA Speech and Natural Language Workshop,* pp. 435–440, 1992.

[SFT94] M. Seligman, L. Fais, and M. Tomokiyo. A bilingual set of communicative act labels for spontaneous dialogues. Technical Report Technical Report TR-IT-0081, ATR Interpreting Telecommunications Laboratories, Kyoto, Japan, 1994.

[Ste94] A. Stenström. *An Introduction to Spoken Interaction.* London: Longman, 1994.

[Tay94] P. Taylor. The Rise/Fall/Connection model of intonation. *Speech Communication*, 15:169–186, 1994.

[TB94] P. Taylor and A. W. Black. Synthesizing conversational intonation from a linguistically rich input. *Proceedings of the ESCA/IEEE Workshop on Speech Synthesis, Mohonk, NY*, pp. 175–178, 1994.

[WC94] C. W. Wightman and W. N. Campbell. Automatic labelling of prosodic structure. Technical Report TR-IT-0061, ATR Interpreting Telecommunications Laboratories, Kyoto, Japan, 1994.

[WH85] G. Ward and J. Hirschberg. Implicating uncertainty: The pragmatics of fall-rise intonation. *Language* 61:747–776, 1985.

10
Effects of Focus on Duration and Vowel Formant Frequency in Japanese

Kikuo Maekawa

ABSTRACT The effect of contrastive focus upon duration and vowel formant frequency was examined. The effect upon duration was found at various levels involving utterance, accentual phrase, and individual segments. The effect upon the formant frequencies was also confirmed; it could be different depending on the prosodic location of the vowel in question. Vowels that were directly focussed became more peripheral in terms of vowel height, while /e/ vowels outside the domain of focus became less peripheral in terms of frontness. It was interesting that the observed effects of focus were mostly omni-directional, i.e., focus influenced not only the temporally preceding constituents but also the following ones in all phonetic parameters examined. This requires certain revision of the treatment of focus in the current phonological theory of intonation.

10.1 Introduction

10.1.1 The Aim of the Study

In experimental studies of Japanese prosody, it is widely recognized that manifestation of prosodic information depends primarily upon the voice fundamental frequency or $F0$ (see [Sug82], among others). However, Japanese prosody cannot be completely analysed by paying attention to pitch alone. It is known that the synthesis component of a text-to-speech conversion system must involve rules that reflect various effects of prosodic boundaries upon duration, such as utterance and phrase final lengthening (e.g., [KTS92], for Japanese). Although in Japanese these effects turn out to be less prominent when compared to those in English ([Kla76]), synthetic Japanese speech without the rules sounds dull and less intelligible. The importance of duration control may increase as we go on to handle more spontaneous speech, whose prominence varies more widely than in laboratory speech. Also, it is expected that wider prominence has some influence on the spectral characteristics of speech as well. See [dJ95] for the effect of stress on articulatory gestures in English. This paper is a preliminary report of a series of experiments that aim at a deeper understanding of the nature of prominence control in Japanese. In the rest of the paper, I examine the effect of contrastive focus in Japanese and show two hitherto unknown facts: the acoustic manifestation of focus in Japanese depends on duration in a very different way from languages like English, and there are changes in the vowel formant frequencies which can be traced back, at least partly, to the presence of focus.

10.1.2 Accent and Focus in Japanese

Tokyo and many other dialects of Japanese use pitch to show lexical contrast. In Tokyo, there are word pairs like *"hashi"* (chopsticks) and *"hashi"* (edge) that contrast solely by means of pitch shapes. If we pay attention to the pitch shift from high to low and represent the fall by an apostrophe, we can represent these words as ha'shi (with fall) and hashi (without fall), respectively. Another pitch phenomenon also occurs at the beginning of a word: the first and the second morae of a word are low- and high-pitched, respectively, unless either a pitch fall is present or the first two morae constitute one heavy syllable. If we symbolize this initial pitch rise by a slash, the pitch contour of hashi (edge) can be represented as ha/shi. Using this notation, we can arrange the nouns of Tokyo Japanese as in Table 10.1. For example, the surface pitch contour of words like aozo'ra begins low, rises to high at the second mora, and falls to low again to the right of the apostrophe; the contour of words like tomodachi begins low, rises to high, and continues to be high to the end of the word; and the contour of words like ko'Hmori begins high, falls to low at the second mora, and continues to be so to the end. The location of pitch shifts—both fall and rise—is sensitive to syllable weight. In Tokyo, a syllable is heavy if it contains two morae and light elsewhere. The morae that can appear as the second mora of a heavy syllable are limited in number and called "special" morae: they involve the moraic nasal /N/, geminate /Q/, diphthong /J/, and long vowel /H/. The high-low pitch fall can occur between the two constituent morae of a heavy syllable but it does not occur between the second mora of a heavy syllable and the following mora. Thus, a word like */koH'mori/, as opposed to /ko'Hmori/, does not exist. As for the low-high pitch rise, it disappears or is strongly reduced when a word begins with a heavy syllable; this is why there is no slash between /ka/ and the moraic nasal /N/ of kaNzi in Table 10.1.

According to Miyata ([Miy27]) and Hattori ([Hat61]), "accent" in Tokyo Japanese is concerned only with the high-low pitch shift, which Hattori called an "accent kernel"; the distinctive features of accent being the existence/absence of the kernel and the location of the kernel if there is any. The word-initial pitch rise was excluded from the distinctive feature specification of accent, because it was predictable from the phonological environment; the rise appears only in the initial position of a prosodic

TABLE 10.1. Surface pitch contour of Tokyo Japanese nouns.

Mora length	1	2	3	4	
Without pitch fall	ha (leaves)	u/shi (cow)	kaNzi (Chinese character)	to/modachi	(friend)
With pitch fall	ki'(tree)	ne'ko (cat)	i'nochi (life)	ko'Hmori	(bat)
		i/nu' (dog)	ko/ko'ro (mind)	mu/ra'saki	(purple)
			ka/gami' (mirror)	a/ozo'ra	(blue sky)
				o/toHto'	(younger brother)

unit that Hattori called "prosodeme", which bears close resemblance to what was called a "minor phrase" or, equivalently, "accentual phrase" by American phonologists (McCawley, [McC68]; Pierrehumbert and Beckman, [PB88]). In the rest of the paper, we will adopt the view of Hattori and delimit the meaning of "accent" to the pitch fall and mark the location of an accent kernel by an apostrophe both in orthography and phonological representation. Hattori's prosodeme will be referred to as an accentual phrase.

At this point, it is necessary to point out that there are two important differences between the accent in Japanese and the pitch-accent of English. For one, Japanese accent has only one phonetic shape and lacks exclusively the paradigmatic contrast of pitch shapes that characterizes English pitch-accent (see Beckman and Pierrehumbert, [BP86]). For another, accent is purely lexical in Japanese, while English pitch-accent is phrasal. It is unlikely that Japanese accent changes its location or is deleted under the effect of various prosodic conditions. Previously reported cases of accent deletion in Japanese turn out to be dubious under fine phonetic analysis [Mae94]. As a consequence of its purely lexical property, the existence of an accent does not signal increased prominence. An accented word is by no means more prominent than unaccented ones. Unlike English, the phonological device used in the control of prominence is not the accent placement. Focus in Japanese is realized by the local expansion of the pitch range in which tones are scaled. From a physical point of view, the local modification of prominence as represented by contrastive, or narrow, focus was primarily concerned with $F0$. Previous experimental studies on focus in Japanese all support this view [FK88, PB88, Kor89b].

However, it is important to note that almost all experimental studies did not pay enough attention to those potential physical correlates of focus like duration, intensity, formant frequencies, and voice quality. The sole exception to this is Kori [Kor89a] which examined the duration, intensity, and $F0$ of unaccented test words in three different focal conditions, viz. post-focus, pre-focus, and under-focus positions. He concluded that duration and intensity played a very limited role in the manifestation of focus. But his experiment has two problems: he examined only the test word that directly bore focus and left the other part of the sentence unexamined; also he did not examine the spectral characteristics at all. In the experiment that follows, we will examine the effect of focus on $F0$, duration, and formant frequencies paying attention not only to the target of focus but also to those constituents that were outside the domain of focus.

10.2 Experimental Setting

The sentence below was used as the speech material. The sentence begins with a time adverbial followed by two noun phrases, which are followed in turn by a verb phrase that ends the sentence. These four syntactic constituents correspond to four accentual phrases all having lexical accent phrase initially.

```
Accentual (Acc. phrase\#1)(Acc. phrase\#2)(Acc. phrase\#3)(Acc. phrase\#4)
phrasing
          ke'sa          X-to         te'rebi-o        mi'ta
[Gloss]   this morning   with X       TV-OBJ           watched
          This morning, (I) watched TV with X.
```

The slot X in the sentence was filled by one of the following five words. They were either kinship terms or names (given as well as family), and share the same phonological configuration except for the membership of the onset consonant and the target long vowel of the first heavy syllable. These words will be referred to as target words and the accentual phrase that consists of the target words and the following particle "to" will be referred to as the target phrase. Also, we will refer to the five accented long vowels in the target words as the target vowels.

1. ji'isan /zi'HsaN/ [dʃi:san] (Grandpa);
2. te'esan /te'HsaN/ [te : san](SurnameTei + san);
3. ka'asan /ka'HsaN/ [ka : san](Mom);
4. to'osan /to'HsaN/ [to : san](Dad);
5. chu'usan /chu'HsaN/ [tʒu:san] (Given name Chuu + san).

The resulting five sentences were pronounced under three different focal conditions on the target words.

(a) No-focus condition; no narrow focus was required on any part of the sentence (Abb. N-focus).

(b) Moderate focus condition; ordinary narrow focus on the target words (M-focus).

(c) Strong-focus condition; extraordinarily strong focus on the target words (S-focus).

The sentences were printed on an index card, and the focal conditions were marked by underscore; no, single, and double underscores beneath the target phrases stood, respectively, for N-focus, M-focus, and S-focus conditions. Two male speakers in their mid-thirties participated in the recording. Speaker one was the present author, and speaker two was a teacher of Japanese as a foreign language who had fine knowledge in pedagogical phonetics but knew nothing about the aim of the experiment. Speakers were instructed to express three degrees of focus by means of the voice pitch, and not to insert a pause into any part of utterance. Speaker two listened to a part of the recorded speech of speaker one prior to his recording session in order to be sure of what was required in the experiment. They read the sentences ten times in random order. The recording was made in a quiet recording room using DAT equipment (Sony TCD-D3 with ECM S220 microphone).

The first five repetitions of the recorded utterances were downsampled with 16-bit quantization at a sampling rate of 16 KHz using a DAT-interface (DAT-LINK) connected to a Sun workstation. The speech files were labelled in the Entropic xwaves environment and then analysed by the "formant" command of the Entropic ESPS signal processing system using an autocorrelation method. The order of analysis was 18 and the analysis step was 0.01 s. As for the target vowels, subsequent formant analysis was performed after the examination of the preliminary results as described in Sec. 3.3.

10.3 Results of Acoustic Analysis

10.3.1 F0 Peaks

Figure 10.1 compares the mean peak $F0$ values of the four accentual phrases as a function of the focal conditions. The data was pooled over all sentences. Phonologically, these peaks correspond to the accentual peak of each initially accented accentual phrase. In Figure 10.1, speaker one shows stair-like descending peaks under N-focus condition; under M-focus condition, the second peak became prominent and all other peaks became less prominent compared to the N-focus condition; under S-focus condition, the second peak became very prominent and other peaks became less prominent than the N- and M-focus conditions. Speaker two's result is by and large similar to that of speaker one, but there are three differences between the two speakers. First, under N-focus condition, the first peak of speaker two was slightly lower than the second peak, thus the overall peak-to-peak configuration did not show a stair-like descending pattern. Second, the peak differences between the M- and S- focus conditions were generally smaller in speaker two's data than in speaker one's. Third, judging from the

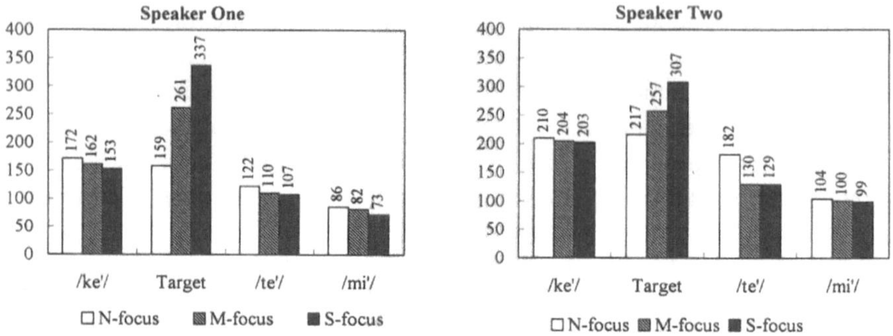

FIGURE 10.1. Averaged peak $F0$ as a function of focal conditions [Hz]. Pooled over the first five repetitions of all sentences.

peak $F0$ values of the second accentual phrase, speaker one used a wider range of prominence in the recording than speaker two.

Table 10.2 summarizes the results of one-way ANOVA on the effect of focus on the peak $F0$ values. The overall effects were statistically highly significant. The sole exception was the first peak of speaker two, which was significant only at the 0.025 level (Note in this paper the expression "highly significant" is used for a significance level higher than 0.001). The results of post hoc tests revealed that only the second peak maintained a significant difference between any pair of the three focal conditions, but all peaks showed significant differences between N- and S-foci. The effect of focus was not localized upon the peak that directly bore focus but was widespread upon both the preceding and following peaks.

TABLE 10.2. Results of ANOVA of the effect of focal condition on $F0$ peaks. Post hoc pairwise comparisons by means of Tukey HSD.

Speaker	Peak	N	Overall effect	Post hoc test by Tukey HSD N vs. M	N vs. S	M vs. S
1	1st	75	<.001	<.051	<.001	<.107
	2nd	75	<.001	<.001	<.001	<.001
	3rd	75	<.001	<.001	<.001	<.396
	4th	75	<.001	<.316	<.001	<.003
2	1st	73	<.025	<.109	<.026	<.793
	2nd	73	<.001	<.001	<.001	<.001
	3rd	73	<.001	<.001	<.001	<.997
	4th	66	<.003	<.022	<.005	<.749

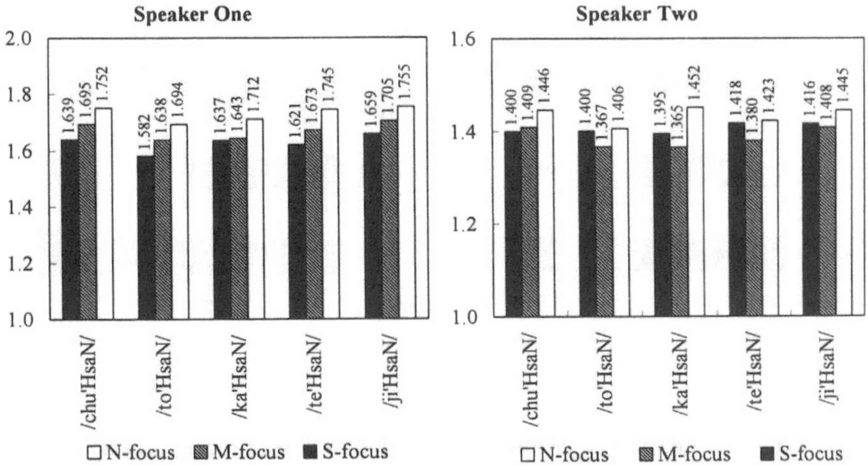

FIGURE 10.2. Total utterance duration as a function of focal conditions [s]. Sentences are distinguished by the target words shown on abscissa.

10.3.2 Utterance Duration

Figure 10.2 shows the total duration of each utterance averaged over the first five repetitions as a function of focal conditions. It is important to note that the total duration differed considerably between the unfocused (N-focus) and focused (M- and S-foci) utterances in both speakers. Closer examination of the figure reveals that the two speakers behaved in slightly different ways: while the utterance duration is inversely correlated to the focus strength in speaker one's data, this is not the case in speaker two. The durations of utterances under S-focus were mostly longer than their M-focus counterparts in speaker two's data. Moreover, in two out of the five sentences of speaker two (i.e., utterances containing /to'HsaN/ and /te'HsaN/), the difference between the durations of N- and S-foci utterances were minimal, although the duration of M-focused utterances showed clear shortening in both cases.

It is also important to note that the duration change caused by focus was not linear over the whole utterance. Figure 10.3 shows the utterance-internal duration change at the level of the accentual phrase. In speaker one's data, the total duration decreased under M- and S-focus conditions, but the duration of the target phrase stayed nearly constant and increased slightly under S-focus. Hence the relative duration of the target phrase became longer when focussed, the ratios of the duration of target phrases

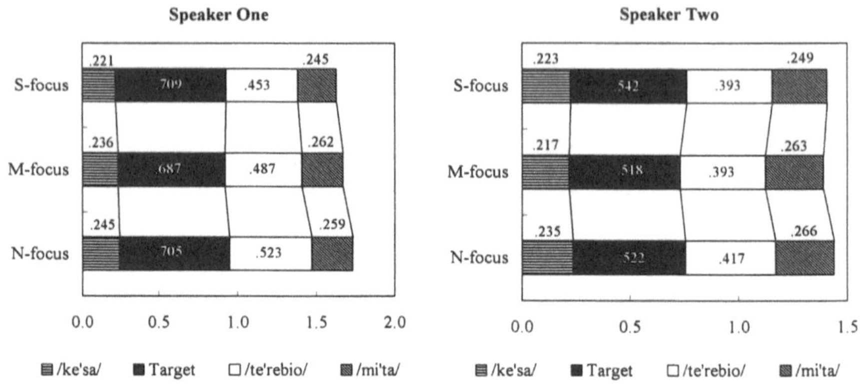

FIGURE 10.3. Accentual phrase duration as function of focal conditions [sec]. Pooled over the first five repetitions of all sentences.

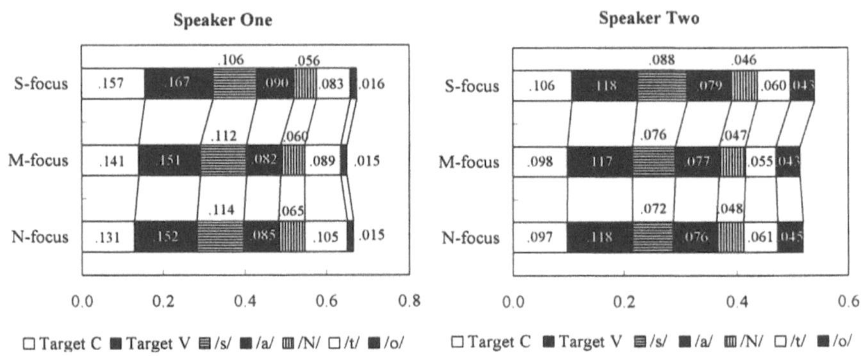

FIGURE 10.4. Averaged duration of segments in the target words [s]. Durations of the first consonant and the following long vowels where averaged across different consonants and vowels. Pooled data over the first five repetitions of all sentences.

relative to the utterance durations being 0.407, 0.411, and 0.435 for N-, M-, and S-foci, respectively.

On the contrary, the non-target accentual phrases were shortened as focus became stronger and contributed to the decrease in the overall utterance durations. The data of speaker two showed the same tendencies, but the relative ratios of the target phrase increased more rapidly than for speaker one: they were 0.362, 0.372, and 0.385 for the three focal conditions.

Figure 10.4 shows the mean durations of the segments contained in the target phrase. Here again, the duration change is not linear. The duration of the target words (i.e., "X-san") increased under S-focus both in speaker one and two's data. The duration change of the grammatical particle "to", however, was speaker dependent. In speaker one's data, the particle duration decreased under M- and S-focus conditions and compensated for the duration change at the level of the accentual phrase, while the particle duration was nearly constant and showed no such compensatory effect in speaker two's data. Finally, the effect of focus on individual vowel segment duration is shown in Table 10.5, which appears in the discussion section. To sum up, Figures 10.2–10.4 and Table 10.5 showed that focus influences the duration of utterance at various levels. They also showed that the influence was greater in speaker one's speech than in speaker two's.

10.3.3 Formant Frequencies

In this section, the effect of focus on vowel formant frequencies is examined. The vowels will be classified into two categories. The first category is called "target vowels" and involves the five accented long vowels in the target words; we will symbolize them as /i'H/, /e'H/, /a'H/, /o'H/, and /u'H/. The second category is called "context vowels" and involves the following: /e/ and /a/ in the time adverbial ke'sa, /a/ in the politeness affix saN, /o/ in particle to, two /e/'s in te'rebi, and /a/ in the verb mi'ta. These context vowels will be symbolized as /ke'/, /sa/, /saN/, /to/, /te'/, /re/, and /ta/, respectively. /saN/ and /to/ were classified as context vowels even though they were in the target phrase, because they were shared by all utterances. Also, their behavior was more similar to the other context vowels than the target vowels as we will see below. Vowels /i/ and /o/ at the end of the third accentual phrase te'rebio were excluded from the analysis because it was impossible to derive a reliable acoustic boundary between the two consecutive vowels. Also, the /i/ of the last accentual phrase mi'ta was excluded because its duration was too short to make consistent segmentation.

As mentioned earlier, the formant estimation was done by autocorrelation method. However, the visual comparison of the DFT spectra and the estimated formant frequencies revealed that the estimation of the focussed target vowels was frequently unsuccessful, presumably because of the high $F0$ of the focussed vowels. In order to remedy this difficulty, all target vowels were reanalysed pitch-synchronously by a novel method of formant estimation described in Ding, Kasuya, and Adachi [DKA95]. The new method simultaneously estimates the parameters of the vocal tract transfer function and the Rosenberg–Klatt voice source model [KK90] based on the ARX (autocorrelation-exogeneous) model. The reanalysis was conducted with a sampling frequency of 10 KHz and 16-bit quantization. The same speech samples used in the earlier analysis were reused after digital downsampling. The formant values obtained by the new method turned out to be generally more reliable than the earlier ones. All subsequent analyses of the target vowels, both focussed and unfocussed ones, are based on the formant data obtained by the reanalysis. Note also that the last five repetitions of

TABLE 10.3. Statistical test on the effect of focus upon the target vowels.

Speaker	Vowels	Focus	N	F1 [Hz] Mean	F1 [Hz] SD	F2 [Hz] Mean	F2 [Hz] SD	Separate ANOVA F1	Separate ANOVA F2	MANOVA F1&F2
1	/i'H/	N	10	356	18	1951	33	F=26.65	F=5.65	F=14.19
		M	9	302	19	1995	42	DF=2,26	DF=2,26	DF=4,48
		S	10	339	13	2005	36	P<0.001	P<0.007	P<0.001
	/e'H/	N	10	455	10	1898	35	F=1.20	F=0.65	F=0.76
		M	10	435	14	1886	40	DF=2,27	DF=2,27	DF=4,50
		S	10	464	73	1929	139	P<0.317	P<0.533	P<0.554
	/a'H/	N	10	719	42	1227	26	F=3.19	F=2.405	F=1.91
		M	10	765	83	1275	63	DF=2,26	DF=2,26	DF=4,48
		S	9	813	108	1278	74	P<0.058	P<0.110	P<0.125
	/o'H/	N	10	437	16	807	27	F=3.98	F=33.3	F=16.21
		M	10	430	30	851	41	DF=2,27	DF=2,27	DF=4,50
		S	10	408	24	933	32	P<0.032	P<0.001	P<0.001
	/u'H/	N	10	340	13	1545	35	F=5.70	F=4.67	F=4.82
		M	10	324	18	1584	44	DF=2,27	DF=2,27	DF=4,50
		S	10	354	27	1604	51	P<0.009	P<0.018	P<0.002
2	/i'H/	N	5	295	15	2247	66	F=14.04	F=0.37	F=7.09
		M	5	256	9	2224	70	DF=2,12	DF=2,12	DF=4,20
		S	5	288	13	2210	72	P<0.001	P<0.698	P<0.001
	/e'H/	N	5	395	20	1897	40	F=0.421	F=0.725	F=0.38
		M	5	380	23	1957	36	DF=2,12	DF=2,12	DF=4,20
		S	5	382	37	1932	128	P<0.666	P<0.504	P<0.818
	/a'H/	N	5	758	19	1302	40	F=1.87	F=7.86	F=3.32
		M	5	808	35	1406	26	DF=2,12	DF=2,12	DF=4,20
		S	5	770	61	1333	56	P<0.196	P<0.007	P<0.031
	/o'H/	N	5	402	27	956	22	F=2.27	F=4.13	F=2.65
		M	5	387	48	879	26	DF=2,12	DF=2,12	DF=4,20
		S	5	351	39	903	67	P<0.145	P<0.044	P<0.064
	/u'H/	N	5	353	28	1488	67	F=16.93	F=0.57	F=8.14
		M	5	282	13	1466	67	DF=2,12	DF=2,12	DF=4,20
		S	5	302	14	1445	57	P<0.001	P<0.578	P<0.001

speaker one's target vowels were analysed in addition to the first five. This is why the number of vowels is greater for speaker one in Table 10.3.

10.3.4 Target Vowels

Figures 10.5 and 10.6 are the F1-F2 scatter plots of the target vowels of speaker one and two. Each data point shows the arithmetic mean over all pitch periods contained in the vowel segment. Data points were classified according to the focal conditions: the digits 0,1, and 2 stand, respectively, for the vowels under N-, M-, and S-focus conditions. Table 10.3 summarizes the means and SD's of F1 and F2 of the target vowels as a function of the focal conditions. In addition, the table shows the results of

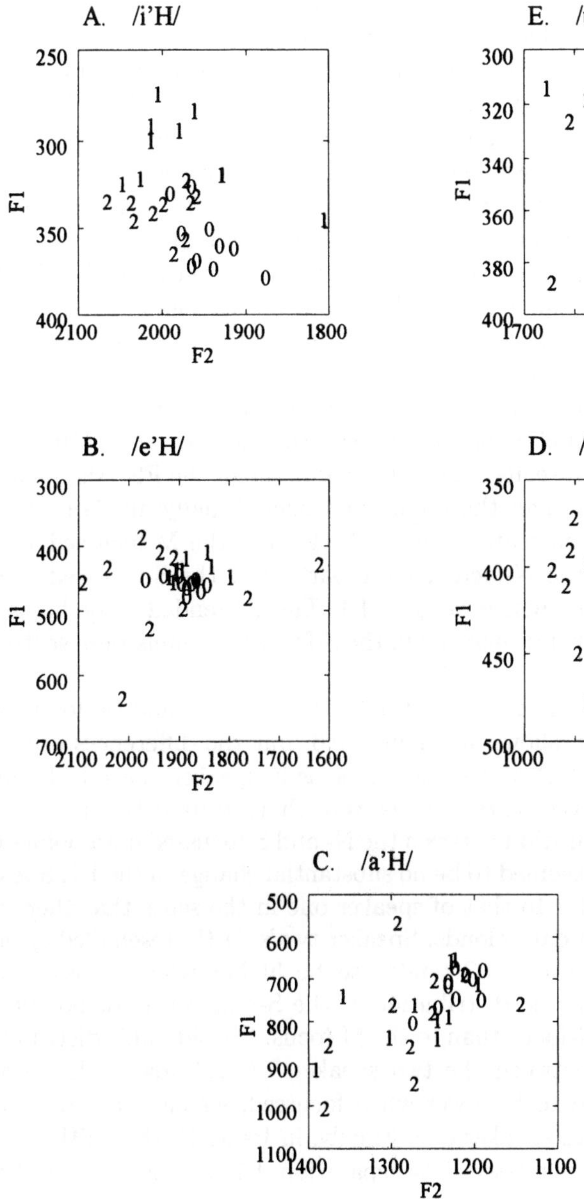

FIGURE 10.5. F1-F2 scatter plots of the five target vowels of speaker one [Hz]. Digits 0, 1, and 2 stand, respectively, for N-, M-, and S-focus conditions.

the statistical tests on the effect of focus. For every vowel, the effects of the three-way difference of focus upon F1 and F2 were tested separately by univariate ANOVA, and two-dimensional MANOVA was applied for the two-dimensional mean vector of F1 and F2. This multidimensional dependent variable is referred to as F1&F2 in the table.

In Panel A of Figure 10.5, both F1 and F2 of the target /i'H/ vowel showed considerable changes according to the focal conditions. The data points under the M-focus condition were characterized by relatively lower F1 and relatively higher F2 when compared to the N-focus counterparts. The data points under the S-focus condition were also characterized by relatively higher F2 but there seems to be no substantial change in the F1. The data clouds of focussed vowels were well separated from the cloud of unfocussed vowels. In Panel B, the /e'H/ vowel did not show any separation of focus-related data clouds. In Panel C, the /a'H/ vowel showed focus-related data separation on both the F1 and F2 axes. Focussed vowels were characterized by higher F1 and F2, but the focus-related separation was not as clear as in Panel A on either axis. In Panel D, the /o'H/ vowel showed clear-cut focus-related data separation along the F2 axis; focussed vowels had higher F2 than the unfocussed ones. Finally, in Panel E, the /u'H/ vowel showed intricate changes. As for F1, the M-focussed vowels had higher F1 than the N-focus counterparts, but the S-focussed vowels were scattered over the whole range of F1. The F2 tended to be higher as focus became stronger, but here again the S-focussed vowels were scattered widely.

The focus-related difference of formant frequencies could be found also in speaker two's data, shown in Figure 10.6, but the differences observed in each vowel were not altogether the same as in speaker one's. In Panel A of the figure, M-focussed /i'H/ vowels were characterized by distinctively lower F1, but the separation between the N- and S-focussed data points was not clear. Also, there seemed to be no substantial change in the F2. Speaker two's /e'H/ was similar to that of speaker one in the sense that there was no clear separation of data clouds. Speaker two's /a'H/ resembled speaker one's in that both F1 and F2 tended to be higher when focussed. But in speaker two's data, the distribution of the S-focussed data points was closer to that of the N-focus than to the M-focus. Considerable distribution difference was seen between the two speakers' /o'H/ vowels. In speaker two's data, F2 tended to be lower when focussed, which is the reverse of the change observed in speaker one. Finally, in Panel E, the /u'H/ vowels showed very clear focus-related data separation; F1 was distinctively lower when focussed.

The results of the statistical tests in Table 10.3 support the impressionistic observations given above. As far as speaker one is concerned, the table indicates that /i'H/, /o'H/, and /u'H/ of speaker one showed statistically significant focus-related formant changes in all three tests; /a'H/ was near the border of the 0.05 level significance, and /e'H/ did not show significance in any one of the three tests. As for speaker two, no vowel achieved significance in all three tests, but /i'H/, /a'H/, /o'H/, and /u'H/ achieved

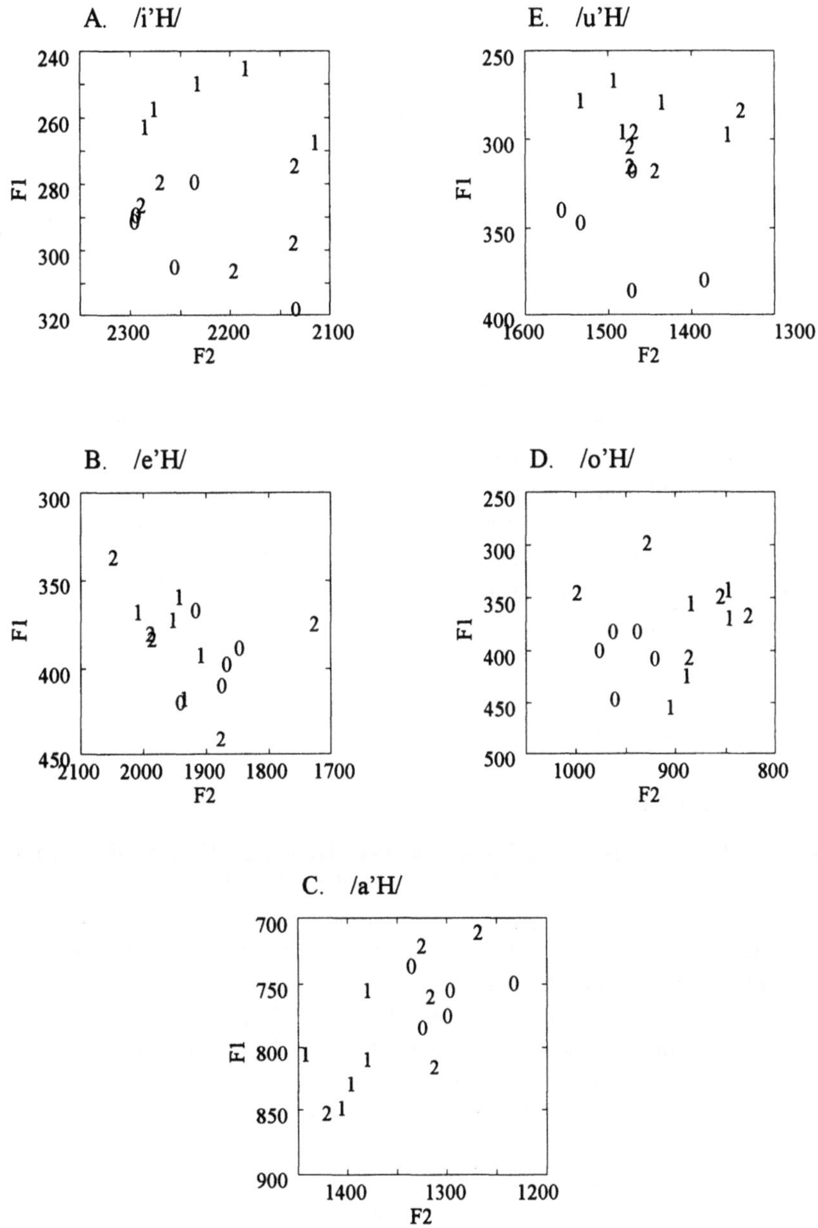

FIGURE 10.6. F1-F2 scatter plots of the five target vowels of speaker two [Hz]. Digits 0, 1, and 2 stand, respectively, for N-, M-, and S-focus conditions.

significance at least in one of the tests. As in speaker one's data, /e'H/ did not show significance.

10.3.5 Context Vowels

Table 10.4 summarizes the results of the context vowels in the same manner of presentation as in Table 10.3. Because context vowels were shared by all sentences, the numbers of observation are larger than the target vowels. Tokens whose estimated pole frequencies were judged to be dubious by a visual comparison to DFT spectra were excluded from the table. This is why some vowels like /saN/ of speaker one have smaller numbers of observation.

The results of the context vowels differ considerably from those of the target vowels. First, many of the context vowels were not influenced by the focal condition in a statistically significant way. Second, although some context vowels were influenced by the focal conditions, the ways they changed were not the same as in the target vowels. For example, /ke'/, /te'/, and /re/ of speaker one continuously lowered their F2's as focus became stronger, while the same speaker did not show any clear focus-related formant change of the target /e'H/ in Table 10.3. These facts will be discussed in detail in Sec. 4.2.

10.4 Discussion

10.4.1 Duration

The results of acoustic analyses revealed that duration could undergo change due to the influence of focus as well as $F0$. This is a new finding in the study of Japanese prosody. When a target phrase was focussed, durations of the preceding and/or following phrases were reduced, while the duration of the target phrase stayed nearly constant. Also, as shown in Table 10.5, focus did not affect the duration of the target vowels in the focussed accentual phrases. This is very different from the duration change in English as reported in Erickson and Lehiste [EL95]. In English, the most remarkable duration change caused by contrastive emphasis consists of a drastic increase of the emphasized constituent per se. One reason that Japanese focus does not lengthen the focussed constituent as in English can be found in the fact that Japanese is a so-called quantity language where the segment duration is a part of the phonological representation of a word. In Japanese all vowels as well as some consonants have a two-way phonological contrast of short vs long. The contrast would not be maintained if a focussed vowel is locally lengthened as in English. In any case, it is important to note that the effect of focus on duration was not localized upon the focussed phrase, but spread over the temporally preceding and following phrases as

TABLE 10.4. Statistical tests on the effect of focus upon the context vowels.

Speaker	Vowels	Focus	N	F1 [Hz] Mean	F1 [Hz] SD	F2 [Hz] Mean	F2 [Hz] SD	Separate ANOVA F1	Separate ANOVA F2	MANOVA F1&F2
1	/ke'/	N	25	497	26	1790	45	F=4.14	F=8.03	F=4.134
		M	25	485	27	1775	43	DF=2,69	DF=2,69	DF=4,134
		S	22	475	22	1740	44	P<0.020	P<0.001	P<0.001
	/sa/	N	25	605	51	1388	96	F=0.875	F=0.336	F=0.609
		M	25	613	44	1404	80	DF=2,72	DF=2,72	DF=4,142
		S	25	596	41	1407	79	P<0.421	P<0.716	P<0.657
	/saN/	N	20	633	48	1295	47	F=0.433	F=1.214	F=0.785
		M	17	628	61	1304	29	DF=2,49	DF=2,49	DF=4,96
		S	15	611	34	1315	32	P<0.651	P<0.306	P<0.538
	/to/	N	24	475	15	1014	59	F=1.408	F=2.624	F=2.579
		M	25	469	11	1038	70	DF=2,71	DF=2,71	DF=4,140
		S	25	470	12	1053	54	P<0.251	P<0.080	P<0.040
	/te'/	N	30	493	11	1861	38	F=2.110	F=38.994	F=17.822
		M	30	489	9	1801	42	DF=2,87	DF=2,87	DF=4,172
		S	30	486	14	1761	51	P<0.127	P<0.001	P<0.001
	/re/	N	25	459	10	1899	50	F=4.881	F=2.772	F=3.063
		M	21	463	13	1876	54	DF=2,61	DF=2,61	DF=4,120
		S	18	470	11	1861	56	P<0.011	P<0.070	P<0.019
	/ta/	N	24	706	38	1587	56	F=1.549	F=2.500	F=1.469
		M	24	696	58	1598	60	DF=2,69	DF=2,69	DF=4,136
		S	24	720	46	1561	63	P<0.220	P<0.089	P<0.215
2	/ke'/	N	25	316	27	1956	99	F=0.463	F=0.963	F=0.947
		M	25	311	28	1920	104	DF=2,71	DF=2,71	DF=4,138
		S	24	318	24	1956	114	P<0.631	P<0.387	P<0.439
	/sa/	N	24	478	132	1404	87	F=0.930	F=1.826	F=1.640
		M	23	503	128	1397	129	DF=2,65	DF=2,65	DF=4,126
		S	21	532	141	1347	100	P<0.400	P<0.169	P<0.168
	/saN/	N	25	220	186	1313	93	F=0.769	F=0.655	F=0.495
		M	24	479	169	1287	88	DF=2,67	DF=2,67	DF=4,130
		S	21	453	198	1284	105	P<0.468	P<0.523	P<0.739
	/to/	N	24	344	31	1187	88	F=1.049	F=3.102	F=1.985
		M	23	356	22	1240	134	DF=2,66	DF=2,66	DF=4,128
		S	22	351	28	1159	108	P<0.356	P<0.052	P<0.101
	/te'/	N	23	330	29	1884	183	F=25.929	F=0.890	F=14.089
		M	25	371	24	1932	206	DF=2,69	DF=2,69	DF=4,134
		S	24	378	22	1854	230	P<0.001	P<0.415	P<0.001
	/re/	N	23	338	21	1837	152	F=11.728	F=0.787	F=9.006
		M	21	363	20	1872	201	DF=2,54	DF=2,64	DF=4,124
		S	23	365	22	1911	237	P<0.001	P<0.459	P<0.001
	/ta/	N	25	542	52	1437	80	F=2.407	F=0.953	F=1.538
		M	25	509	61	1413	51	DF=2,72	DF=2,72	DF=4,140
		S	25	508	72	1425	51	P<0.097	P<0.390	P<0.195

was the case with the $F0$ peaks. This point is relevant to the discussion in Sec. 4.4.

10.4.2 Target Vowels

Figure 10.7 compares the mean F1 and F2 frequencies of the target vowels under the three focal conditions; Panels A and B compare the data under N- and M-foci, and Panels C and D compare the data under N- and S-foci. Panels A and B show that, when focussed, F1-F2 formant space was enlarged along the vertical axis by a decrease in F1 of closed vowels –/i'H/ and /u'H/– and an increase in F1 of open vowel /a'H/. Nearly the same tendency could be observed in Panels C and D, with the sole exception of the speaker one's F1 of /u'H/ under S-focus, which was higher than the N-focus counterpart. These observations suggest that the effect of focus upon the target vowels was manifested primarily by the adjustment of jaw height. Target vowels that bear direct focus become more peripheral, or "hyper-articulated", in vowel height: close vowels become more closed, and open vowels become more open. See de Jong [dJ95] for the stress-related hyper-articulation in English.

However, it is unlikely that the peripherality of the jaw position is the only factor in formant changes. There are at least two more factors. The first is the change in the vertical larynx position. As is well known, the larynx changes its vertical position as $F0$ goes up and down: the larynx is relatively high when $F0$ is high and, conversely, low when $F0$ is low. Since a higher larynx position should result in lower pharynx volume, it is expected that focussed vowels have higher F1 compared to their unfocussed counterpart.

The second factor is the physiological interaction between the tongue and the larynx. Since the tongue and the larynx are anatomically coupled via the hyoid bone, it is expected that the raised larynx pushes the tongue forward, and, conversely, the tongue is pulled backward as larynx goes down. The MRI data presented in Honda *et al.* [HHD94] showed clear differences of the overall tongue shape among the /a/ vowels uttered with various $F0$ levels and contours. Hirai [Hir95a] analysed the effect of various $F0$ levels and changing $F0$ contours upon the formant frequencies of vowels uttered by three male subjects and found that F2 increased proportionally to $F0$ in all vowels and subjects. This tendency could be found in the data of speaker one in Table 10.3, where all target vowels had higher F2 under S-focus than under N-focus. However, the tendency was not clear in the data of speaker two in the same table.

There seems to be two mutually related reasons for the inconsistency between Hirai's experiment and the present one. For one, the two experiments were conducted under very different pitch ranges. In Hirai's experiment, $F0$ was controled in relatively narrow pitch range of 100–200Hz, while the upper limit of the pitch range used in our experiment exceeded 300 Hz. For another, and more importantly, the verbal behaviours required in the two experiments were qualitatively very different. The subjects of the current experiment were instructed to "emphasize" (put focus upon)

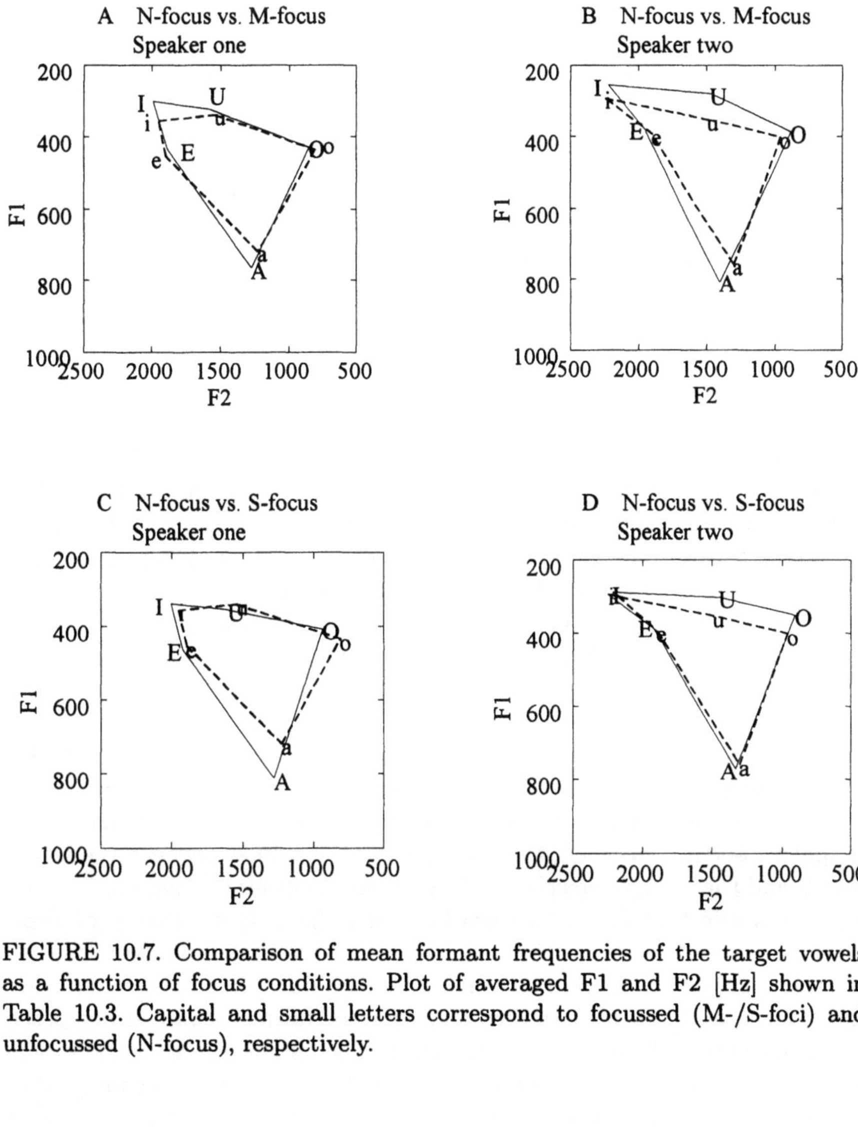

FIGURE 10.7. Comparison of mean formant frequencies of the target vowels as a function of focus conditions. Plot of averaged F1 and F2 [Hz] shown in Table 10.3. Capital and small letters correspond to focussed (M-/S-foci) and unfocussed (N-focus), respectively.

the designated element of meaningful utterances, while in Hirai's experiment subjects were requested to utter several nonsense vowel sequences like /aia/ while changing the $F0$. As Hirai noted at the end of his paper, the implementation of linguistic emphasis may contain positive controls of articulatory gestures planned at the higher level of speech production process in addition to the low level tongue-larynx interaction. It may be that the expanded formant space found in the current experiment is one of the instances of such positive "linguistic" control.

10.4.3 Context Vowels

The formant changes observed in the context vowels showed considerable inter-speaker variability. Because speaker one was by far more sensitive to focal conditions than speaker two, I will mostly concentrate upon the data of this speaker. Panels A, B, and C of Figure 10.8 show the F1-F2 scatter plots of speaker one's /ke'/, /te'/, and /re/. As mentioned earlier, the focussed vowels had lower F2 than their unfocussed counterparts. This change is qualitatively different from the one observed in the target /e'H/ vowel of the same speaker in three respects. First, focus influenced mid vowel /e/ which was insensitive to the focus conditions in the target vowel position in both speakers. Second, it was the F2 rather than the F1 that changed. From an articulatory point of view, the change seems to be concerned with the frontness of tongue constriction rather than the jaw height. Third, and most important, the change in the F2 of /e/ so happened that the vowel became less peripheral in terms of frontness. The focus-related F2 changes observed in these vowels were statistically highly significant as shown in Table 10.4. At the same time, it is equally important that focus-related formant change was not observed in all context vowels. Panels D, E, and F of Figure 10.8 show the F1-F2 scatter plots of /sa/, /saN/, and /ta/ of speaker one. Here, it is impossible to see any separation of data clouds in relation to the focal conditions.

Although it is impossible to provide a full-fledged explanation for the variablitiy of the context vowels at this stage of inquiry, a tentative explanation can be presented. Table 10.5 shows the mean vowel durations as a function of the focal conditions and the results of one-way ANOVA. As far as speaker one is concerned, the results of ANOVA differ between the target vowels and the context vowels. Whie all context vowels showed statistically significant effect of focus on duration with the only exception of /sa/, the target vowels were not influenced by focus with the single exception of /u'H/. Moreover, it is interesting that the durations of speaker one's /ke'/, /te'/, and /re/ that showed focus-related formant changes decreased monotonically as focus became stronger. The monotonical decrease was observed also in this speaker's /sa/ and /to/. (Note, in passing, that speaker two's result was similar to speaker one's in that all target vowels were not influenced by focus.) Next, Table 10.6 shows the Pearson correlation coefficients calculated between the vowel durations and the F1, F2 frequencies. Here again the context vowels showed higher correlation than the target vowels. The data shown in Tables 10.5 and 10.6 suggest the interpretation that focus-related formant changes of context vowels were evoked mainly as a result of the duration change: namely, the lower F2 of /ke'/, /te'/, and /re/ could be regarded as the undershooting of formant.

If duration is the principal factor of focus-related formant change in the context vowels, it is expected that the duration of vowels that showed focus-

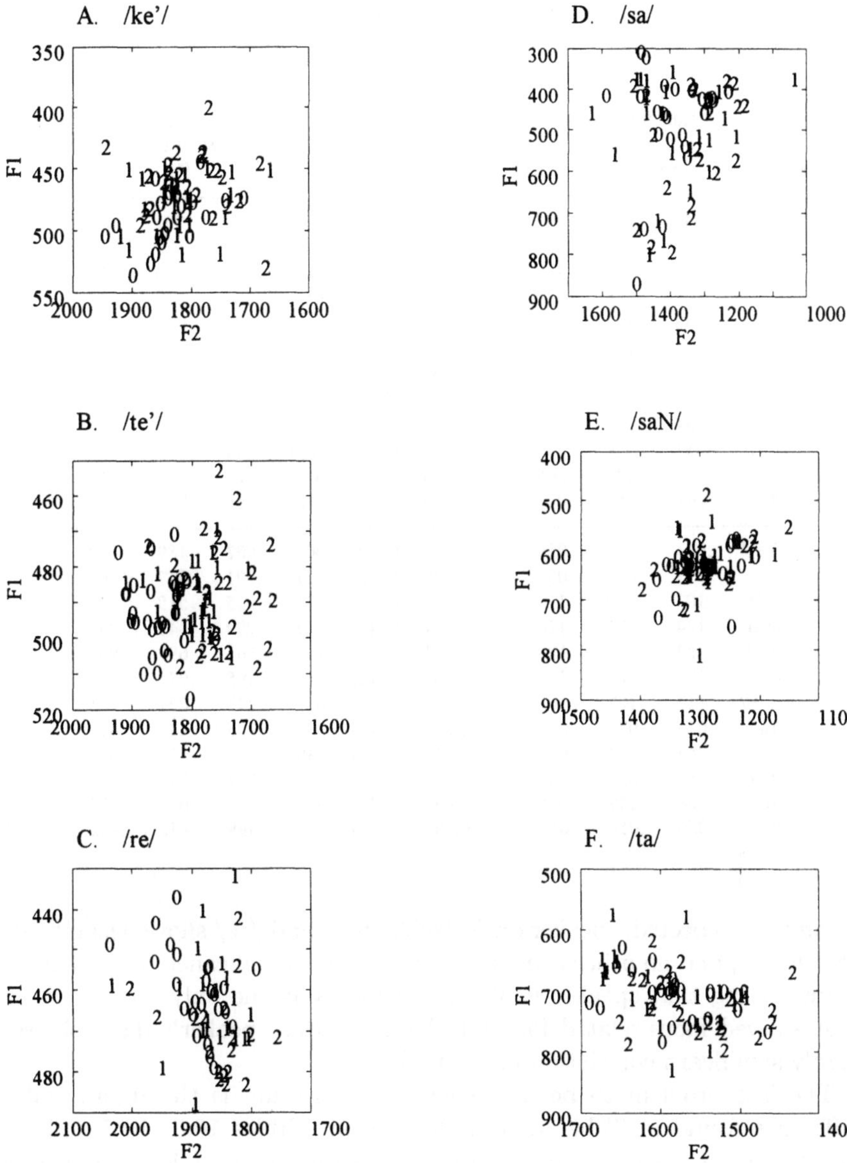

FIGURE 10.8. F1-F2 scatter plots of speaker one's context vowels [Hz]. Data points were classified according to the focal conditions as in Figures 10.5 and 10.6.

related formant changes correlate also with the peak $F0$ of the accentual phrase, because focus is manifested in $F0$ in any way. Table 10.7 shows the Pearson correlation coefficients calculated between the vowel durations and the peak $F0$ values of the accentual phrases to which the vowels

TABLE 10.5. Mean vowel durations as a function of focal conditions. See text for the symbols used for each vowel.

		Mean duration [sec] under			ANOVA			Tukey HSD		
		N-	M-	S-				N vs.M	N vs.S	M vs.S
Speaker	Vowel	focus	focus	focus	F	DF	P<	P<	P<	P<
1	/ke'/	.040	.037	.033	14.622	(2,72)	.001	.070	.001	.007
	/sa/	.075	.072	.067	1.662	(2,71)	.197	.676	.170	.606
	/i'H/	.162	.152	.167	1.314	(2,12)	.305	.526	.888	.292
	/e'H/	.162	.154	.176	3.677	(2,12)	.057	.663	.217	.051
	/a'H/	.148	.162	.177	2.985	(2,12)	.089	.218	.088	.843
	/o'H/	.156	.147	.162	1.839	(2,12)	.201	.513	.726	.179
	/u'H/	.134	.139	.151	4.133	(2,12)	.043	.682	.040	.166
	/saN/	.087	.084	.090	5.585	(2,71)	.006	.183	.263	.004
	/to/	.054	.051	.050	4.791	(2,70)	.011	.122	.009	.555
	/te'/	.054	.054	.049	5.796	(2,72)	.005	.944	.019	.008
	/re/	.077	.071	.065	12.737	(2,70)	.001	.027	.001	.052
	/ta/	.041	.050	.041	4.906	(2,72)	.010	.056	.807	.011
2	/ke'/	.063	.054	.059	6.216	(2,72)	.003	.002	.189	.189
	/sa/	.071	.064	.067	2.270	(2,72)	.111	.091	.517	.561
	/i'H/	.107	.109	.113	0.660	(2,12)	.535	.935	.519	.727
	/e'H/	.124	.124	.128	0.277	(2,12)	.277	.999	.341	.341
	/a'H/	.122	.126	.128	0.351	(2,12)	.711	.861	.693	.951
	/o'H/	.127	.122	.118	2.881	(2,12)	.095	.395	.081	.567
	/u'H/	.112	.107	.105	3.095	(2,12)	.082	.242	.077	.761
	/saN/	.076	.077	.079	1.003	(2,72)	.372	.996	.417	.470
	/to/	.045	.043	.043	0.328	(2,72)	.238	.362	.264	.978
	/te'/	.052	.052	.055	2.337	(2,72)	.104	.954	.209	.119
	/re/	.068	.059	.056	19.972	(2,72)	.001	.001	.001	.355
	/ta/	.056	.056	.052	0.362	(2,72)	.362	.999	.432	.432

belong. As expected, speaker one's /ke'/, /te'/, and /re/ showed significant correlation, and no other context vowels showed significance. In passing, it is interesting that speaker two's /re/, which was the only context vowel that showed focus-related formant change in this speaker's data, showed highly significant correlation in the table.

The last problem to be discussed in this section is the inter-speaker difference found in Table 10.4. Judging from Tables 10.5- 10.7 and the duration data presented in Sec. 10.3.2, it is obvious that focus influences the duration of an utterance at various levels in both speakers' speech. But the effect of focus on the formant frequencies of context vowels differed considerably for each speaker. Perhaps, this is related to the magnitude with which each speaker changed the duration of his speech under the effect of focus. As mentioned earlier in Sec. 10.3.2, the effect of focus upon duration was less evident in speaker two's speech. Most probably, this is a direct consequence of the fact that the range of prominence recorded in the current experiment was wider in speaker one's data than in speaker two's. It may be that speaker two would show clearer focus-oriented formant

TABLE 10.6. Pearson correlation coefficients between the vowel duration and F1, F2.

		Correlation of duration and			
Speaker	Vowel	F1	P<	F2	P<
1	/ke'/	.342	.003	.403	.001
	/sa/	-.221	.059	.486	.001
	/i'H/	.211	.450	-.114	.686
	/e'H/	-.076	.787	.666	.007
	/a'H/	.035	.900	.154	.583
	/o'H/	.310	.261	.179	.524
	/u'H/	-.098	.728	.291	.292
	/saN/	-.039	.742	.160	.178
	/to/	.194	.099	-.222	.059
	/te'/	.130	.265	.239	.039
	/re/	-.150	.213	.446	.001
	/ta/	.563	.001	-.382	.001
2	/ke'/	.273	.020	.345	.003
	/sa/	-.455	.001	-.066	.588
	/i'H/	.034	.904	.244	.381
	/e'H/	-.352	.198	-.065	.819
	/a'H/	-.004	.988	.430	.109
	/o'H/	.498	.059	.289	.297
	/u'H/	.508	.053	-.190	.498
	/saN/	-.039	.741	.157	.178
	/to/	.082	.485	-.096	.413
	/te'/	.163	.161	-.032	.784
	/re/	-.366	.002	-.106	.376
	/ta/	.265	.022	.099	.396

change in his context vowels if he varied the range of prominence as widely as speaker one.

Concluding Remarks

The interpretation presented in the previous sections gives rise to two issues in the phonology of prosody. First, the interpretation depends crucially upon the existence of an omnidirectional effect of focus. In all the phonetic dimensions examined in this study, i.e., $F0$, duration, and formant frequency, focus influenced not only the target but also both the

TABLE 10.7. Pearson correlation coefficients between $F0$ and context vowel furation.

Speaker	Vowel	N	Corr.	P<
1	/ke'/	72	.469	.001
	/sa/	72	.230	.049
	/saN/	74	.217	.064
	/to/	75	.138	.238
	/te'/	75	.275	.017
	/re/	73	.658	.001
	/ta/	75	.033	.781
2	/ke'/	73	.018	.880
	/sa/	73	-.077	.516
	/saN/	73	.114	.336
	/to/	73	-.177	.135
	/te'/	73	-.130	.272
	/re/	73	.570	.001
	/ta/	75	.246	.047

preceding and following constituents. In Figure 10.1 and Table 10.2, focus reduced the $F0$ peak of both the preceding and following accentual phrases in speaker one's data. Omnidirectional reduction was observed also in the duration of accentual phrases in both speakers (see Figure 10.3). And, as a consequence of the duration reduction, presumably, context vowels /ke'/ in the first accentual phrase and /te'/ in the third both showed considerable F2 reduction in speaker one.

These observations are in contradiction to the assumption held in the current theory of intonation as represented by Pierrehumbert and Beckman [PB88]; the theory presumes that the effect of focus does not affect the part of utterance that precedes the focussed constituent. It may be that the theory needs revision with regard to the treatment of focus, because there are independent studies like Fujisaki and Kawai [FK88] and Kori [Kor89b] that report the existence of the regressive effect of focus on $F0$ in Japanese. Also, see Grønnum [Grø95] for the regressive effect of focus on $F0$ in Dutch, and Erickson and Lehiste [EL95] for clear regressive effect of focus on duration in English. As for the progressive effect of focus on temporally following constituents, it is not clear if Pierrehumbert and Beckman [PB88] presumes the effect. But the effect does exist and makes substantial contribution to the realization of linguistic information like the distinction between WH and Yes-No questions ([Mae91, Mae94]).

The next issue is concerned with the effect of focus within a single accentual phrase; it is an interesting question to know whether the effect of focus can be different depending on the location within an accentual phrase. The duration data presented in Figure 10.4 suggests that the effect of focus is not uniformly distributed within a phrase. In this respect, Hattori's claim that the initial segments of an accentual phrase (i.e., his "prosodeme") were stronger in intensity and clearer in articulation than the following segments ([Hat61]) is of particular interest, because in the current experiment, the vowels that showed focus-related formant changes were located accentual phrase initially with the sole exception of the context vowel /re/. From this, it can be stipulated that the phrase-initial position is more sensitive to a change in prominence than the other positions. However, it is possible to propose an alternative interpretation for the coincidence. Because the vowels that showed formant changes were all accented ones, again with the exception of /re/, it is equally possible to claim that a tonally linked vowel, i.e., a vowel that is associated with a phonological tone—an accent in this case, was more sensitive to a change in prominence than a tonally unlinked vowel. Unfortunately, it is impossible to evaluate the validity of these interpretations at the present, because the current speech material involves only initially-accented accentual phrases. The evaluation should be the objective of a further study in which accentual phrases varying in the presence and/or location of accent would be examined.

Acknowledgments

The author is very grateful for the courtesy of Hideki Kasuya and Wen Ding of Utsunomiya University, who, upon the request by the author, conducted the acoustic analysis by their novel method of formant and voice source parameter estimation. His gratitude also goes to Donna Erickson of the Ohio State University and an anonymous referee who gave very helpful comments to an earlier manuscript of this study.

References

[BP86] M. Beckman and J. Pierrehumbert. Intonational structure in Japanese and English. *Phonology Yearbook 3*, pp. 255–309. 1986.

[dJ95] K. de Jong. The supraglottal articulation of prominence in English: Linguistic stress as localized hyperarticulation. *J. Acoust. Soc. Am.*, 97:491–504, 1995.

[DKA95] W. Ding, H. Kasuya, and S. Adachi. Simultaneous estimation of vocal tract and voice source parameters based on an ARX model.

IEICE Transactions on Information and Systems, E78-D, 6:738–743, 1995.

[EL95] D. Erickson and I. Lehiste. Contrastive emphasis in elicited dialogue: durational compensation. In *Proceedings of the 13th International Congress of Phonetic Sciences, Stockholm, Sweden*, Vol. 4, pp. 352–355, 1995.

[FK88] H. Fujisaki and H. Kawai. Realization of linguistic information in the voice fundamental frequency contour of the spoken Japanese. *Ann. Bull. RILP*, 22:183–191, 1988.

[Grø95] N. Grønnum. Superposition and subordination in intonation: A non-linear approach. In *Proceedings of the 13th International Congress of Phonetic Sciences, Stockholm, Sweden*, Vol. 2, pp. 124–131, 1995.

[Hat61] S. Hattori. Prosodeme, syllable structure and laryngeal phonemes. *Bulletin of the Summer Institute of Linguistics*, Vol. 1: Studies in Descriptive and Applied Linguistics pp. 1–27. Tokyo: International Christian University. 1961.

[HHD94] K. Honda, H. Hirai, and J. Dang. A physiological model of speech production and the implication of tongue-larynx interaction. In *Proceedings of the International Conference on Spoken Language Processing, Yokohama, Japan*, Vol. 1, pp. 175–178, 1994.

[Hir95a] H. Hirai. F_0 henkani tomonau boinno forumanto shuuhasuuno sen'i (changes of vowel formant frequency due to F_0 change). Technical Report SP94-102, IEICE, 1995.

[KK90] D. H. Klatt and L. Klatt. Analysis, synthesis, and perception of voice quality variations among female and male talkers. *J. Acoust. Soc. Am.*, 87:820–857, 1990.

[Kla76] D. H. Klatt. Linguistic uses of segmental duration in English: Acoustic and perceptual evidence. *J. Acoust. Soc. Am.*, 59:1208–1221, 1976.

[Kor89a] S. Kori. Fookasu jitsugenni okeru onseino tsuyosa, jizokujikan, F_0 no eikyoo (acoustic manifestation of focus in Tokyo Japanese: the role of intensity, duration and F_0). *Onsei Gengo (Studies in phonetics and speech communication)*, 3:29–38, 1989.

[Kor89b] S. Kori. Kyoochoo to intoneeshon, (focus and intonation). In M. Sugito, editor, *Nihongono Onsei on'in*, Vol. 2 of *Kooza Nihongoto Nihongo-kyooiku*. Tokyo: Meijishoin, 1989.

[KTS92] N. Kaiki, K. Takeda, and Y. Sagisaka. Linguistic properties in the control of segmental duration for speech synthesis. In G. Bailly, C. Benoît, and T. R. Sawallis, editors, *Talking Machines: Theories, Models, and Designs*, pp. 255–263. Amsterdam: Elsevier Science, 1992.

[Mae91] K. Maekawa. Perception of intonational characteristics of wh and non-wh questions in Tokyo Japanese. In *Proceedings of the 12th International Congress of Phonetic Sciences, Aix-en-Provence, France*, Vol. 4, pp. 202–205, 1991.

[Mae94] K. Maekawa. Is there "dephrasing" of the accentual phrase in japanese? In *Working Papers in Linguistics*, Vol. 44, pp. 146–165. The Ohio State University, 1994.

[McC68] J. D. McCawley. *The Phonological Component of a Grammar of Japanese.* Mouton, The Hague, 1968.

[Miy27] K. Miyata. Atarashii akusentokanto akusentohyookihoo (a novel theory of accent and accent notation). *Study of Sounds*, 1, pp. 18–22, Tokyo: Phonetic Society of Japan, 1927.

[PB88] J. B. Pierrehumbert and M. E. Beckman. *Japanese Tone Structure.* Cambridge, MA: MIT Press, 1988.

[Sug82] M. Sugito. Nihongo akusentono kenkyu (studies in Japanese accent). Tokyo, Sanseido, 1982.

Part III

Prosody in Speech Synthesis

11
Introduction to Part III

Gérard Bailly

11.1 No Future for Comprehensive Models of Intonation?

The general problem of speech synthesis is to produce the most intelligible and natural audio (and visual) stimuli from a set of linguistic, paralinguistic, or emotional instructions. While traditional text-to-speech systems are intrinsically limited by the automatic processing of text to the rendering of lexical, syntactic, and some semantic information, the insertion of speech synthesis in more sophisticated person-machine communication systems—including person-person interpreted dialogue—requires the generation of far variable speaking styles than so far studied. The prosodic generators described in this paper all use automatic learning: the prosodic parameters most likely to be associated with the instructions mentioned above are generated by an associator. This associator, whether regression tree, multinomial regression, neural network, etc., is learned through examples of natural associations. Does this mean that prosodic research will be restricted only to identifying the most appropriate instruction descriptors and should therefore focus only on labelling?

11.2 Learning from Examples

Empirical studies share common methodological problems. We will attempt here to elucidate some of these problems with reference to some of the points made in the papers in this section.

11.2.1 The Reference Corpus

The first methodological problem is how to obtain relevant training material. A speech segment is subject to the interaction of five main components: (a) motor control and production constraints, often referred to as coarticulation effects; (b) perceptual constraints; (c) structure of the language; (d) intrinsic environmental conditions—speaker's commitment, mood, and listener's expectations; and (e) extrinsic environmental conditions—such as signal-to-noise ratio. The essential feature of speech is its ability to adapt to these constraints in order to ensure the most efficient—economical in all terms—communication. In order for a model to adequately learn the strategy followed by a particular speaker in a certain situation, the reference corpus must have a sufficient statistical coverage of the constraints mentioned above. As mentioned by Jan van Santen, an unconstrained recording of spontaneous speech will often result in a corpus with severe sparcity of factors. The lexicon daily used by speakers is often limited to a few thousand words: in the phonetic dictionary for French "De A à Zut" [BT92a], 102137 lexical items and 304752 sounds from various sets of conversations have been segmented by hand. The number of occurrences `N` for each word is a drastic function of its rank ($R : N = 10.75 - 1.25\log(R)$), e.g., the most frequent word "De" (of) occurs 3305 times whereas the word "appeler" (to call) with rank 826 only occurs 10 times, and more than 4000 words only occur once. Three solutions may resolve this sparcity: (a) introducing experimental procedures to directly or indirectly control the speaker's production: from read to real spontaneous speech the span of possible experimental designs is large, (b) using the generalization abilities of statistical models such as regression trees (see also Hirai et al.), neural networks (see Fujio et al.) or multinomial regressions (see van Santen); or (c) using models which can be trained separately in order to avoid the combinatory explosion (see the use of a superposition model in Hirai et al. and the criticism of two-stage models for duration control in van Santen). But the problem of sparcity is perhaps ill-posed, as pointed out by Campbell: If the reference corpus reflects the intended characteristics of the target speech, then the phonetic balance represented in the corpus is adapted to the degrees-of-freedom imposed by the application; and the synthesis can thus be considered as an information retrieval process which selects the reference speech segments that most closely match the target characteristics. This argument is completed by naturalness constraints: concatenative synthesis outperforms complex rule-generated synthetic speech because prosodic deformation of natural samples does not introduce disruptive artifacts such as those introduced by speech production modelling or stylization procedures. The best way to preserve the intrinsic naturalness of the sounds of the reference database is to leave most parts of the speech untouched.

11.2.2 Labelling the Corpus

Once the database has been collected according to some given experimental settings, the audio-visual signals and the hypothesised constraints which the speaker obeyed during his oral production have to be explicitly associated in the reference database. These symbolic descriptions range from subjective general description of the emotional state of the speaker, and environmental conditions, to phonological descriptions closely linked with the actual realizations. The limit between the descriptions of the intended speech acts and their actual realization is fuzzy and often considered as information best expressed in terms of features—thus subject to interpretation—vs measurable variables.

11.2.2.1 Pauses

Pause insertion is one such phenomenon, which is often treated at both symbolic and numerical levels: pause insertion is often considered as part of the phonological description, i.e., the pause is treated as part of the phonetic string, delimits and determines largely phonological constituents, and is predicted from linguistic structure (see Fujio *et al.*). Whereas the absence/presence of a pause is then used as a feature to determine duration and melodic contours, the generation of its own duration is often not considered with the same care.

However, the work done by Grosjean and colleagues [GG83, MG93] has shown that pause durations encode fine details of the linguistic structure and suggests that the pause and the rhyme of the preceding syllable function as a coherent rhythmical unit (see also the similarities between pause and prosodic phrase boundary locations in Fujio *et al.*). A computational model has been proposed in [BB96], showing that it is possible to compute both pause insertion and duration from boundary strength and speech rate without any featural interface. This example shows that not only the choice and ordering of prosodic predictors may affect the performance of automatic learning but also the judicious use of theoretical models of prosodic structure.

11.2.2.2 Discourse Structure

Campbell's paper in this section (cf., also Black's earlier) is concerned with the nature of high level discourse information that could be relevant for describing speaker–listener interactions for an interpreted dialog. The advantage of such synthesis applications is their intrinsic rich description of the synthesis task: in order to generate an adequate part-of-speech the dialog system must have a complete and accurate view of the current state of the dialog. How far do the features predicted by the dialog system describe the actual intentions of a real speaker? Each aspect of the speech enriches the shared knowledge by the speaker and the listener in order to further the purpose of the interaction. The prosodic structure may only carry a small part of this information and the speaker makes at each moment important choices according to what is most relevant for him and what he supposes the listener already knows and has understood. Thus analysis of such dialogs should face sparcity, underspecification and uncompleteness.

11.2.3 The Sub-Symbolic Paradigm: Training an Associator

Once the reference corpus has been designed, recorded, and labelled, diverse learning paradigms have been proposed, including CART techniques, multinomial interpolators, or neural networks. These tools share a common feature: they produce a least-square fit of reference numerical values with predicted values. As previously argued by van Santen [vS92], such an optimization procedure should be preceded by a selection of features affecting particular subsets of sounds. Following this procedure, prosodic trajectories are thus predicted on a sound-by-sound basis. The cohesion of the temporal structure of speech events and the kinematics of the trajectories is thus only maintained through the phonological relationship these sounds entertain with higher levels. When phonological and phonotactic information is given with sufficient precision, such systems might be able to give an accurate prediction. However, since prediction errors are rarely null, it should be perhaps interesting to evaluate the perceptual relevance of these errors. The appropriate answer to this question should be to incorporate more perceptual knowledge in the processing of reference prosodic data. Various intonation models for automatically simplifying the melodic curve have been proposed to date [tB93a, dM95] including the one used by Campbell [HNE91]. Similarly, we may question whether sound durations contribute equally to the perception of momentary tempo. For example, perceptual experiments described in [BB94] showed that listeners are more sensitive to errors in relative timing of vocalic onsets than in errors uniformly distributed among segments, and the paper by Kato in his section shows that we are sensitive to durational variation in units larger than a single segment, arguing for criteria that include human perceptual factors when evaluating prediction models.

11.2.4 The Morphological Paradigm

These functional models of intonation may adequately describe the prosody used by reference speakers. They do not however supply any comprehensive model of the speaker–listener interaction. One could claim that this lack of theoretical background is the key to the success of sub-symbolic approaches in speech technology. Synthetic speech may have prosody as planes have wings: wings assume some common function but need not be as rapidly moving as those of birds (although the astute reader may argue that birds do not need airports and can land on electric cables). In the present case prosody is charged with carrying information on the human's concrete and abstract world: prosodic patterns such as pitch accents, phrase accents, and boundary tones are phonological objects on which humans belonging to a certain linguistic community have agreed. It is thus quite legitimate to strongly base our prosodic models on the way people do use prosody to encode "what they think and how they feel about what and when they say" [Bol89]. Our speech communication system is robust because it is redundant: coarticulation, and anticipatory patterns are effectively used by the perceptual system. Each sound thus contributes at more than one prosodic level; all melodic descriptions include at least a *phrase* level, to control utterance chunking, as well as signalling *word* accents. How many such prosodic elements overlap in time? What are their phonetic characteristics and how do they combine? These questions are still open: however some morphological decompositions of the $F0$ curve have already been proposed: Thorsen [Tho83] proposes a model of Danish intonation where word accent are added to an Overlap-and-Add of declination lines. A similar approach was also proposed by O'Shaughnessy and Allen for English [OA83], Gårding for Swedish [Går91], and of course by Fujisaki [FS71a] for Japanese. A more generic approach involving an Overlap-and-Add of intonation contours has been proposed by Aubergé [Aub92] and is actually further developed (see[MBA95]).

Conclusions

The experimental approaches put forward in this chapter make claims for three main research efforts.

First, both the use of statistical methods to automatically learn phonetic models and the search for morphological descriptors claim for the creation of large prosodically labelled corpora of speech data. The use of phonological descriptors should not obscure the concrete prosodic phenomena and should be used as filters to enable speech researchers to extract lawful variability from prosodic signals: discourse intonation not only encodes salience and segmentation but also more subtle attitudes and affects with different *illocutory forces*, and the predominance of "prominence" in current prosodic phonology may hide more global patterns.

Second, much care should also be taken when defining and designing these research corpora: the development of experimental physics at the beginning of the 18th century demonstrated that the ultimate structure of nature may be captured and clever ideas validated only by using well-controlled experiments. Recording predefined text with different attitudes or emotions [Moz95] while reading or by memorising it, or describing figures appearing on a computer screen [SC92] are just two examples of such research paradigms. Many more experimental settings have to be invented.

Third, as the number of instructions which the synthetic prosodic generator should obey increases, the combinatory explosion of interacting factors will have to be mastered. The coherence and the statistical significance of the data needed to train associators or to store prototypical templates will be difficult to maintain as long as comprehensive models of rhythm and intonation are not able to propose phonetic models where segments and prosody can be processed in parallel [PMK93] and where discourse structure and phrasal levels are well identified. Superpositional models which combine prosodic contours with specific operators have already been used in synthesis [FS71a] as well as in recognition experiments [NS94, JMDL94]. Linear models combined with non-overlapping prototypical prosodic templates have the advantage of simplifying the composition/decomposition problem and may drastically limit the amount of learning data required: user-specific templates may be learned separately for each representation level (discourse, turn, paragraph, sentence, proposition, group ...) as micro-prosody contours are often added to a melody free of segmental effects.

I claim that a better understanding of the phonetic representation used to encode the various functions of prosody is the right way to understand strategies used by speakers even within spontaneous corpora. Psycholinguistic experiments often use constraints or perturbations together with control experiments to examine the robustness of human biological systems. Arguments against the use of carefully controlled data always focus on differences between read speech and freer speaking styles [Bla95]. We should also focus on their similarities; linguistic and prosodic strategies often preserve local syntactic or prosodic structures, as already observed by Fónagy and others:

> *"The opinion that any utterance is an instantaneous production ...applies only to the domain of modern poetry. Except for this limited domain, our utterances are made of large prestored pieces... "* (Translated from [FBF84, p.182])

References

[Aub92] V. Aubergé. Developing a structured lexicon for synthesis of prosody. In C. Benoît, G. Bailly, and T. R. Sawallis, editors, *Talking Machines: Theories, Models, and Designs*, pp. 307–321. Amsterdam: Elsevier Science, 1992.

[BB94] P. Barbosa and G. Bailly. Characterization of rhythmic patterns for text-to-speech synthesis. *Speech Communication* 15: pp. 127–137, 1994.

[BB96] P. Barbosa and G. Bailly. Generation of pauses within the z-score model. In J. P. H. van Santen, R. W. Sproat, J. P. Olive, and J. Hirschberg, editors, *Progress in Speech Synthesis*. New York: Springer-Verlag, 1997.

[Bla95] E. Blaauw. On the Perceptual Classification of Spontaneous and Read Speech. Ph.D. thesis, OTS Dissertation Series, Utrecht University. ISBN 90-5434-045-2, 1995.

[Bol89] D. L. Bolinger. *Intonation and its Uses.* London: Edward Arnold, 1989.

[BT92a] L. J. Boë and J. P. Tubach. De A à Zut: dictionnaire phonétique du français parlé. *ELLUG—Université Stendhal*, 1992.

[dM95] C. d'Allessandro and P. Mertens. Automatic pitch contour stylization using a model of tonal perception. *Computer Speech and Language* 9:257–288, 1995.

[FBF84] I. Fónagy, E. Bérard, and J. Fónagy. Clichés mélodiques. *Folia Linguistica* 17:153–185, 1984.

[FS71a] H. Fujisaki and H. Sudo. A generative model for the prosody of connected speech in Japanese. *Annual Report of Engineering Research Institute* 30:75–80, 1971.

[Går91] E. Gårding. Intonation parameters in production and perception. In *Proceedings of the XIIème International Congress of Phonetic Sciences, Aix-en-Provence, France*, Vol. 1, pp. 300–304, 1991.

[GG83] J. P. Gee and F. Grosjean. Performance structures: a psycholinguistic and linguistic appraisal. *Cognitive Psychology*, 15:418–458, 1983.

[HNE91] D. Hirst, P. Nicolas, and R. Espesser. Coding the F_0 of a continuous text in French. In *Proceedings of the XIIème International Congress of Phonetic Sciences, Aix-en-Provence, France*, Vol. 5, pp. 234–237, 1991.

[JMDL94] U. Jensen, R. Moore, P. Dalsgaard, and B. Lindberg. Modelling intonation contours at the phrase level using continuous density hidden Markov models. *Computer Speech and Language* 8:247–260, 1994.

[MBA95] Y. Morlec, G. Bailly, and V. Aubergé. Synthesis and evaluation of intonation with a superposition model. *Proceedings of the European Conference on Speech Communication and Technology, Madrid, Spain*, Vol. 3; pp. 2043–2046, 1995.

[MG93] P. Monnin and F. Grosjean. Les structures de performance en français: caractérisation et prédiction. *L'Année Psychologique*, 93:9–30, 1993.

[Moz95] S. Mozziconacci. Pitch variations and emotions in speech. In *Proceedings of the 13th International Congress of Phonetic Sciences, Stockholm, Sweden*, Vol. 1, pp. 178–181, 1995.

[NS94] M. Nakai and H. Shimodaira. Accent phrase segmentation by finding N-best sequences of pitch pattern templates. In *Proceedings of the International Conference on Spoken Language Processing, Yokohama, Japan*, Vol. 1, pp. 347–350, 1994.

[OA83] D. O'Shaughnessy and J. Allen. Linguistic modality effects on fundamental frequency in speech. *J. Acoust. Soc. Am.*, 74:1155–1171, 1983.

[PMK93] V. Pasdeloup, J. Morais, and R. Kolinsky. Are stress and phonemic string processed separately? Evidence from speech illusions. *Proceedings of the European Conference on Speech Communication and Technology, Berlin, Germany*, 2:775–778, 1993.

[SC92] M. G. J. Swerts and R. Collier. On the controlled elicitation of spontaneous speech. *Speech Communication* 11:463–468, 1992.

[tB93a] L. ten Bosch. Algorithmic classification of pitch movement. *Working Papers 41, Proceedings of the ESCA Workshop on Prosody, Lund University, Sweden*, pp. 242–245, 1993.

[Tho83] N. G. Thorsen. Standard Danish sentence intonation—Phonetic data and their representation. *Folia Linguistica*, 17:187–220, 1983.

[vS92] J. P. H. van Santen. Contextual effects on vowel duration. *Speech Communication*, 11:513–546, 1992.

12
Synthesizing Spontaneous Speech

W. N. Campbell

ABSTRACT This paper addresses the issue of producing synthetic speech for an interpreted dialogue where the attitudinal colouring of an original utterance is to be preserved; it describes differences in speaking style between read and spontaneous speech from the viewpoint of synthesis research and discusses the design of a synthesis system incorporating labels to encode the prosodic and segmental variation simultaneously. Spontaneous speech confronts us with phenomena that were not encountered in corpora of prepared or read speech, and to account for these we have to identify increasingly higher-level factors of discourse structure and speaker involvement. The paper makes three specific claims: (a) that it is better to label the distinctive characteristics of speech through higher-level context dependencies, and to select units for synthesis from appropriate contexts, rather than attempt to predict and modify fine phonetic detail; (b) that the labelling of segmental and prosodic characteristics can be done adequately for speech synthesis using automatic techniques, leaving the human labeller free to identify higher-level discourse-related aspects of the speech; and (c) that instead of minimizing the size of the source database of speech units, we should rather be concerned to maximize its variety and to efficiently select from it the units that most closely express the characteristics of the target speech. The CHATR resynthesis toolkit performs many of these tasks.

12.1 Introduction

Speech synthesis is not spontaneous, nor can it be. However, there are applications of synthesis where modelling of the spontaneous characteristics of natural speech is required, such as in an interpreted dialogue where speakers talk in their own language and the speech is then automatically converted into the language of the listener. In such a dialogue the prosodic attributes, such as speed of speaking, degree of segmental reduction, tone-of-voice, etc., carry information that signals among other things speech-act type, stage of the discourse, the speaker's mood, and her commitment to the utterance. For the successful interpretation of such para-linguistic information, the system must be capable of recognizing and expressing quite subtle prosodic and pragmatic voice-quality changes.

It is probable that when using such a system, speakers will be more careful than usual to control their style of speaking, and it is perhaps even questionable whether the synthesised translation should be required to sound completely natural at all (if that were possible), because of accountability issues, but we are instead concerned here with the still theoretical issues of how to automatically identify and label such stylistic and discoursal speech information and with the techniques of synthesis best used to encode it.

12.1.1 Synthesizing Speech

There are three primary methods of synthesizing speech: (a) articulatory synthesis, which produces a speech waveform by modelling the physiological characteristics and excitation of the human vocal tract; (b) formant synthesis, which directly models the acoustics of the speech waveform; and (c) concatenative synthesis, which uses pre-recorded segments of real speech to construct a novel utterance. For the manipulation of intonation and timing in the synthesized speech, (b) offers the most flexibility, and (a) the most natural built-in constraints, but (c), while producing the most natural-sounding speech, is the most difficult.

Because concatenative synthesis employs digitized segments of recorded speech, it is in principle capable of reproducing all the fine variation of detail that is still too complex or too subtle for the other methods to model. However, in manipulating the intonation of a concatenated sequence of segments, we must resort to signal processing and encoding techniques such as PSOLA [MC90] that inevitably introduce some distortion and thereby reduce the naturalness. The more the prosody in the synthesis is varied from that of the original recording, the more the waveform is distorted from its natural shape, and the more artifacts are introduced by the processing. This may be the reason why even concatenative speech synthesis still sounds more like a machine talking than a human being. However, by increasing the size and variety of the segment inventory, this problem with concatenative synthesis can be greatly reduced.

Although concatenative synthesis is the most natural-sounding of the three methods, we have yet to hear automatically generated synthetic speech that can consistently be confused with a human original; i.e., speech synthesis has yet to pass its turing test. A likely reason for this is that although the source segments (*units*) for concatenation were originally produced by a human speaker, they have typically been excised from recordings of carefully prepared read speech, and although they may be *phonemically representative* of the sound combinations of a given language, they are *prosodically constrained* and invariant, i.e., they form a set of typical sound sequences that represent the phonetic-context-dependent allophones required to reproduce a spoken language, but they fail to adequately model the range of prosodic-context-dependent variation that occurs when speech is produced in various natural contexts. In warping them to a different prosodic configuration, as is required in the synthesis, the original naturalness is lost.

A major point of the work being presented here is to show that units for synthesis can instead be excised from a corpus of speech produced under less constrained circumstances, which therefore includes more natural prosodic variation typical of different styles of speech. Because of the variation in such a corpus though, the accuracy of the labelling becomes much more important as it then becomes necessary to identify and select source units not just from an appropriate phonemic environment but also with respect to the prosodic and voice-quality[1] dimensions as well. If by dint of improved labelling we can extract the units for concatenation from a context that is similar to the target in all significant dimensions, then we can reduce the amount of signal processing that will be required to produce the appropriate intonation, and therefore maintain a level of synthesis that is closer to the quality of real human speech. In this way we shift the main task of synthesis research away from the *modelling* of speech and in the direction of its characterization (or *labelling*) instead, identifying superpositional levels of influence and taking advantage of their consequences.

Most of the corpora so far studied for speech synthesis have been of read speech, and there is already a considerable body of experience in the automatic or assisted labelling of segmental and prosodic aspects of such corpora ([TW94, CS92, Cam93a, Cam92b, WC94, KKN+95a, Koh95a, BGG+96]). However, for the synthesis of dialogue or conversational speech, such additional aspects as voice quality, hesitations, and speaking style will also need to be identified as additional features. However, rather than resulting in a proliferation of the number of labels that are required, this actually reduces the labelling load, by incorporating superpositional information. We will see below that the labelling can be performed in a simple hierarchical way, with each level inheriting features from higher-level descriptors, so that rather than describing (and having to identify) fine variations in articulatory characteristics we can predict their occurrence instead. That is, by labelling the higher-level features of a spoken utterance we are thereby able to predict the circumstances under which the dependent lower-level characteristics change.

12.1.2 Natural Speech

Speech is "natural", but not all speech is similarly natural, and recorded speech, especially when produced in a controlled environment as a source for synthesis units, can be highly constrained. In its natural form, speech is inter-personal and often functionally goal-directed, but in recordings of "lab

[1]i.e., "Prosodic" in a Firthian sense.

speech", where the listener is replaced by a microphone, the act becomes production-based rather than listener-oriented. There are significant and perceptually relevant differences between a spontaneous natural utterance and a prepared one that mimics it even though the text of what is said may be identical [Bla95]. As a consequence, the materials that are typically collected for analysis and as training data for rule-based prediction, may not be representative of what people actually *do* when they speak normally.

Furthermore, because the reader producing a source-unit database for synthesis is faced with the daunting task of carefully having to read into a microphone long lists of unconnected sentences (or worse, nonsense words) to produce all the required sound combinations, there is a high probability of boredom or fatigue having effects on the voice quality, naturalness, and vitality of such recordings[2].

To obtain source units for the simulation of lively natural speech, it must be preferable to replace production controls at the recording stage with statistical controls in a later analytical stage, and to use these to process instead large representative corpora of spontaneously produced spoken material. Such corpora are now becoming more widely available but the tools for their analysis were developed for a more restricted speaking style, when read speech was the main source of data. The differences between such read corpora and "real speech" can be so great that models based on the analysis of lab speech may no longer be of use in the prediction of natural speech phenomena.

12.2 Spontaneous Speech

The range of prosodic variation increases as the speech becomes more spontaneous. As an illustration of the contrasts between read and spontaneous speech in British English, we will examine here some segmental duration characteristics, shown in Figures 12:1–12:4 which plot mean segmental duration against the coefficient of variance (i.e., the standard deviation of the durations expressed relative to the mean) for each phone class for each speaking style. These figures show how speakers clearly discriminate between different phone types when they have less context to provide redundancy of information, and show well how this clarity of production reduces as contextual information becomes more reliable. In the case of isolated words, when there is no continuity from one word to the next, durational discrimination is maximal, but when the same words are produced in a meaningful sentence, the differences are blurred, and in the extreme case, when the listener has the whole of the previous discourse as contextual

---

[2]Santen (this volume) for example mentions recording 2000 repetitions of the type "Now I know C.V.X" from a single speaker as data for his experiments

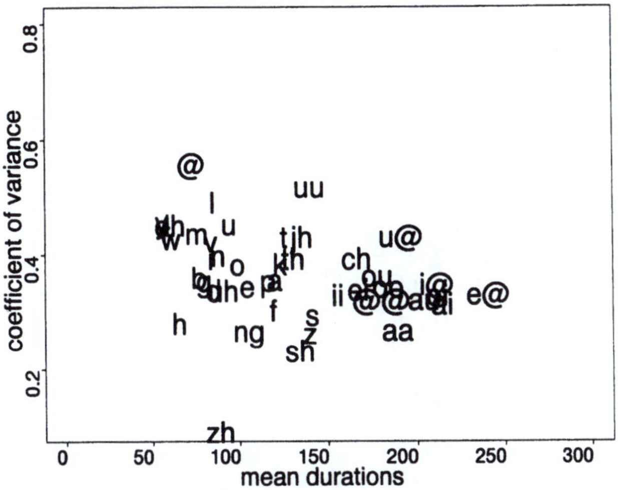

FIGURE 12.1. Segment durations in isolated words.

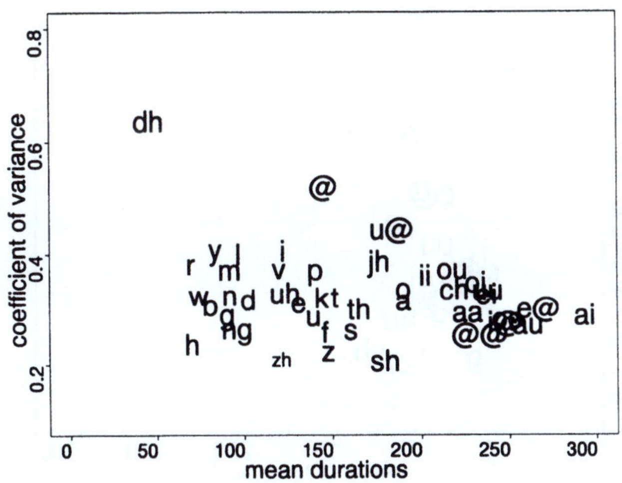

FIGURE 12.2. Segment durations in isolated word sentences

information, there is very little difference in durational characteristics between phones of very different types.

The data examined in this section come from four related corpora. The first contains citation-form readings of 5000 English words; the second, a subset of these words in the form of 200 meaningful sentences but read as isolated words; the third, the same sentences read in connected form as meaningful sentences; and the fourth, 20 min of spontaneous interactive monologue (i.e., dialogue with a passive partner). They are of British

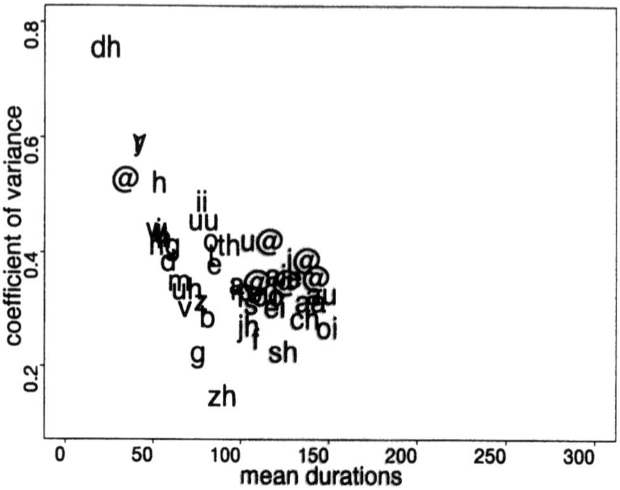

FIGURE 12.3. Segment durations in continuous sentences.

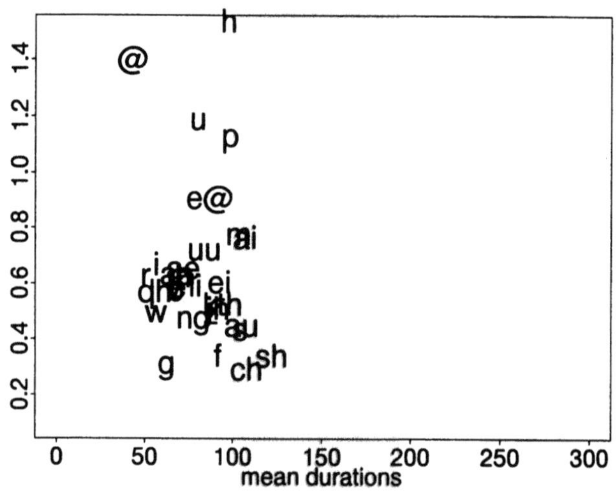

FIGURE 12.4. Segment durations in spontaneous speech.

English from a young adult female speaker, and show a wide range of production variation according to speaking style. The phone labels used in these figures are the Edinburgh University MRPA machine-readable phonemic symbols; "@" represents a schwa, "@@" a long central vowel, and so on.

We can see from Figures 12.1 and 12.2 that in the isolated-word citation-form readings, there is a good dispersion in the mean durations for each phone class (as represented by the horizontal spread), and relatively constant variance in their durations (as represented by the close vertical clustering). Figure 12.3 shows the opposite to be the case for the exact same sequence of words as for Figure 12.2 but read in continuous sentences. Here the variance increases and segments undergo considerable shortening, with the result that they are no longer as distinct in their mean durations. Separate examination of segments in word-initial and word-medial position confirms that this is not just a result of more phrase-final lengthening (isolated words being also complete phrases), rather that the articulation of the citation-form words was generally slower and more distinct.

When the speech contains little contextual information, and the speaker is only concerned to be clearly distinct, then segmental durations appear to be maximally separated, and can be well predicted by simple statistical models. However, as the speech becomes more situated, not only do the words become more predictable but the listener can also rely on prosodic phrasing as an aid to its interpretation. All words in the connected speech data tend to be shorter overall, but with much more variance within each phone type. The prediction of segmental durations for such speech will require far more detailed knowledge of the discourse environment than is currently being incorporated in synthesis systems.

The spontaneous monologue from the same speaker (Figure 12.4) shows the same trends more clearly: We find not only that the mean durations for all segment types are low and uniform, but that the variances are huge—notice especially that the vertical scale of Figure 12.4 is *twice* that of the other three figures. When a listener is present, the speaker has a better idea of the extent of their mutual understanding, and can hurry through some parts of the speech, and linger before others. There is much greater range of variance in the natural speech than was found in the more carefully prepared "lab" speech. As Mehta and Cutler have pointed out [MC88], prosodic prediction may be of limited applicability with spontaneous input.

If we try to predict the prosodic characteristics of such utterances using training data based on read speech we will only be able to predict a small fraction of the variance observed, since many of the factors now coming into play will not have been considered. If, on the other hand, we are able to predict different local speaking styles, marking areas of confidence or high mutual understanding, then we stand a much better chance of determining the durational (and other articulatory) characteristics of the segments they include.

12.2.1 Spectral Correlates of Prosodic Variation

To confirm that this difference of style is not unique to durations, nor specific to the prosody of one speaker, we can compare the above differences with the changes that take place in spectral tilt associated with prominence and focus in read vs. interactive speech. We see significant correlations between changes in the prosodic dimensions of emphasis and focus, and the acoustic features of segmental articulation. This suggests the need for a multi-tiered, superpositional, labelling system for describing the variation that occurs in natural speech, so that these fine phonetic differences can be captured by description of the prosodic characteristics that correlate so closely with them.

The data reported next are taken from a corpus of 300 focus-shifting sentences, produced by a young female American speaker, which illustrate the effects of contrastive focus. Three sets of 100 sentences selected from a larger corpus contained syntactically and semantically identical word sequences but differed only in the focus given to each. The sentences were produced in three utterance styles: (a) read in grouped order with focus shifting from earlier words to later words within groups of identically worded sentences. This emphasized the contrast and increased the articulatory emphasis, (b) the same sentences were then read in randomized order so that emphasis would not be forced, and (c) the different emphasis renditions of each sentence were produced spontaneously as a result of elicitation in interactive discourse. Shifts of emphasis in the read speech were controlled by use of capitalization to signal different interpretations, and elicited in the interactive discourse by (deliberate) misinterpretations on the listener's part.

Using normalized segmental duration and energy as cues for automatic prominence detection. (described more fully in [Cam92b, Cam95]), we found that it was much easier to recognize the focussed words in style (a) (read in groups), for which we achieved 92% correct detection of the focussed word from among those detected as prominent, than in style (b) (78%) or (c) (72%). The elicited corrections of style (c) resulted in a perceptually clearer articulation than style (b), but because the durational organization of the more spontaneous speaking style was much more varied, prominence and focus were not as easy to detect automatically using duration and energy alone as cues.

A follow-up study including spectral tilt information ([Cam95] after [SvH93, Slu95b]) increased the detection accuracy and confirmed that speakers appear to change their phonation according to the discourse context and the type of information they impart. The detection algorithm using both duration and spectral tilt (measured by the relative amount of energy in the mid-third of an ERB-scaled spectrum between 2 kHz and 4 kHz, normalised by the overall energy within each frame) showed the correlations given in Table 12.2.

It is interesting to note that although the durational cues to prominence were weakened by greater variance in the interactive speech, the spectral measure was apparently strengthened, as Table 12.2 shows. We can suppose (like Lindblom [Lin90]) that this trade off is not coincidental, and that the speaker varies her production according to the needs of the discourse context.

TABLE 12.1. ± prominent spectral tilt.

	Student's t	df
(a) Read grouped	35.63	7676
(b) Read randomized	19.01	6110
(c) Interactive	42.76	6974

Showing the separation in mean spectral tilt between prominent and non-prominent syllable peaks.

12.3 Labelling Speech

Traditional phonetics labels speech according to segmental content alone, and while allowing the use of diacritics to describe prosodic variation, typically regards this as very much a secondary feature. I argue the contrary: that in order to describe a speech segment sufficiently to use it in concatenative synthesis where the goal is to reproduce the characteristics of natural speech, we have to label *both* the segmental *and* the prosodic attributes equally. This is because simple phonetic labels do not sufficiently describe the location on the scale of hypo- hyper- articulation of segment sequences with supposedly identical phonemic structure.

A serious consequence of this under-labelling is that units with the same phonetic labels excised for synthesis from naturally occurring speech may not be similar enough to concatenate without a noticeable discontinuity. This explains why such care has been taken to reduce the variation in conventional source databases for concatenative synthesis, but it is a problem that must be solved if we are to synthesize natural-sounding speech that carries a clear pragmatic message. Lindblom [Lin90] has shown that the phonatory characteristics of articulation vary according to speaking style and speaker familiarity. Kohler has similarly described a cognitively based reduction coefficient [Koh95a, Koh96], under the control of the speaker, that governs reduction and elision causing scalar variation in the articulation of a given sequence of phones in different contexts. Mechanisms for these effects have been accounted for in articulatory phonology as simple consequences of overlapping constituents [Col92a] but described by psycholinguists as being used intentionally [Wha90]. Since they are largely predictable from higher-level features of the discourse, such as speaking rate and familiarity, then it should in general be sufficient to know just the speaking style (and its prosodic correlates) and context in order to describe the degree of reduction on any given segment in an utterance.

For identifying such higher-level features in a large corpus, canonical segment labels can first be automatically aligned from an orthographic transcription in order to provide access to discrete portions of the speech waveform. From these we can then extract prosodic information, which in turn will be used to detect and label the higher-level structural and stylistic features which can then be used to account for the finer articulatory differences that are then predictable from context.

12.3.1 Automated Segmental Labelling

What can be predicted does not need to be explicitly labelled. Kohler (this volume) argues that a linear segmental representation of canonical citation forms can account well for the phonological reorganization of speech, and shows that although a segment may be elided or deleted in the production of fluent speech, a non-segmental residue remains to colour the articulation of the remaining segments. This supports our contention that rather than attempt a fine labelling of the surface representation of sounds in an utterance, it is preferable to label only the underlying canonical segment sequences but to relate them to their prosodic environment separately in a multi-tiered description. Similarly in synthesis, rather than attempt to predict the fine microsegmental variation, we can use selection according to prosodic environment to bypass this difficult task. A canonical representation of the phone sequence is easily accessible from a machine-readable pronunciation dictionary, so given an orthographic transcription of a speech corpus, segmental labelling can be automated to a large extent by using speech recognition technology to predict and align a default phone sequence. This is then complemented by an encoding of the prosodic structure of each utterance to capture the interactions.

By training single-phone hidden Markov models (HMMs) corresponding to the set of phonetic labels in a machine-readable pronunciation dictionary, and generating networks of default pronunciations for each word in the orthographic transcription, we can obtain a first-pass estimate of the segmental realization of each utterance. Separate lexical sub-entries must be included for some particularly different pronunciation variants such as "gonna" for "going to", but in general a single pronunciation for each word will suffice.[3] A finer segmental alignment can then be achieved after a second pass by using Baum–Welsh re-estimation (as in the HTK toolkit [Ent93]) to retrain the HMM models specifically for each corpus, using the transcription derived from the orthography to constrain the alignments. We can thereby achieve segmentation accuracy comparable to a human transcription (see, e.g., [TW94]).

A criticism of this blind transcription technique is that without human intervention we do not know for certain what pronunciation a word was given in a particular utterance, and that it is possible for example that

[3]In the synthesis stage, only one pronunciation for any word will be generated, but its actual phonemic/phonetic realization will depend on its prosodic context.

the consonant cluster in "handbag" although actually pronounced as [mb] was force-aligned (from the canonical form) as an [ndb] sequence. The counter to this criticism is that given the supplementary information about the prosodic environment, such knowledge is no longer required; the same sequence in a markedly prominent or contrastively focussed environment is likely to be given one pronunciation, and in a normal or reduced environment the other, or something in-between. The claim being made in this paper is that this type of segmental reduction is to a large extent predictable from the prominence (or lack of it) realized on the sequence, in conjunction with the speaking style, rate, and boundary information.

12.3.2 Automating Prosodic Labelling

Whereas prosodic variation is scalar and multi-dimensional, including at least fundamental-frequency, segmental duration, and amplitude changes, prosodic *structure* can as a first approximation be represented as binary and in two dimensions, by a combination of the higher-level labels of prominence and phrase finality, as in the ToBI system of prosodic transcription [SBP$^+$92, PBH94]. In read speech at least, phrasal boundaries and prominences appear to be the most basic elements marking prosodic structure, and we can predict much about the phonatory (acoustic) characteristics of a segment from knowledge of its place in the syllable and of that syllable's position with respect to its neighbours at the various levels of prosodic phrasing and prominence.

Taking segmental duration as an example, a syllable immediately before a prosodic phrase boundary is likely to be lengthened, with amplitude low and decaying, and it may exhibit vocal fry in the rhyme. The lengthening is likely to be greater with increasing strength of phrase break, and a pause is likely to follow if utterance final. There will also be lengthening observed in a prominent (or nuclear accented) syllable, but in this case it is likely to be more marked on the onset segments [dJ95, Cam93b] and there may be more aspiration after plosives in the onset, increases in spectral tilt resulting from changes in vocal effort [PT92, CB95, GS89, SvH93] and differences in supraglottal phonation arising from local hyperarticulation [Lin90].

Figure 12.5 (from [Cam92a] but see [Cam93b] for more detail) illustrates three types of prosodic context that interact to govern the duration of a syllable: In terms of lengthening, the effects of prominence are biased more towards early segments (in the onset and peak) and those of phrase finality on later ones (in the offset or rhyme). Rate-related lengthening affects segments more uniformly across the syllable, warping each segment similarly in terms of its distributional characteristics (as fitted by a two-parameter Gamma probability model).

In labelling the source database, each syllable is therefore tagged according to the following features to determine its prosodic environment: (a) $\pm$*prominent* (a binary indicator),

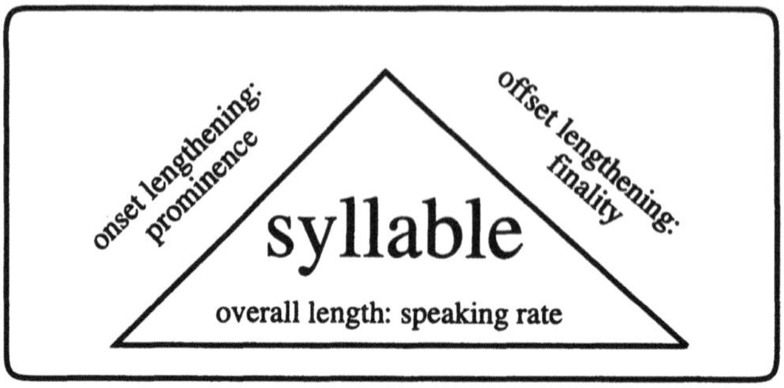

FIGURE 12.5. The prosodic lengthening effects on a syllable.

(b) ±*phrase-final* (binary at three levels of phrasing),
where "prominence" on a syllable is perhaps best defined perceptually as "having been uttered with a greater degree of vocal effort than surrounding syllables"[4], and "prosodic phrase finality" is defined at three levels: (i) the accentual (minor) phrase; (ii) the intonational (major) phrase; and (iii) the utterance-final major-phrase variant. Higher levels of chunking are already required, e.g., at the paragraph level for read text, and to mark disfluencies and turns in more natural speech, but cannot yet be performed automatically.

Wightman and Campbell [WC94] were able to correctly predict most of the hand-labelled prominences and intonation boundaries in a corpus of 45 min of professionally read American radio-news speech. Section f2b of the Boston University Radio News Corpus [OPSH95b] contains news broadcasts produced by one adult female speaker that a consistent marked style typical of professional announcer speech. The corpus had been prosodically labelled by hand according to the ToBI conventions to differentiate high and low tones at intonational boundaries and on prominent syllables, and to mark the degree of prosodic discontinuity at junctions by break indices between each pair of words.

A model incorporating acoustic, lexical, and segmental features derivable from the phone labels, the dictionary, and the speech waveform, was trained, and achieved automatic detection of 86% of hand-labelled prominences, 83% of intonation boundaries, and 88% correct estimation of break indices (at ±1). The acoustic features extracted from the speech waveform for the autolabelling of prosody include (in order of predictive strength) silence duration, duration of the syllable rhyme, the maximum pitch target[5], the mean pitch of the word, intensity at the fundamental, and spectral

[4]Prominence thus defined frequently but not necessarily co-occurs with lexical stress, but should not be confused with "intrinsic vowel length" or absence of schwa-reduction.

[5]Pitch targets were calculated using Daniel Hirst's quadratic spline smoothing to estimate the underlying contour from the actual $f0$[Hir80].

tilt (calculated from the harmonic ratio). Non-acoustic features included end-of-word status, polysyllabicity, lexical stress potential, position of the syllable in the word, and word class (function or content only). These latter were all derivable directly from the dictionary used in the aligning.

12.3.3 Labelling Interactive Speech

Although much of the labelling of significant levels of information can now be performed to a large extent automatically and requires only minimal hand correction for corpora of read speech, these techniques do not extend easily to the processing of dialogue speech or to spontaneous monologue. Here we realize the need for extra levels of information to describe the structuring of discourse events that cannot yet be achieved automatically. Whereas the read speech was highly predictable, the unplanned (spontaneous) speech is characterized by bursts of faster and slower sections where the speaker displays switches in speaking style [Bar95], and by much greater variation in $f0$ range and pausing as she expresses different degrees of confidence, hesitation, involvement, and uncertainty.

In order to compare the speech of one individual, in a highly restricted domain, under a variety of interaction styles, we recorded a native speaker of American English taking one side in a series of 20 task-related instruction-giving dialogues. These were performed in a multi-modal environment, alternatively with and without a view of the interlocutor's face [Fai94].

Transcribing the orthography of such spontaneous speech required more than just the skills of an audio-typist, and to allow auto-segmentation, decisions had to be made about marking disfluencies and repairs. To include this information in our labelling, two extra tiers of information were added to the basic ToBI transcription. One, after Nakatani and Shriberg [NS93, Shr94], which extended the miscellaneous tier of the ToBI transcription to describe interruptions in the speech flow, and one after Stenström [Ste94]), to label illocutionary force type (IFT) speech-act information. The following set of speech-act labels was used:

> *inform, expressive, good_wishes_response, apology_response, invite, vocative, suggest, instruct, promise, good_wishes, yn_question, do_you_understand_question, wh_question, yes, no, permission_request, acknowledge, thank, thanks_response, alert, offer, offer_follow_up, action_request, laugh, greet, farewell, apology, temporize, hesitation, confirmation.*

Hirshberg [Hir92, Hir95b] has noted that the major differences between lab speech and spontaneous speech appear to be prosodic (concerning speaking rate and choice of intonation contour), but also acknowledges significant segmental differences. She notes, for example, that some disfluencies in spontaneous speech are marked by characteristic phonetic effects, such as interruption glottalization (which is acoustically distinct from articulatorily similar laryngealization). In labelling speech to include these characteristics, we need not just the prosodic and segmental information derived from HMM alignment, but also an indication of the broader context in which they are uttered, including fluency and even extending to the state-of-mind of the speaker. That is, we need to label not just what the speaker says, but what she is *doing* in saying it, and how she feels about what she is saying.

As an example, in the dialogue corpus the word "okay" was said 140 times. It was variously labelled as "acknowledge", "confirmation", "offer_follow_up", "accept", and "do_you_understand_qn.", etc., 12 categories in all. The intonation, duration, and articulation varied considerably; sometimes short, sharp, and rising, on a high tone, sometimes slow and drawn out on a falling tone. Since we were able to find significant correlations between the intonation and the speech act label for most of these cases (see [BC95] for details), we continue in our assumption that instead of trying to predict and model the lower-level acoustic variations, we should instead be accessing them through higher-level labels.

Spontaneous speech appears to be most marked in terms of its rhythmic structuring, exhibiting greater ranges of variation with corresponding differences in phonation style. These prosodic changes appear to have clear correlates in the speech-act labels that we are now using. However, to more fully describe them, we need also to formalise a measure of the speaker's commitment to her utterance. Impressionistic comments such as "she's thinking ahead", "her mind's not on what she's saying", "she's said this many times before", and "she doesn't quite know how to put this" are triggered by such differences in speaking style, but none of the labels we have considered so far are sufficient to mark such differences. The next step in this work is to determine the appropriate labels, in order to categorize their prosodic and articulatory correlates. Since human listeners can respond consistently to such subtle speaking-style changes, then the clues must be present somewhere in the speech signal but rather than search at the acoustic level, we will continue to explore higher levels of labelling in an attempt to capture them.

12.4 Synthesis in CHATR

CHATR[6] [Cam92d, BT94b, Cam94b] is a set of tools that process an arbitrary speech corpus, with its orthographic transcription, and automatically generate from these a labelled database of segments with derived features for synthesis. Selection from this database is then performed through a

[6]Collective hacks from ATR (pronounced "chatter" for obvious reasons).

weighted combination of the segmental and prosodic features to satisfy a target utterance specification.

Using simple waveform concatenation, the method is speaker-independent. It is also language-independent since the target description for any novel utterance must be completely specifiable in terms of the segmental and prosodic labels of the database from which it is to be generated. The language-specific processing required to predict the appropriate representation of the phone sequence and prosody for a text-to-speech synthesiser is not addressed in this chapter, as we take such a representation as basic input to the synthesis module.

From a database prepared as described above, CHATR can now extract and concatenate speech segments to reproduce the voice and speaking style of any sufficient labelled speech corpus. Preparing a corpus for a new speaker (e.g., from 40 min of phonetically balanced utterances) has been completed in less than a day, from initial recording, through segmentation and weight training, to eventual synthesis.[7] See [Cam94b, CB96, HB96] for details of the weight-training procedures.

It has been argued (van Santen, personal communication) that the combinatrics of prosodically labelled units is formidable, and would logically require hundreds of hours of recorded speech. However, whereas in theory any of the 70,000 English triphones can occur in any prosodic context, in practice the collocations of the language constrain the distributions such that a small number occur very often in idiomatic (possibly prestored) word sequences. In our defined-domain speech synthesis applications, we find that it is of greater benefit to be able to model these fluent articulations well first. Then, in the less frequent case of an uncommon sound sequence, as for example in a foreign personal name, the CHATR synthesiser will automatically select shorter (non-uniform but typically phone-sized) more prototypical clearly articulated units. This fallback technique simulates the care with which a human speaker would also pronounce the less-familiar word. For natural-sounding synthetic speech it is at least as important to model the reductions of fluent articulation in the carrier dialogue as it is to model the careful articulation of its information-rich sub-sections.

The method is *not* speaking-style independent, and we can only model the style(s) of speech found in the source corpus—i.e., news speech always sounds like news speech—but this can be an advantage: with enough disk space, we can now reproduce the characteristics of any speaker or speaking style, given a sufficient source corpus.

Many previous methods of speech synthesis were limited by machine and memory size, and so were constrained to modelling intelligibility rather

[7]We have currently tested this process with corpora from 12 speakers of Japanese, five of English, two of German, and (without requiring any changes to the c-code) one of Korean.

than naturalness. However, with the advent of multi-media computing, many more resources have become available. The recently agreed magnetic-optical standard of 4.7 gigabytes for a single "floppy" disk allows sufficient room for more adventurous techniques of speech production, since even a high quality recording (without compression) requires only about a megabyte of memory per minute of speech, and for non-interactive speaking (read-speech), 20 min currently seems to be an adequate minimum size.

Once the prosodic and segmental features are labelled for a given database, training of the weights to determine the strength of contribution of any given feature in a specific database is performed automatically by jack-knife substitution, removing each utterance of the original database in turn and synthesizing an approximation of it using the segments remaining in the database according to a range of different weight settings to produce a measure of the Euclidean cepstral distance for each [BC95, HB96].

Campbell and Black [Cam94b] reported results using the BU Radio News corpus as the basis for a resynthesis test of the assumption that labels of prosodic and canonical segmental context suffice to encode the lower-level spectral and articulation characteristics, employing the CHATR speech synthesis toolkit to select segments from a labelled corpus for concatenative synthesis as described above. Using similar jack-knife substitution, we resynthesized each utterance by concatenation of segment sequences selected from the remaining utterances according to suitability of their prosodic environment, with no signal processing performed on the concatenated sequences. Measures of Euclidean cepstral distance between target and synthesized utterances confirmed that the use of prosodic features in the selection resulted in a closer match between the spectra of target and synthesized utterances. When equivalent tokens from the same segmental sequences were selected from less appropriate prosodic environments, ignoring the weights on the non-segmental features, the resulting synthetic speech showed considerable degradation. Table 12.2 shows similar results for a database of Japanese speech.

Because the source corpus typically includes natural non-speech noises, these can also appear in the synthesis if in an appropriate context for selection. It frequently happens that a sequence of segments across a prosodic phrase boundary is resynthesized using tokens selected from pre- and post-pausal locations such that the "silence" between them includes an appropriate sharp intake of breath. Such noises coming from a synthesizer make the resulting speech sound even more "natural".

12.5 Summary

To summarize the main points of this paper, I have argued that concatenative synthesis currently offers the best method of generating synthetic

TABLE 12.2. Mean Euclidean cepstral difference for different selection methods.

Selection based on equal weights	1.9349
Selection using weighted features	1.6700
Theoretical minimum	1.5456

Notes: (a) "equal weights" is equivalent to selection using only phonemic environment and provides a measure of the dispersion in the spectra of phonemically identical units in the corpus. (b) "weighted features" shows the reduction in this distortion that can be achieved by including prosodic descriptors in the selection. (c) the "theoretical minimum" is defined by selection based on cepstral targets which although impossible to predict in synthesis, nonetheless allow us to measure the optimality of available units in a given corpus.

speech by rule, and that although in the past few years the quality of output has been steadily improving, the technique is inherently limited by the nature of the source units, typically few in number and lacking in the necessary variety to generate human-sounding speech. Naturally occurring speech offers a richer source of units for synthesis than specially recorded databases, but the success of selecting units from a natural-speech database crucially depends on the labelling of the corpus.

For the efficient characterization of speech sounds, it is preferable to label superpositionally; not requiring detection of fine phonetic features explicitly nor numerical quantification of their prosodic attributes, but taking advantage of the natural consequences of the higher-level structuring of the discourse in which they occur. By labelling a large corpus of natural speech as a source of units for concatenative synthesis and selecting non-uniform-sized segments by a weighted combination of segmental and prosodic characteristics, we have been able to reduce the need for disruptive warping to contort a given waveform segment into a predicted context, and can therefore maintain a higher level of naturalness in the resultant synthetic speech.

For non-interactive or read speech, knowing the phonemic context of a segment, its position within the syllable, and whether that syllable is prominent, prosodic-phrase-final, or both, allows us to predict much about its lengthening characteristics, its energy profile, its manner of phonation, and whether it will elide, assimilate, or remain robust. In the case of interactive speech, however, a significant part of the message lies in the *interpretation* of *how* it was said, and to encode sufficient information about such aspects of the utterance as phonation style and speaking style, we need also to design labels for the discourse and communication strategies that allow the listener to estimate the attitudinal colouring of an utterance and the speaker's commitment to its content.

The remaining challenge is therefore to label large corpora of real speech with a small but sufficiently descriptive set of higher-level features so that more of the relevant variations can be indexed and retrieved. This requires a definition of the *perceptually salient* characteristics of speech and of the *contexts* that contribute to their variation.

If a large and sufficiently representative corpus could be labelled in terms of the higher-level factors that govern phonemic, phrasal, prosodic, speech-act, etc., variation, then it would perhaps no longer be necessary to attempt to predict or modify such fine details as segmental duration or articulation variation at all; it would be sufficient just to select a unit from a context with the appropriate labels in order to characterize the desired target speech. The durations and other relevant acoustic features would be contextually appropriate and natural by default.

Finally, much of the previous research on speech synthesis has been performed on small computers. If we compare the resources currently available to, e.g., image processing with those available for speech processing, we see a tremendous mismatch. I maintain that speech is no less complex a signal than image and that if we are to model it accurately, then we need to devote much more processing power and disk space than we are currently considering. In compensation, we see that the currently popular multi-media computing devices are increasingly being equipped with just such facitities.

Acknowledgments

This paper includes material first presented at the ATR 1995 International Workshop on Computational Modelling of Prosody for Spontaneous Speech Processing, and later expanded upon in the Symposium on Speaking Styles at the XIIIth Congress of the Phonetic Sciences in Stockholm. I am grateful to many colleagues, in particular to Jan van Santen and Klaus Kohler, for their helpful suggestions and comments.

References

[Bar95] W. J. Barry. Phonetics and phonology in speaking styles. In *Proceedings of the 13th International Congress of Phonetic Sciences, Stockholm, Sweden*, 1995.

[BC95] A. W. Black and W. N. Campbell. Predicting the intonation of discourse segments from examples in dialogue speech. *Proceed-*

ings of the ESCA Workshop on Spoken Dialogue, Hanstholm, Denmark, 1995.

[BGG+96] G. Bruce, B. Granström, K. Gustafson, M. Horne, D. House, and P. Touati. On the analysis of prosody in interaction. In *Computing Prosody: Approaches to a Computational Analysis of the Prosody of Spontaneous Speech.* New York: Springer-Verlag, 1997.

[Bla95] E. Blaauw. On the perceptual classification of spontaneous and read speech. Ph.D. thesis, OTS Dissertation Series, Utrecht University. ISBN 90-5434-045-2, 1995.

[BT94b] A. W. Black and P. Taylor. CHATR: A generic speech synthesis system. *Proceedings of COLING-94*, II:983–986, 1994.

[Cam92a] W. N. Campbell. Multi-level timing in speech. PhD thesis, University of Sussex, Department of Experimental Psychology, 1992. Available as ATR Technical Report TR-IT-0035.

[Cam92b] W. N. Campbell. Prosodic encoding of English speech. In *Proceedings of the International Conference on Spoken Language Processing, Banff, Canada*, pp. 663–666, 1992.

[Cam92d] W. N. Campbell. Synthesis units for natural English speech. Technical Report SP 91-129, IEICE, 1992.

[Cam93a] W. N. Campbell. Automatic detection of prosodic boundaries in speech. *Speech Communication*, 13:343–354, 1993.

[Cam93b] W. N. Campbell. Predicting segmental durations for accommodation within a syllable-level timing framework. *Proceedings of the European Conference on Speech Communication and Technology, Berlin, Germany*, pp. 1081–1084, 1993.

[Cam94b] W. N. Campbell. Prosody and the selection of source units for concatenative synthesis. *Proceedings of the ESCA/IEEE Workshop on Speech Synthesis, Mohonk, NY*, pp. 61–64, 1994.

[Cam95] W. N. Campbell. Loudness, spectral tilt, and perceived prominence in dialogues. In *Proceedings of the 13th International Congress of Phonetic Sciences, Stockholm, Sweden*, 1995.

[CB95] W. N. Campbell and M. Beckman. Stress, loudness, and spectral tilt. *Proceedings of the Acoustical Society Japan, Spring Meeting*, 3-4-3, 1995.

[CB96] W. N. Campbell and A. W. Black. Prosody and the selection of source units for concatenative synthesis. In *Progress in Speech Synthesis.* Berlin: Springer-Verlag, 1996.

[Col92a] J. C. Coleman. The phonetic interpretation of headed phonological structures containing overlapping constituents. In *Phonetics Yearbook 9*, pp. 1–44. New York: Academic, 1992.

[CS92] W. N. Campbell and Y. Sagisaka. Automatic annotation of speech corpora. *Proceedings of the SST92 Queensland, Australia*, pp. 686–691, 1992.

[dJ95] K. de Jong. The supraglottal articulation of prominence in English: Linguistic stress as localized hyperarticulation. *J. Acoust. Society Am.*, 97:491–504, 1995.

[Ent93] Entropic Research Laboratory, 600 Pennsylvania Avenue, Washington DC 20003. *HTK - Hidden Markov Model Toolkit*, 1993.

[Fai94] L. Fais. Conversation as collaboration: some syntactic evidence. *Speech Communication*, 15:230–242, 1994.

[GS89] J. Gauffin and J. Sundberg. Spectral correlates of glottal voice source waveform characteristics. *JSHR*, 32:556–565, 1989.

[HB96] A. Hunt and A. Black. Unit selection in a concatenative speech synthesis system using a large speech database. *Proceedings of the International Conference on Acoustics, Speech, and Signal Processes*, 1996.

[Hir80] D. Hirst. Automatic modelling of fundamental frequency using a quadratic spline function. *Travaux de l'Institut de Phonétique 15, Aix en Provence*, pp. 71–85, 1980.

[Hir92] J. Hirschberg. Using discourse context to guide pitch accent decisions in synthetic speech. In G. Bailly, C. Benoît, and T. R. Sawallis, editors, *Talking Machines: Theories, Models, and Designs*, pp. 367–376. Amsterdam: Elsevier Science, 1992.

[Hir95b] J. Hirschberg. Acoustic and prosodic cues to speaking style in spontaneous and read speech. *Proceedings of the 13th International Congress of Phonetic Sciences, Stockholm, Sweden*, Vol. 2, pp. 36–43, 1995. Symposium on speaking styles.

[KKN+95a] A. Kiessling, R. Kompe, H. Niemann, E. Nöth, and A. Batlinger. Detection of phrase boundaries and accents. *Progress and Prospects of Speech Research and Technology: Proceedings of the CRIM/ORWISS Workshop, Sankt Augustin*, pp. 266–269, 1995.

[Koh95a] K. J. Kohler. Articulatory reduction in different speaking styles. *Proceedings of the 13th International Congress of Phonetic Sciences, Stockholm, Sweden*, Vol. 2, pp. 12–19, 1995. Symposium on speaking styles.

[Koh96] K. J. Kohler. Modelling prosody in spontaneous speech. In *Computing Prosody: Approaches to a Computational Analysis of the Prosody of Spontaneous Speech*. New York: Springer-Verlag, 1997.

[Lin90] B. E. F. Lindblom. Explaining phonetic variation: A sketch of the H&H theory. In H. J. Hardcastle and A. Marchal, editors, *Speech Production and Speech Modelling*, pp. 403–409. Dordrecht: Kluwer, 1990.

[MC88] G. Mehta and A. Cutler. Detection of target phonemes in spontaneous and read speech. *Language and Speech*, 31:135–156, 1988.

[MC90] E. Moulines and F. Charpentier. Pitch-synchronous waveform processing techniques for text-to-speech synthesis using diphone. *Speech Communication*, 9:453–467, 1990.

[NS93] C. Nakatani and L. Shriberg. Draft proposal for labelling disfluencies in ToBI. Paper presented at 3rd ToBI Labelling Workshop, Ohio, 1993.

[OPSH95b] M. Ostendorf, P. J. Price, and S. Shattuck-Hufnagel. The Boston University Radio News Corpus. Technical Report ECS-95-001, Boston University ECS Dept., 1995.

[PBH94] J. Pitrelli, M. E. Beckman, and J. Hirschberg. Evaluation of prosodic transcription labelling reliability in the ToBI framework. In *Proceedings of the International Conference on Spoken Language Processing, Yokohama, Japan*, Vol. 1, pp. 123–126, 1994.

[PT92] J. B. Pierrehumbert and D. Talkin. Lenition of /h/ and glottal stop. In G. Doherty and D. R. Ladd, editors, *Papers in Laboratory phonology 2*, pp. 90–127. Cambridge, UK: Cambridge University Press, 1992.

[SBP+92] K. Silverman, M. Beckman, J. Pitrelli, M. Ostendorf, C. Wightman, P. Price, J. Pierrehumbert, and J. Hirschberg. ToBI: a standard for labelling English prosody. In *Proceedings of the International Conference on Spoken Language Processing, Banff, Canada*, Vol. 2, pp. 867–870, 1992.

[Shr94] L. Shriberg. Preliminaries to a theory of disfluencies. Ph.D. thesis, University of California at Berkeley, 1994.

[Slu95b] A. C. M. Sluijter. *Phonetic correlates of stress and accent.* Holland Institute of General Linguistics, 1995.

[Ste94] A. Stenström. *An Introduction to Spoken Interaction.* London: Longman, 1994.

[SvH93] A. Sluijter and V. J. van Heuven. Perceptual cues of linguistic stress: intensity revisited. *Working papers 41, Proceedings of the ESCA Workshop on Prosody, Lund University, Sweden,* pp. 246–249, 1993.

[TW94] D. Talkin and C. W. Wightman. The aligner: text-to-speech alignment using Markov models and a pronunciation dictionary. *Proceedings of the ESCA/IEEE Workshop on Speech Synthesis, Mohonk, NY,* pp. 89–92, 1994.

[WC94] C. W. Wightman and W. N. Campbell. Automatic labelling of prosodic structure. Technical Report TR-IT-0061, ATR Interpreting Telecommunications Laboratories, Kyoto, Japan, 1994.

[Wha90] D. Whalen. Coarticulation is largely planned. *Journal of Phonetics,* 18:3–35, 1990.

13
Modelling Prosody in Spontaneous Speech

Klaus J. Kohler

ABSTRACT Following on from general considerations of requirements for prosodic modelling of spontaneous speech, this paper outlines a prosody model for German, its incorporation in a TTS system as a prosody research tool, the model-based development of a prosodic labelling system for the application to spontaneous speech, and the use of the resulting prosodic label files as input to the TTS system for transcription verification and model elaboration.

13.1 Introduction

Whereas speech analysis and modelling have traditionally focussed on scripted speech (logatomes and words in isolation or in standard sentence frames, connected speech in sentences and texts), spontaneous speech is now receiving increased attention, also in the area of prosody. But—at least initially—work in this new field of research proceeds on the assumption that the theoretical categories and operations on them, established for scripted speech, can simply be transferred to spontaneous speech. This will most certainly require adjusting in two ways: some categories will no longer be adequate (declination operating over time being a case in point), and new categories and operations will have to be added (e.g., in connection with dysfluencies and repairs).

Furthermore, the modelling of prosody will have to take the following points into account.

(1) Prosodic universals. The study of prosody has grown out of dealing with individual languages, especially with English, more than with any other language. Categories and operations (e.g., prosodic rules) are—to a large extent—determined by the particular linguistic structures. What we need for a general prosodic theory, however, are independently motivated categories and operations. Candidates are pitch direction (falling, rising) and synchronization of pitch "peaks" and "valleys" with syllable timing, in each case independently of the functional use they may be put to in individual languages (e.g.,

tone or intonation), further, the prominence-lending features of pitch movement and segment duration for the functional use of sentence stress (focus), and timing factors at various levels (global speech rate, utterance-final lengthening, stress/syllable timing).

(2) A unified theory integrating segmental and prosodic aspects, as well as intonation and timing among the latter. We have become used to thinking in and dealing with dichotomies in the study of speech: segmentals vs prosody, intonation vs timing. It is quite clear that all these levels of description form an intricate mutually conditioning network. Prosody (e.g., stress, timing, especially speech rate) provides conditioning factors for articulatory reduction; on the other hand segmental structures determine the manifestation of prosodic categories (different synchronization of pitch "peaks" and "valleys" in high/low or short/long vowels, curtailing of falls in pitch peaks before voiceless consonants). Global utterance timing has not only an influence on individual segment durations, but also on their qualitative realization and on the manifestation of pitch patterns (upward scaling of $F0$ and reduction of $F0$ range in increased speed). Contrarywise, segmental features determine their timing in global utterance speed: vowel and consonant durations are adjusted differently, not by a uniform proportionate factor across all segments; depending on the segmental type and context, shortening in fast speech also involves assimilation and elision of articulatory gestures over and above their changes in timing.

(3) A prosodic phonology as an interlevel between syntax/semantics/pragmatics and the phonetic signal: phonetic substance—phonological form—linguistic function. The phonetic-semantic relationship is not direct in the sense that the measured values themselves represent syntactic or semantic categories, but that the link operates via formal elements that, on the one hand, are related to features of meaning, but are, on the other hand, defined by phonetic ranges. This phonetic substantiation of phonological categories is just as essential as the recognition of structure in phonetic substance. Both phonetic substance and phonetic structure (or signal measures and phonological form) are required for an adequate description of the phonetic-syntactic/semantic relationship, and consequently for prosodic modelling.

(4) Integration of prosodic modelling into a linguistic environment. It follows from (3) that prosodic modelling requires strong links with syntactic, semantic, and, particularly in the case of spontaneous speech, also pragmatic levels.

This contribution outlines

(1) a prosody model that has been developed for German in accordance with the four requirements specified for prosodic modelling: the Kiel Intonation Model (KIM);

(2) its incorporation in a TTS system (RULSYS/INFOVOX) as a prosody research tool;

(3) the model-based development of a prosodic labelling system (PROLAB) to create prosodic label files of spontaneous speech, which may in turn be used as input into the TTS system for transcription verification and model elaboration.

13.2 A Prosodic Phonology of German: The Kiel Intonation Model (KIM)

13.2.1 The Categories of the Model and its General Structure

This chapter gives a summary of Kohler ([Koh91a, Koh91b, Koh96]), supplemented by the most recent additions that became necessary to cope with spontaneous speech phenomena encountered in prosodic labelling. The following domains have to be recognized in a prosodic model of German:

(1) lexical stress;

(2) sentence stress;

(3) intonation:

  (a) categories of pitch "peaks" and "valleys" as well as their combinations at each sentence-stress position;

  (b) types of pitch category concatenation;

  (c) pitch of pre-head preceding the first sentence stress in a prosodic phrase;

(4) synchronization of pitch "peaks" and "valleys" with stressed syllables;

(5) "downstep" of successive pitch "peaks"/"valleys" and pitch reset;

(6) "upstep" of successive pitch "peaks"/"valleys";

(7) prosodic boundaries (degrees of cohesion);

(8) overall speech rate (changes) between the utterance beginning and successive prosodic boundaries;

(9) register change;

(10) dysfluencies: pauses, breathing, hesitations, breakoffs, and resumptions.

A system of prosodic distinctive features is used to specify the abstract phonological categories in these domains, and they enter into sets of ordered symbolic rules. The features are either graded or binary and determine the parametric value spaces activated by parametric phonetic rules following the symbolic ones. The prosodic features are attributed to phonological units, which are either segmental (vowels and consonants) or non-segmental (morphological and phrase boundaries). Attached to vowels is the fundamental distinction whithin the German prosodic system, viz. stress and intonation. In separating symbolic rules from subsequent parametric ones for generating acoustic output, KIM recognizes two levels in prosodic modelling:

(1) the defining of phonology-controlled prosodic patterns by a small number of significant $F0$ points (macroprosody);

(2) the output of continuous $F0$ contours influenced by articulation-related modifications (microprosody, for further details see [Koh96]).

KIM is integrated into a pragmatic, semantic, and syntactic environment. The input into the model are symbolic strings in phonetic notation with additional pragmatic, semantic, and syntactic markers. The pragmatic and semantic markers trigger, e.g., the pragmatically or semantically conditioned use of "peak" and "valley" types or of sentence focus. Lexical stress position can largely be derived by rule, and syntactic structure rules mark deaccentuation and emphasis in word, phrase, clause, and sentence construction. Phrasal accentuations are thus derived from the syntactic component preceding the prosodic model, and are given special symbolizations in the input strings to the model (see [Koh91a, Koh91b] for further details).

13.2.2 Lexical and Sentence Stress

At the abstract level of phonological specifications in the lexicon, every German word has at least one vowel that has to be marked as potentially stressable, as being able to attract the feature specifications of sentence stress. This lexical stress is thus not a distinctive stress feature, it only marks a position that can attract such a feature at the sentence level. In non-compounds as well as in at least one compound element of compounds there is one vowel with such "primary" lexical stress; other compound elements have one vowel each with "secondary" lexical stress. All remaining vowels are lexically "unstressed".

Sentence stress is attributed to a word as a whole, and manifests itself phonetically at the lexical stress position. By default a non-function word receives the category "accented". Deviations from this are either in the direction of emphatic "reinforcement", or of "deaccentuation", which may be "partial" or "complete". Function words are by default "unaccented" (= "completely deaccented"). Deviations are "partially (de)accented", "accented" or "reinforced".

Vowels receive combinations of the stress features <+/–FSTRESS> and <+/–DSTRESS> (referring to the association of sentence stress with the two important parameter domains of $F0$ and duration). In sentence-stressed words, the vowel with "primary" lexical stress is <+FSTRESS,+DSTRESS>; in "completely deaccented" content words it is <–FSTRESS,+DSTRESS>. Vowels with "secondary" lexical stress are also <–FSTRESS,+DSTRESS>, irrespective of sentence stress.

Finally, in "unaccented" function words as well as in lexically "unstressed" syllables, the combination is <–FSTRESS,–DSTRESS>. In "partially deaccented" sentence stresses <+DEACC> is added to the two positive stress features, all other vowels are <–DEACC>. Words that are to get additional emphasis receive the feature <+EMPH> in their lexical stress position, all other vowels <–EMPH>.

Whether <+DSTRESS>, responsible for longer duration, is associated with <+FSTRESS>, marking the vowel as the recipient of intonation features ("peak" and "valley" contours), or as <–FSTRESS>, not providing the vowel with this potential, depends on the rules of grammar and context of situation in speech communication, which allocate sentence stress digit markings in the input string to the prosodic model. They have to be supplied by the linguistic environment of the prosodic phonology (see [Koh91a, Koh91b]). The same applies to the attribution of <+DEACC>.

To distinguish degrees of emphasis, <+EMPH> vowels may be given the graded stress level feature <@STRLEV>, with @ = 1, 2, ... , 7; <–EMPH> vowels are <0STRLEV>. These vowels are made the more prominent, the higher the stress level. In "peak" contours, this greater prominence is achieved by raising the $F0$ maximum, and, if the "peak" is non-final in a "peak" series, by having a faster descent as well as by lowering the $F0$ minimum between "peaks", proportionally to stress level. In the case of $F0$ "valley" contours, the final $F0$ point is raised in accordance with stress level. Emphasis is used to put words and phrases within sentences in focus, particularly when the expansion of intonation contours on certain structural elements is coupled with the deaccentuation of others. <+EMPH> and <@STRLEV> associated with <+FSTRESS> do not automatically change the duration linked to <+DSTRESS>. The parametric variation of <+DSTRESS> may be controlled independently of the other stress features; in the model this is captured by the categories of speech rate and hesitation lengthening (2.5, 2.7).

In summary, the following distinctive sentence-stress features are proposed for a comprehensive contrastive categorization in the prosodic phonology of German:

<+/–FSTRESS>
<+/–DSTRESS>
<+/–DEACC>
<+/–EMPH>
<@STRLEV>, with @ = 0, 1, ... , 7.
The feature pair <+/–EMPH> and the graded feature <@STRLEV> constitute the link with the intonation features. The following tree graph represents the hierarchical relationship between the various sentence-stress features.

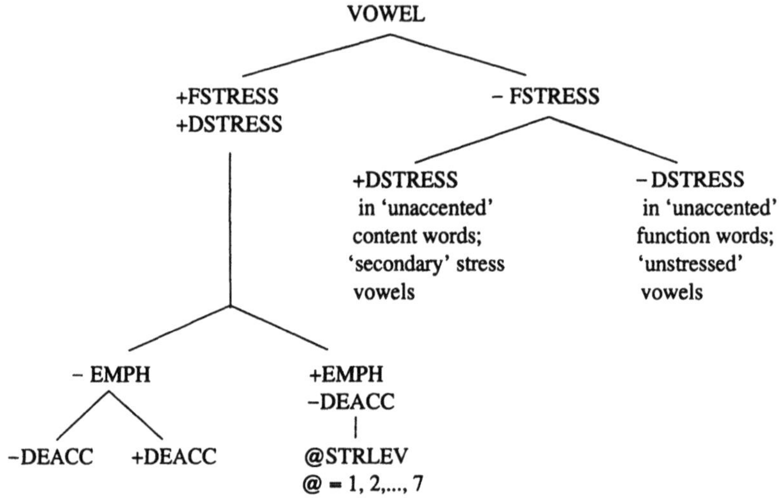

13.2.3 Intonation

13.2.3.1 Pitch Categories at Sentence Stresses

All vowels with "accented" or "reinforced" or "partially deaccented" sentence stress, i.e., with the feature specification <+FSTRESS,+DSTRESS> receive intonation features, which may be either "valleys" or "peaks", specified as <+/–VALLEY>, and in the case of "peaks" (<–VALLEY>), they may contain a unidirectional $F0$ fall, classified as <+TERMIN>, or rise again at the end, resulting in a (rise-) fall-rise, categorized as <–TERMIN>. <+VALLEY> is <–TERMIN> by definition. <–TERMIN> may have a low, narrow rise, to indicate, e.g., continuation, or a high, wide rise, used, e.g., in questions, with the specifications <+/–QUEST>. All "peaks" and "valleys" may have their turning points ($F0$ maximum in "peaks" or $F0$ minimum in "valleys") early or later with reference to the onset of <+VOK,+FSTRESS>, categorized as <+/–EARLY>, and finally, for "peaks" <–EARLY> may be around the stressed vowel centre or towards its end, classified by the feature opposition <+/–LATE>. The categorization of <–VALLEY> into <+EARLY> and <–EARLY>, with a further subdivision of the latter into <+/–LATE>, captures the grouping of "late" and "medial" vs "early peaks", as it showed up in perceptual experiments with stepwise "peak" shift from left to right ([Koh90a]).

Thus the distinctive intonation features needed in the prosodic phonology of German are:
<+/–VALLEY>
<+/–TERMIN>
<+/–QUEST>
<+/–EARLY>
<+/–LATE>
<+/–EMPH>
<@STRLEV>, with @ = 0, 1, ... , 7.

Their hierarchical relationships, linked to sentence stress, are represented by the following tree graph.

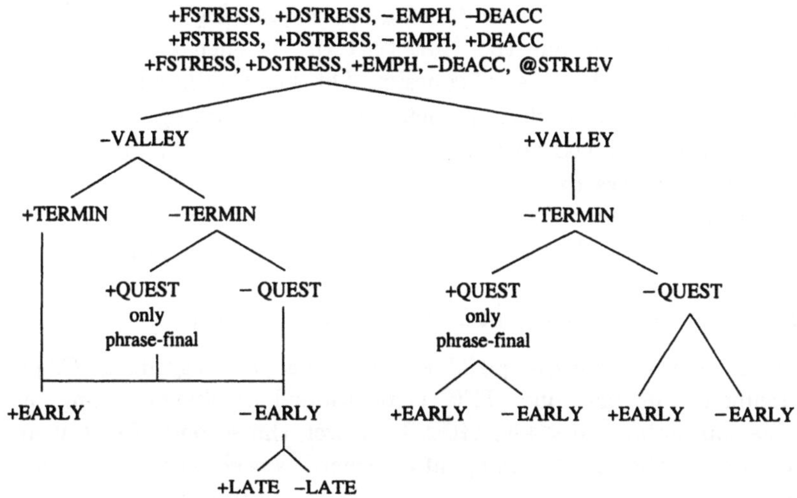

The resulting feature combinations for the intonation categories are illustrated in Figure 3 of the appendix:

(a1) <–VALLEY, +TERMIN, +EARLY>: "early peak"
(a2,3) <–VALLEY, +TERMIN, –EARLY, –/+LATE>: "medial/late peak"
(b1,2) <–VALLEY, –TERMIN, –/+QUEST, +EARLY>: "low/high non-terminal early peak"
(b3,4,5,6) <–VALLEY, –TERMIN, –/+QUEST, –EARLY, –/+LATE>: "low/high non-terminal medial/late peak"
(c1,2,3,4) <+VALLEY, –TERMIN, –/+QUEST, +/–EARLY>: "early/non-early low/high valley"

Which feature combinations are to be activated in the prosodic model again depends on the rules of grammar and context of situation in speech communication, which allocate the intonation markings in the input string. They have to be supplied by the linguistic environment of the prosodic phonology (see [Koh91a, Koh91b]).

13.2.3.2 Pitch category concatenation and pre-head

In a concatenation of pitch "peaks" without prosodic boundaries between them (see Sec. 2.4), $F0$ may fall to a low or an intermediate level and then rise again for the next "peak". There is also the boundary case of the absence of an $F0$ descent between "peaks", which results in a "hat pattern" [tHCC90]. The slight fall due to "downstep" between the "peaks" (see 13.2.3.4) justifies subsuming this boundary case under the intonation feature <+TERMIN>. In such an intonation structure an "early peak" is not possible initially, and a "late" one is excluded non-initially. The boundary case of zero peak-descent is also applied to a high level $F0$ before a phrase boundary (see Figure 13.3 **(d1)**). In <–TERMIN> "peaks" the same differentiation between full and intermediate $F0$ descents must be made (see Figures 13.3 **(d2,3)**).

When prosodic boundaries intervene any sequencing of "peaks" and/or "valleys" is possible, but the "hat pattern" is then excluded since it represents a very high degree of cohesion. On the other hand, a "late peak" with a full $F0$ descent marks a dissociation from a following "peak" and will then normally be linked with a prosodic boundary, i.e., final lengthening and $F0$ reset afterwards.

Unstressed syllables preceding the first sentence stress in a prosodic phrase may be either low or high: they are different types of pre-head.

13.2.3.3 Temporal Alignment of "Peaks" and "Valleys"

Taking the default, "medial peak" as a reference, two significant $F0$ points are defined. The first one, TF0, is positioned at the beginning of the syllable containing the <+FSTRESS> vowel, the second, T2F0, near the vowel centre, the exact timing after voiced vowel onset depending on vowel quantity, vowel height, number of following unstressed syllables and position in the utterance. The calculation of the time point T2F0 after vowel onset is carried out on the basis of the segmental duration rules for German. They have adopted the principle proposed by [Kla79]) for the rule synthesis of English (see also [Koh88]), defining different classes of segments (e.g., diphthongs vs long vs short vowels, low vs high vowels) by different pairs of values for intrinsic duration (Di) and for minimal duration (Dmin) and geneyrating actual segment durations in various segmental, prosodic and syntactic contexts by the application of the following rules:

(1) <DUR> ^ <(Di-Dmin)*PRCNT/100+Dmin>
(2) <PRCNT> ^ <PRCNT*PRCNT1/100>.

In (1), PRCNT = 100 initially; the rules then change the PRCNT values successively by introducing a rule-specific PRCNT1 value into (2). This way all the factors influencing segmental durations (tempo, position in the word, and sentence, stress, segmental context) can be captured in specific rules by inserting a new PRCNT1 value each time. This model assumes that all the factors affecting duration operate independently of each other and that it is only the amount exceeding the minimal duration of a segment that is adjusted by these factors. The two assumptions provide a good approximation of segment timing in languages like German and English, and certainly result in prosodically acceptable speech synthesis.

T2F0 for "medial peaks" is now derived from the basic vowel-type related duration. The only percentage factor that enters the calculation is the one referring to speech rate; it is normally set at 100, a speeding up lowers, a slowing down increases the factor, i.e., it is essentially the intrinsic vowel duration that determines the point in time after <+VOK,+FSTRESS> onset (T2F0) where the "medial peak" is positioned. But this has to be adjusted in the case of aspiration. On the one hand, aspiration lengthens the total vowel duration, compared with vowels in non-aspirated contexts, but this increase is not as large as the total aspiration phase; on the other hand, it shortens the stop closure duration compared with unaspirated cases, but again not by the total amount. So the larger part of the aspiration (AH) should be added to the vowel, but some of it attached to the plosive, and the *F*0 "peak" placement has to take this ambivalence into account:

(3) <+VOK,+FSTRESS> ^ <T2F0=((Di-Dmin)*PRCNT/100+Dmin)
*0.6+TLAH*0.75>,

i.e., three quarters of the period up to the last aspiration time point are added to T2F0, shifting it further to the right by this amount.

Sentence-final "medial peaks" receive a third *F*0 point, T3F0, at 150 ms after the "peak" maximum in a medium speech rate (see Sec. 2.5); in all non-final cases, the default treatment of a descending *F*0 is that the "peak" summit of one <+FSTRESS> connects with the left-base point of the next <+FSTRESS>. Other possibilities are that the low *F*0 point between "peaks" occurs at any other intervening word, or that there is a relatively level "dip" in between.

As the absolute *F*0 "peak" position is not affected by vowel duration modifications due to voiced/voiceless context, number of syllables in the word, sentence position, etc., its relative position changes with vowel shortening or lengthening, moving closer towards or further away from, the end. This way the microprosodic *F*0 truncation before voiceless obstruents is automatically built into the rules. That no longer applies to rising *F*0. The intended high value of a "valley" always has to be physically reached.

An "early peak" has its maximum value at the <+FSTRESS> syllable onset, TF0 100 ms before, and T3F0—in sentence-final position—in an area where the "medial peak" has its maximum. A "late peak" has TF0 at the same point as a "medial peak", then an additional low *F*0 point T2F0 is inserted at vowel onset, and the late summit (T3F0) occurs 100 ms after the point where a "medial peak" has its centre, or at the end of the last voiced segment in a non-final monosyllabic word if this distance is less than 100 ms. If there is an unstressed syllable following, the summit coincides with the unstressed vowel voice onset. In utterance-final position, a fourth $F0$ point, T4F0, occurs 100 ms after the summit. In monosyllables without final voiced consonants, T3F0 has to occur at least 30 ms before vowel offset to signal the $F0$ descent to T4F0.

"Valleys" have their left and centre $F0$ points at the same positions as TF0 and T2F0 in "medial peaks" (except for "early low valleys" where T2F0 as in "early peak"); in an "early valley" the left point is the lowest, whereas in a "non-early valley" it is the centre point. In both cases the right high point is located at the end of the last voiced segment.

13.2.3.4 "Downstep" and Pitch Reset

Declination, i.e., the temporally fixed decline of $F0$, has been replaced by downstepping in KIM, i.e., a structurally determined pitch lowering from sentence stress to sentence stress, independent of the time that elapses between them. The downstepping values used in KIM are 6% from "peak" to "peak" (starting from 130 Hz in a male voice), and 18% from a "peak" to the next base. In "valleys" both the low and the high $F0$ value are downstepped by 6%. Downstepping can be interrupted at any point by the feature <+EMPH> or by resetting. Prosodic boundaries (see Sec. 2.4) are usually associated with pitch reset but they need not be. Reset can also occur at other points than phrase boundaries.

13.2.3.5 "Upstep"

Besides the interruption of automatic "downstep" at any point by a controlled restart of the downstepping pattern (pitch reset), we also have to take another systematic deviation from default into account, namely the step-wise upward trend of "peak" or "valley" sequences: "upstep". It is treated as a global superpositional feature in KIM and in its TTS implementation (see 3.). The upstepping values used in KIM and in its TTS implementation are comparable to the ones for downstepping: 6% up from "peak" to "peak", and 12% down from a "peak" to the next base. In "valleys" both the low and the high $F0$ value are upstepped by 6%.

13.2.4 Prosodic Boundaries

One of the functions of prosody is the sequential structuring of utterances and discourse, i.e., the signalling of prosodic boundaries and—at least partially—their hierarchical organization. To decode the syntagmatic chunking of messages in accordance with the speaker"s intention the listener requires signals that index degrees of cohesion or separation, respectively, between phrases, clauses, utterances and turns. The parameters that achieve this are pause duration, phrase-final segmental lengthening and scaling of $F0$ end points at the respective boundaries. They can be controlled by parametric rules in the prosodic model upon appropriate symbolic input.

As at this stage the linguistically and phonetically relevant categorization of these boundaries is not well understood the modelling cannot reduce the categories in this domain to the same small number as in the other areas of prosody discussed so far, but has to allow sufficient degrees of freedom for experimentation with data modelling in a development system, which will be discussed in Sec. 3. Two categories of phrasing have been extracted so far with this very flexible device: [PG1] corresponding to prosodic clauses and [PG2] related to prosodic phrases. Both are always phonetically signalled by lengthening before them, and usually by $F0$ resetting after them. $F0$ resets may occur at other points than the phrasing markers [PG1,2]. [PG1] also coincides with high syntactic structure nodes, whereas [PG2] does not. Both may be further strengthened by the incidence of pauses and intonation patterns. Full $F0$ "peak" descents are particularly frequent with [PG1], and <–TERMIN, +QUEST> is only associated with this phrasing marker.

13.2.5 Speech Rate

The modelling of different absolute speech rates—"slow", "medium", "fast"—within the same speaker, discussed in [Koh96] also includes articulatory reduction and elaboration, and recognizes a fourth degree, viz. "reduced" at an otherwise medium rate (see also [Koh90b]). This level will probably have to be subdivided into subcategories according to the degree of formality and spontaneity of speaking, comprising different rule modules for the respective degrees. A good deal more research into spontaneous speech is necessary before an adequate categorization can be set up in the model.

This modelling of speech rate and degrees of reduction is extremely useful at the development level, using the TTS research tool (see Sec. 3). As a basis for the description of spontaneous speech data and their labelling, however, it is more helpful to set up relative categories of speech rate change, indicating slowing down or speeding up with regard to a preceding stretch of speech (see Sec. 4). Eventually the two approaches should be combined, for instance in such a way that for a particular speaker"s speech production (e.g., a dialogue turn) an initial absolute rate evaluation is provided, upon which slowing down and speeding up operate iteratively.

13.2.6 Register Change

A prosodic model also has to incorporate the category of register (change), in the sense that a speaker may observe equivalent contrastive phonological pitch relations on different pitch levels, with semantic and pragmatic implications, e.g., a lowering of pitch level for asides, and insertions into main arguments, or a raising of pitch level for putting whole stretches of utterance into perspective. It is an open research question at this stage as to how many register levels are needed in the modelling of spontaneous speech. The research tool of Sec. 3 uses three: default—raised—lowered. The values used for register in KIM and in its TTS implementation (see Sec. 3) are 20% up for the raised level and 20% down for the lowered one, compared with the $F0$ points in default register.

13.2.7 Dysfluencies

At the segmental level, pauses, breathing, hesitation particles, laughing, clicks, etc. (see [Koh96]) need to be indicated as elements of utterance structuring and dysfluency. At the prosodic level, hesitation lengthening is to be differentiated from automatic phrase-final lengthening. Break-offs with and without repairs, inside words, and at word boundaries are additional dysfluency categories, characteristic of spontaneous speech, with the potential of phonetic exponents (see [Koh96]).

13.3 A TTS Implementation of the Model as a Prosody Research Tool

The incorporation of KIM in the RULSYS/INFOVOX TTS system for German has been described in more detail in [Koh96]) (see also [CGH90]). This environment constitutes a powerful research tool in the analysis and modelling of German prosody. This particularly applies to the investigation into prosodic phrasing (2.4), amount of $F0$ descent in "peaks", especially in "peak" concatenation (2.3.2), and into "downstep and "pitch reset" (2.3.4).

To test these aspects the development system uses a prosodic boundary (cohesion) marker [p:], which is put after the word at which boundary indices occur. It is preceded by two digits, ranging from [0] to [2], the second of which refers to pause length, the first to utterance-final lengthening ([0] standing for absence in both cases). In the case of pitch "peaks", there is a third boundary-related digit to the left of these two, referring to the scaling of the $F0$ end point, again ranging from [0] to [2]: [2] refers to a descent to the bottom of the speaker"s voice range, [1] to an intermediate fall and [0] to the absence of an $F0$ dip, e.g., between the "peaks" of a "hat pattern". This digit string is preceded by a further digit, ranging from [0] to [3] to mark four degrees of speech rate from fast to slow. The "high pre-head" index is also linked to the prosodic boundary marker in TTS: [p:=]. As our knowledge of prosodic boundary marking increases the degrees of freedom can be reduced by establishing constraints between the parameters in the signalling of the necessary and sufficient number of phonologically relevant distinctions. The need to retain a high degree of flexibility in the prosodic development system is also the reason why, in the implementation of KIM in TTS synthesis, prosodic boundary markings are always preceded by [+], if there is pitch reset, or they remain unspecified, if there is not.

"Upstep" and three levels of register have also been incorporated in KIM-based TTS, for a simulation of natural, especially spontaneous speech, but apart from pauses and hesitation lengthening, dysfluency indices have not been successfully implemented yet. The extension of "upstep" is marked by indexing the two sentence-stressed words where it begins and ends. The devices used are [$+] and [+$] after the sentence-stress digit [2].

Figures 1-5 of the Appendix illustrate the TTS parameters $F0$ and duration for the various KIM-based prosodic categories in this development system (RULSYS). They provide $F0$ (in Hz) for the significant points (square parameter) as well as cosine interpolation, and the TTS phonetic transcription aligned to the time scale (segment durations in cs).

13.4 The Analysis of Spontaneous Speech

13.4.1 PROLAB: A KIM-based Labelling System

The categories of KIM outlined in Sec. 2 have also been the basis for the development of a symbolic system for consistent, systematic, and efficient prosodic labelling of recorded spontaneous speech data: PROLAB. This system meets the following requirements:

(1) unequivocal representation of the categories of the prosodic phonology;

(2) integration into 7 bit ASCII segmental label files;

(3) integration into ASCII orthographic files of German text;

(4) clear typographic separation from the segmental labelling allowing prosodic notations on the same tier for convenient cross-reference between segmental and prosodic aspects of speech;

(5) mnemonic suitability for easy learning and use;

(6) easy retrieval of prosodic phenomena in data bank searches.

The application of these guiding principles to prosodic labelling of data bases has resulted in the standardization of the following repertoire and conventions [Koh96] for insertion in orthographic text or segmental phonetic files.

(1) ['],["] for lexical stress are put in front of the "primary" or "secondary" stress vowel; unmarked vowels are "unstressed". In a segmental label file these stress markers are linked to the vowel symbol, in an orthographic file they are inserted in sequential order before, and

on the same time mark as, the vowel. Function words, identified with suffixed [+], by default do not get a lexical stress symbol; if they receive sentence stress, ["] is inserted before the vowel of the appropriate syllable.

(2) All sentence-prosodic markers are preceded by & to separate them unequivocally from non-prosodic labels, e.g., grammatically determined punctuation marks. The latter are taken over from the orthography and kept as such beside the inserted and &-prefixed punctuations, which refer to intonation categories. Prosodic punctuations follow orthographic ones in their sequential ordering.

(3) Digits [&3],[&2],[&1],[&0], when not combined with punctuation marks, refer to sentence stress. They are put in sequential order before words that receive the "reinforced", "accented", "partially" or "completely deaccented" sentence-stress category. The lexical stress position then determines where $F0$ contours have to be hooked.

(4) Punctuation marks [&.],[&,],[&?] refer to pitch "peaks", "low" and "high" rising "valleys", and the character sequences [&.,] and [&.?] to the corresponding fall-rises. They are put in sequential order before a prosodic boundary or before the next sentence-stress label [&<digit> $\geq$ 1].
[&,] and [&.,] occur before the next word after the $F0$ maximum, i.e., they are also possible before the sentence-stress label [&0]. [&(.)?] can only occur before a prosodic boundary.

(5) Parentheses [)],[(] refer to "early" and "late peaks", the corresponding brackets []] and [[] to "early" and "non-early valleys"; they are put after the sentence-stress digit, e.g., [&2(]; the "medial peak" is also positively marked by [^], which differs from the default implication in 13.2.3.1 and in the TTS implementation. It allows easier access in data bank retrieval. The same applies to the differentiation between parentheses and brackets for "peaks" and "valleys". Sentence-stress digit and pitch synchronization marker form a prosodic label unit.

(6) The pitch movement between successive "peaks" or between a "peak" and a boundary may be a full or an intermediate $F0$ descent or a level $F0$, symbolized by digits before [.]: [&2.],[&1.],[&0.]. Digit and punctuation mark form a prosodic label unit.

(7) A high prehead is marked by [&HP] at the beginning of a prosodic phrase, but in sequential order after a phrasing marker.

(8) "Downstep" is not marked. $F0$ reset is implied by a prosodic boundary; in the case of its absence, [=] is prefixed to the phrasing marker. If reset occurs at other points than boundaries, [+] is prefixed

to the sentence-stress digit, where the reset occurs. In both cases the character forms a label unit with the prosodic symbol it is prefixed to.

(9) "Upstep" is marked by [|], prefixed to each relevant sentence-stress digit, with which it forms a prosodic label unit.

(10) Prosodic phrasing markers [&PG1] and [&PG2] are put after punctuation marks at the appropriate places. A phrasing marker that is associated with breakoffs and resumptions, [/-] or [/+], is indexed as [&PG/]. Asides and insertions into main clauses are indicated by bracketed [&PG1<]...[&PG1>].

(11) Only speech rate changes in relation to the speed in the preceding prosodic phrasing unit are marked: [RP] and [RM] (= "rate plus"/ "rate minus") are put after [PG1/2] (and before [HP]). An absolute rate judgment at the utterance onset may be added at a later labelling stage.

(12) Register is not marked yet in PROLAB.

(13) Dysfluency markers are

 (a) [z:] for hesitation lengthening at the end or inside of a word

 (b) [/-] or [=/-] for break-offs, and [/+] or [=/+] for break-offs and resumptions at word boundaries and within words, respectively.

(14) Markers for segmental phrase-level units are [p:], [h:] (= pause, breathing), [l:], [s:] etc. (= laughing, clicks etc.) (see [Koh96]).

(15) All non-segmental prosodic markers are without duration; they are put on the same time mark as the beginning of the next segmental unit (usually word beginning) or as the end of the last segmental unit in the speech file.

A labelling platform has been created at IPDS by M. Pätzold on an AT, running under X11 and equipped with a sound card, which accepts segmental and/or orthographic label files and *F*0 analysis data as input, allows the display of the speech wave form, of *F*0 contours and labels (see Figure 6 in the appendix), as well as the insertion, deletion, and change of prosodic labels under auditory and visual control. The manual labelling proceeds in cycles dealing with one prosodic domain after another, in the progression from phrasing to sentence stress, to intonation patterns (peaks, valleys), to their alignment, to speech rate changes and to the other variables listed above. The output is a label file that integrates prosodic labels into the segmental and/or orthographic strings. In this cyclical progression from broad to narrow the labelling can stop at any degree of delicacy, defined by the purpose the resulting label files are to be put to. PROLAB is a comprehensive labelling frame for all prosodic variables, but it is at the same time very flexible to allow a wide range of detail and complexity for different applications.

This PROLAB labelling platform, which also includes a formal check program for the correct syntax of labels in label files, has been applied to a large data base of spontaneous dialogues—The Kiel Corpus of Spontaneous Speech [Kie95]. The 31 dialogues on CD-ROM#2 have been provided with prosodic label files in addition to the segmental label files on the CD itself.[1] The following orthographic transcript with prosodic annotations (rather than a complete label file, to reduce the amount of information and for greater ease of intelligibility) provides an illustration of the prosodic labelling of a spontaneous dialogue turn from this corpus.

g071a004 TIS004:
&2 <ähm> &PG2 &2(D´ienstag &0 würde+ &0 mir+ &0 g´ut &1. &2) p´assen &2. &PG1 &2 <ähm> &PG2 &0 das+ &2] h´eißt &, &PG2 p: &2^ Mom´ent &1. &PG2 &2] ´allerdings &, &2] ´erst z: &, &PG2 &2(n´achm"ittags h: &2. &PG1 &RP &HP &0 das+ &0 wird+ &0 dann+ &2^ wahrsch´einlich &0 ´n+ &0 b´ißchen &0. &2) schw´ierig &2. &PG1 &2^ D´ienstag &1. &|2^ m´ittwochs z: &1. &PG2 &0 <äh> &PG2 p: &0 is=/+ &PG/ &0 s´ieht &0 das+ &0 bei+ &2^ m´´ir+ z: &0 sch=/+ &0. &PG/ &2^ schw´ierig &0 ´aus &2. &PG1 &0 da+ &0 hab" &0 ich+ &2^ tags´über &1. &2^ Term´ine &1. &PG1 &RM h: &2 <ähm> &PG2 &HP &0 wie+ &0 s´ieht &0 das+ &0 bei+ &2^ ´´Ihnen+ &0 am+ &1. &3^ D´onnerstag &0 ´aus &2. &PG1

Figure 6 of the appendix gives the labelled speech waves and $F0$ contours of the first and the last prosodic clause of this dialogue turn as examples of the PROLAB platform output, demonstrating the application of sentence stress and intonation categories to spontaneous speech.

13.4.2 Transcription Verification and Model Elaboration

Prosodic label files can now be the input to the RULSYS/INFOVOX TTS system for German (see Sec. 3) to test the adequacy of the manual labelling by comparing its rule synthesis with the original. Not all aspects of spontaneous speech in the area of dysfluencies have been successfully implemented in TTS. Work on them is in progress. But even so the synthesis of spontaneous speech from PROLAB files already produces very convincing, natural sounding results that can be evaluated against the original human production. This comparative assessment also provides feedback for improved prosodic modelling of spontaneous speech. Figure 7

[1]The CD-ROM as well as the prosodic label files may be obtained from IPDS Kiel.

of the appendix provides TTS parameters and speech wave output, as well as its $F0$ analysis for the same PROLAB representations as in Figure 6. In TTS, the microprosodic $F0$ is largely controlled by a separate parameter, not shown in the $F0$ displays of significant points.

Prosodic modelling, its TTS implementation for model testing, prosodic labelling on the basis of the model, prosodic resynthesis of these prosodic label files for transcription verification and renewed model testing and elaboration thus form an integrated framework of prosodic research at IPDS Kiel. The prosodic categories, being related to human sound production beyond particular language phenomena found in German, should also be transferable to the description of other languages, and the portable PROLAB platform be of more general interest in the prosodic labelling of a wide variety of language data.

Acknowledgments

This paper is a revised and expanded version of a plenary paper "Modelling intonation, timing, and segmental reduction in spontaneous speech" which I presented at the ATR International Workshop on Computational Modelling of Prosody for Spontaneous Speech Processing in April 1995. My special thanks are due to the organizers for their kind invitation and generous support. Part of the spontaneous data recording and labelling was carried out with funding from the German Ministry of Education, Science, Research, and Technology (BMBF) under VERBMOBIL Contract 01IV 101 M7.

Appendix

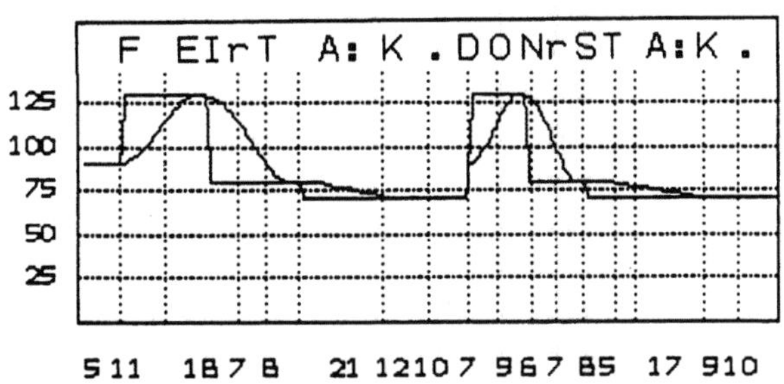

F'eier-t"ag. D´onnerstag.

FIGURE 13.1. Lexical stress and compounding

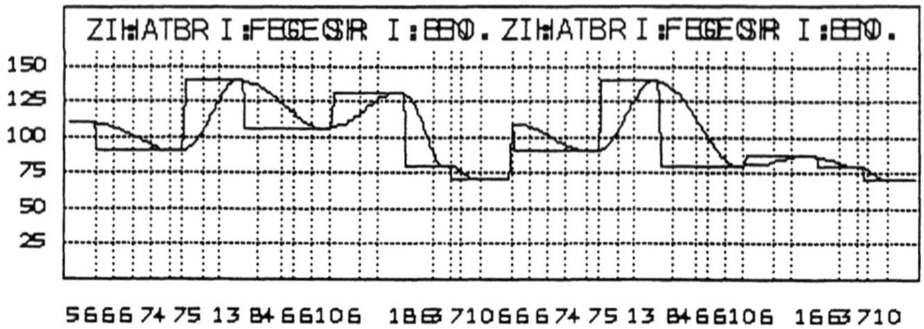

"accented": Sie hat Briefe #2# geschrieben.
"partially deaccented": Sie hat Briefe #1# geschrieben.

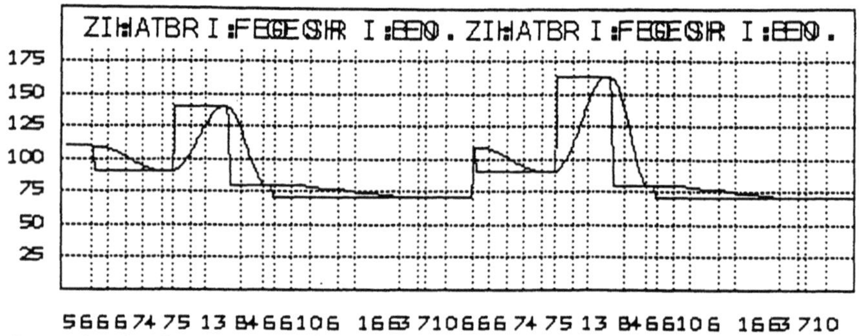

"completely deaccented": Sie hat Briefe #0# geschrieben.
"reinforced": Sie hat #3# Briefe #0# geschrieben.

FIGURE 13.2. Sentence stress.

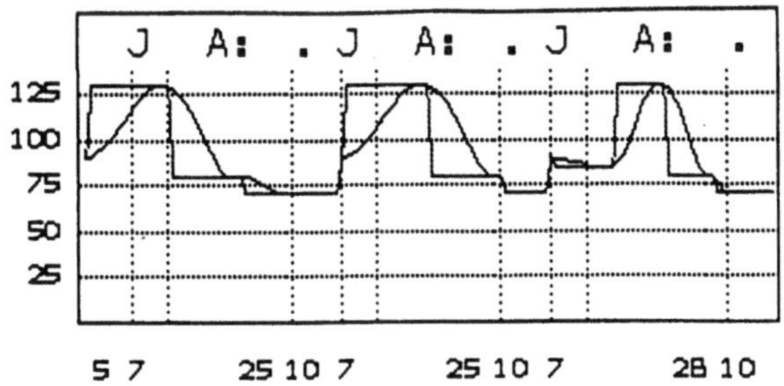

(a1,2,3) "early/medial/late peak": #)#ja. ja. #(#ja.

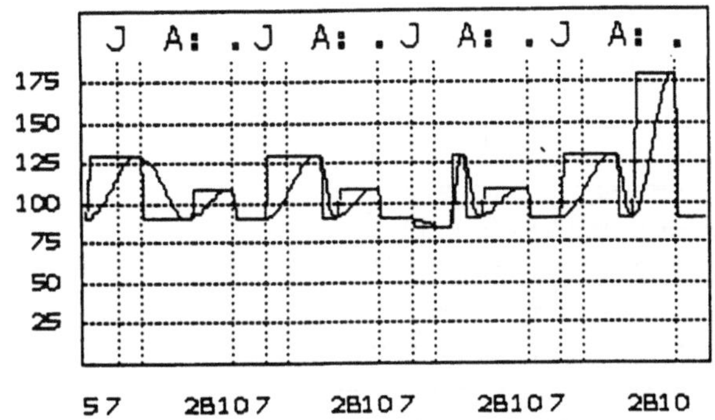

(b1,3,4) "low non-terminal early/medial/late peak": #)#ja., ja., #(#ja., bf (b5) "hight non-terminal medial peak": ja.?

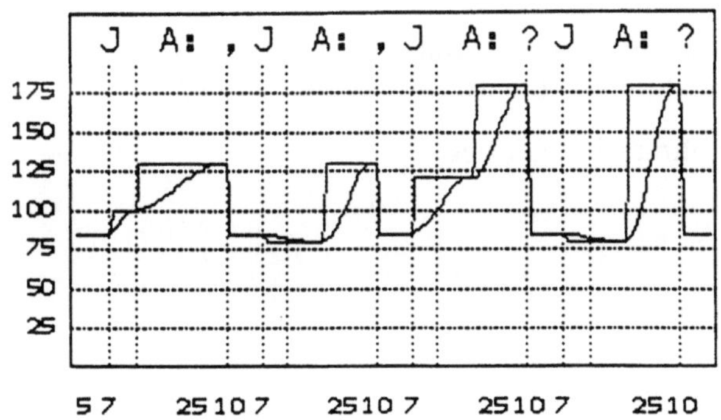

(c1,2,3,4) "early/non-early low/high valley": #)# ja, #(# ja, #)# ja? #(# ja?

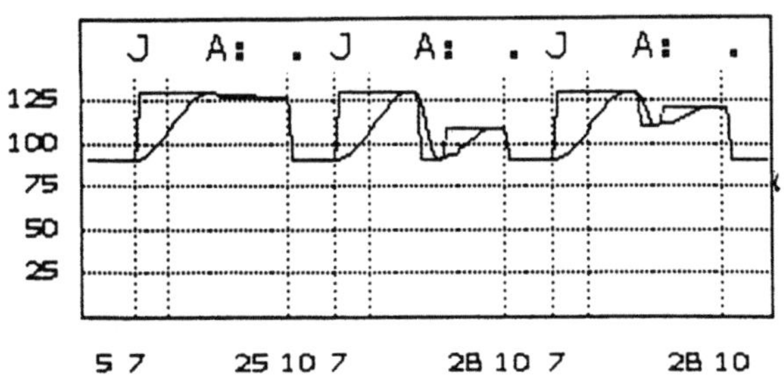

(d1) "medial peak with zero descent": ja 0. **(d2,3)** "full/intermediate descent in <–TERMIN> "peak": ja 2., ja 1.,

FIGURE 13.3. Intonation (cf. Sec. 2.2.1.)

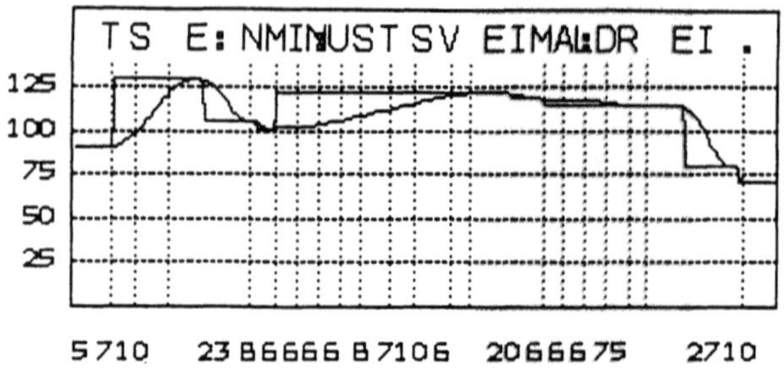

"10 – 2 × 3":

#2# zehn #110p:# minus #2# zwei #000p:# mal #2# drei.

dipped *F*0 pattern between "peaks" followed by "hat pattern"

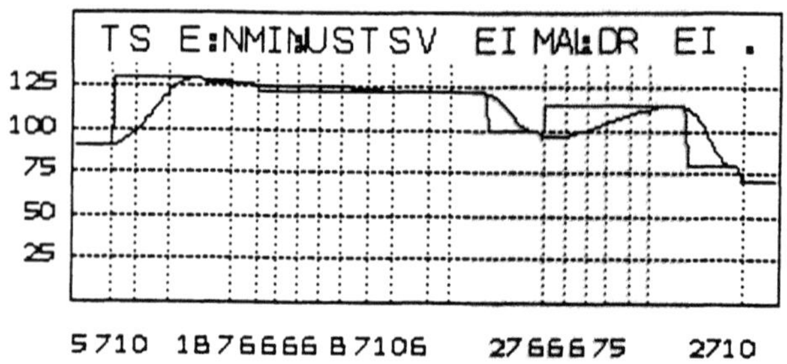

"(10 – 2) × 3":

#2# zehn #000p:# minus #2# zwei #110p:# mal #2# drei.

"hat pattern" followed by dipped *F*0 pattern between "peaks"

FIGURE 13.4. Prosodic phrase boundaries.

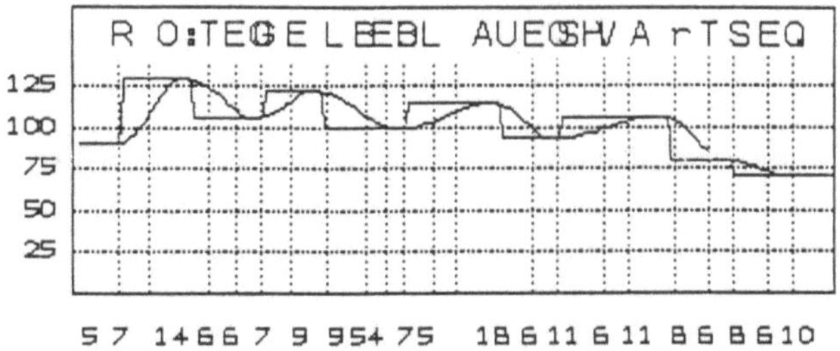

Downstep default: rote gelbe blaue schwarze #212p:#.

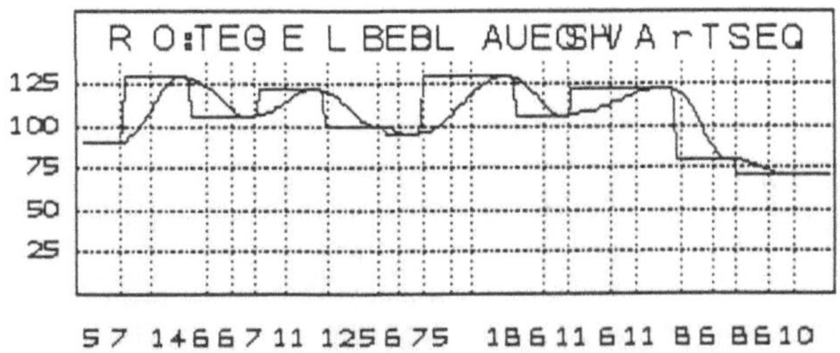

Pitch reset: rote gelbe #+110p:# blaue schwarze #212p:#.

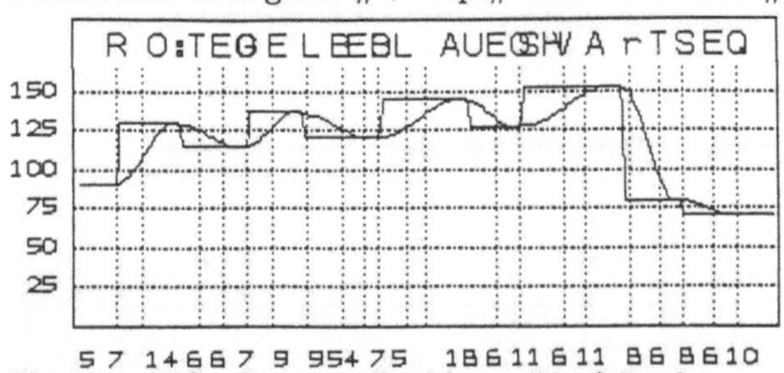

Upstep: #2$+# rote gelbe blaue #2+$# schwarze #212p:#.

FIGURE 13.5. Downstep, reset, upstep.

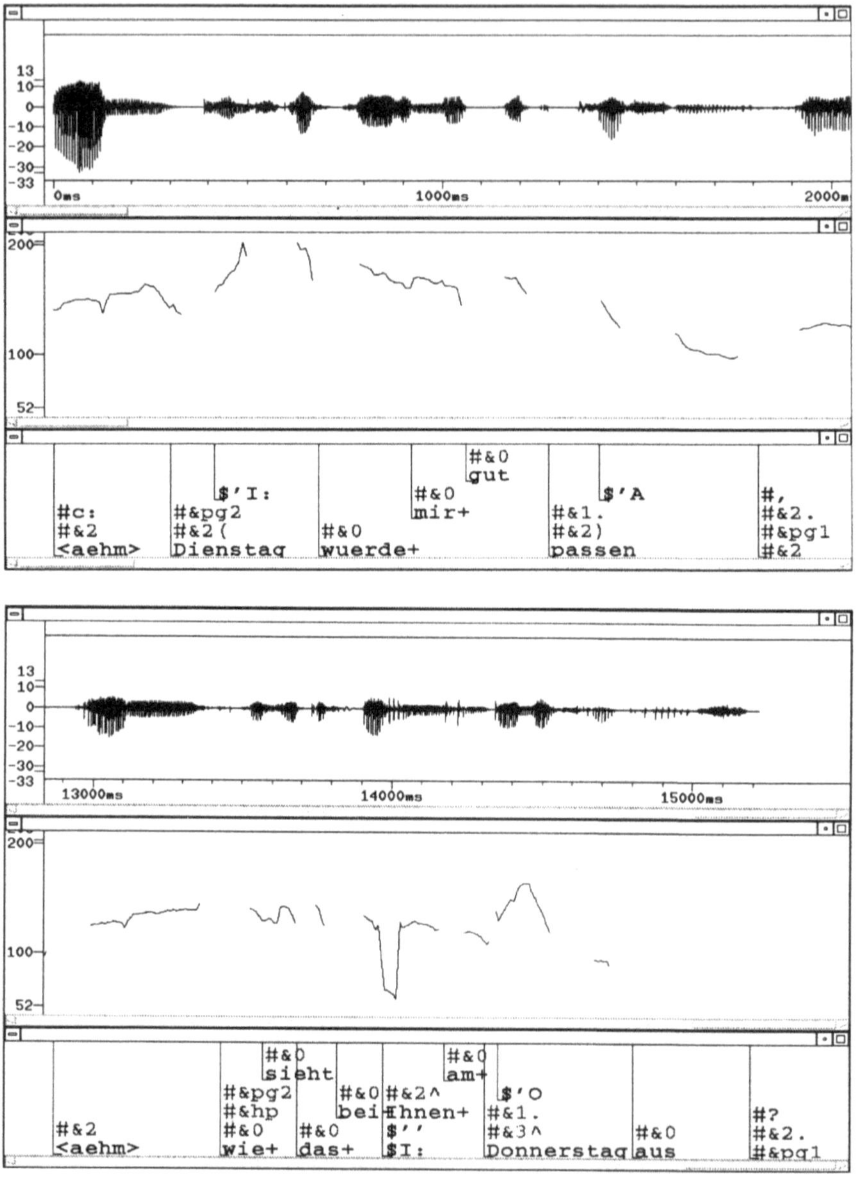

FIGURE 13.6. Labelled speech wave and $F0$ contour of the first and the last prosodic clause in dialogue turn g071a004.

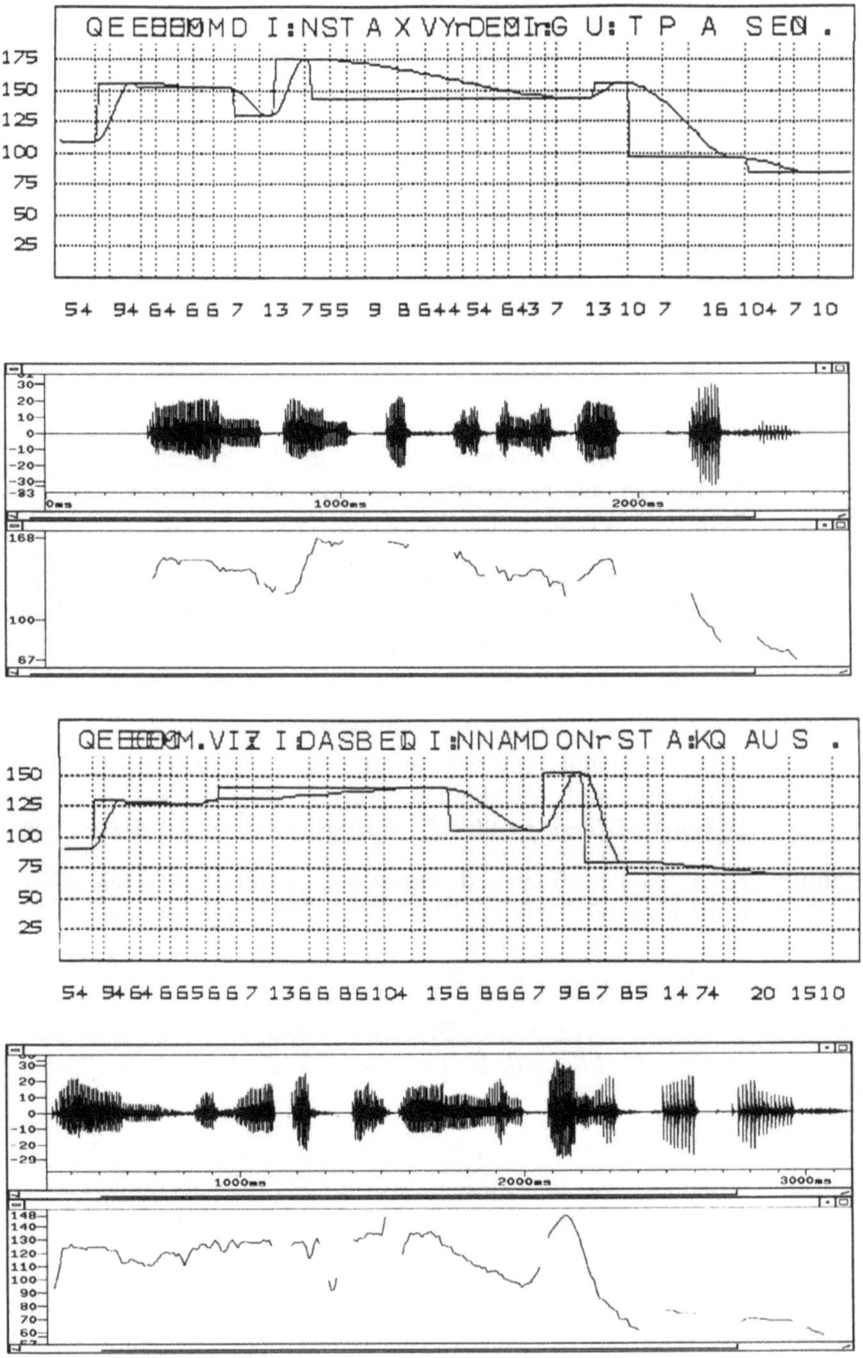

FIGURE 13.7. TTS parameters ($F0$, duration), speech wave output, and its $F0$ analysis for the PROLAB input of Figure 13.6

References

[CGH90] R. Carlson, B. Granström, and S. Hunnicutt. Multi-lingual text-to-speech development and applications. In W.A. Ainsworth, editor, *Advances in Speech, Hearing and Language Processing* pp. 269–296. London: JAI Press.

[Kie95] Kiel, IPDS. *CD-ROM#2: The Kiel Corpus of Spontaneous Speech*, 1995.

[Kla79] D. H. Klatt. Synthesis by rule of segmental durations in English sentences. In B. Lindblom and S. Öhman, editor, *Frontiers of Speech Communication Research*, pp. 287–299. New York: Academic, 1979.

[Koh88] K. J. Kohler. Zeitstrukturierung in der Sprachsynthese. In A. Lacroix, editor, *Digitale Sprachverarbeitung*, pp. 165–170. Berlin: ITG-Tagung, Bad Nauheim, 1988.

[Koh90a] K. J. Kohler. Macro and micro F_0 in the synthesis of intonation. In J. Kingston and M. E. Beckman, editors, *Papers in Laboratory Phonology I*, pp. 115–138. Cambridge, UK: Cambridge University Press, 1990.

[Koh90b] K. J. Kohler. Segmental reduction in connected speech in German: phonological facts and phonetic explanations. In W. J. Hardcastle and A. Marchal, editors, *Speech Production and Speech Modelling*, pp. 69–92. Dordrecht: Kluwer Academic, 1990.

[Koh91a] K. J. Kohler. Terminal intonation patterns in single-accent utterances of German: Phonetics, phonology, and semantics. *Arbeitsberichte des Instituts für Phonetik der Universität Kiel (AIPUK)* 25:115–185, 1991.

[Koh91b] K. J. Kohler. A model of German intonation. *Arbeitsberichte des Instituts für Phonetik der Universität Kiel (AIPUK)* 25:295–360, 1991.

[Koh96] K. J. Kohler. Parametric control of prosodic variables by symbolic input in TTS synthesis. In J. P. H. van Santen, et al. *Progress in Speech Synthesis*. Berlin: Springer-Verlag, 1995.

[tHCC90] J. 't Hart, R. Collier, and A. Cohen. *A Perceptual Study of Intonation*. Cambridge, UK: Cambridge University Press, 1990.

14

Comparison of $F0$ Control Rules Derived from Multiple Speech Databases

Toshio Hirai
Norio Higuchi
Yoshinori Sagisaka

ABSTRACT
In this paper we describe how computational models of $F0$ were derived from four different speech corpora and how their control characteristics were compared to find the possibilities of prosody conversion for speech synthesis. A superpositional $F0$ control model was employed to reduce comptational complexities and a statistical optimization method was used to determine the dominant factors for $F0$ control in each speech corpus efficiently. The analyses showed the invariance of some dominant control parameters and the differences due to speaking styles. These preliminary results also confirmed the usefulness of superpositional $F0$ control for prosody conversion.

14.1 Introduction

In speech synthesis technology, research efforts have traditionally been devoted to synthesizing natural sounding speech of one standard type and not much attention has been paid to variety in speaking styles. Recent improvements in the technology of voice conversion show the feasibility of modelling a speaker's characteristics and speech quality [ES95], but this conversion technology has only been applied to the mapping of a speaker's segmental characteristics, and prosodic characteristics have not yet been well controlled in this scheme. For prosody, only average values are modified according to the statistics of a source speaker and a target speaker. To convert a speaker's prosodic characteristics from one speaker to the other, or to change prosody from a standard style to a specific speaking style without degrading naturalness of the resultant synthetic speech, the prosody control rules themselves should be converted according to the target change.

To adapt prosody control from one style to another in a non-parametric fashion, statistical modelling [KS92b, HHS96] is expected to be useful, as in the conversion of segmental characteristics. For a non-parametric conversion of prosodic rules, a common control framework is to be assumed among different styles. In this paper, multiple models are built for $F0$ control from different speech corpora using the same statistical scheme. They are compared with respect to the effect of different input linguistic factors so as to confirm the effectiveness of the prosody conversion. For this purpose, four different speech corpora were used to evaluate the difference between three speakers and two speech rates.

14.2 Derivation of $F0$ Control Rules and Their Comparison

14.2.1 Overview of the Rule Derivation Procedure

Figure 14.1 shows the three step procedure for the derivation of $F0$ control rules and their comparison. Each $F0$ contour extracted from a speech waveform is decomposed into two components and parametrized using Fujisaki's production process model [FH84] (step (1)), for $F0$ control. The relationship between the linguistic information and the $F0$ control model parameters is analysed using multiple split regression (MSR), a statistical method, in order to predict the value of each model parameter from linguistic information (step (2)). Comparison of the $F0$ control rules generated from speech databases of various speaking styles or from different speakers helps to clarify the differences between them (step (3)). The first two steps are described in the following subsections.

14.2.2 $F0$ Contour Decomposition

Reducing the number of parameters for $F0$ control is useful for constructing a statistical computational model of $F0$ control. For this reduction, Fujisaki's $F0$ control model is used. In this model, a log scaled $F0$ contour is decomposed into phrase commands and accent commands. Each command has only three or four parameters.

In Fujisaki's formation [FH84], an $F0$ contour is described as follows:

$$\begin{aligned}\mathrm{lnF0}(t) &= \ln F_{\min} \\ &\quad + \sum_{i=1}^{I} A_{pi} G_{pi}(t - T_{0i}) \\ &\quad + \sum_{j=1}^{J} A_{aj} \{ G_{aj}(t - T_{1j}) - G_{aj}(t - T_{2j}) \}\end{aligned}$$

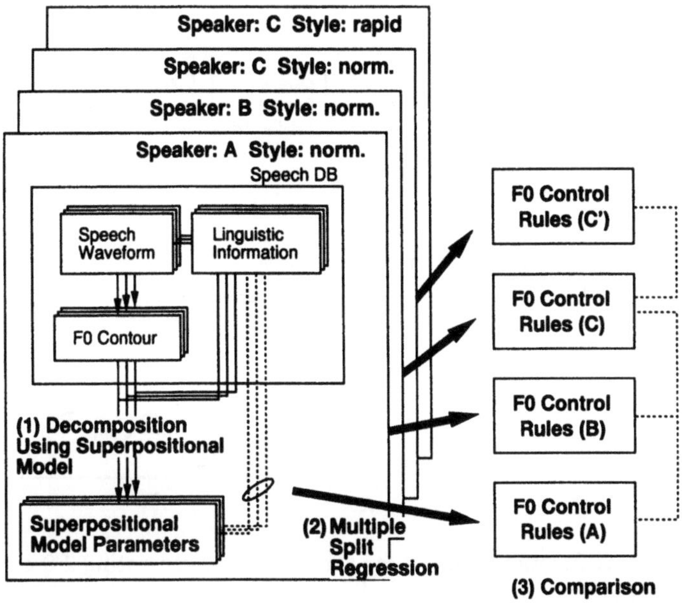

FIGURE 14.1. $F0$ control rule derivation and comparison.

where

$$G_{pi}(t) = \begin{cases} \alpha_i^2 t \exp(-\alpha_i t) & \text{for } t \geq 0 \\ 0 & \text{for } t < 0 \end{cases},$$

$$G_{aj}(t) = \begin{cases} \mathrm{Min}[1 - (1 + \beta_j t)\exp(-\beta_j t), \theta_j] & \text{for } t \geq 0 \\ 0, & \text{for } t < 0 \end{cases}.$$

In these equations, the following control parameters are employed.

$F_{\min}$: base level upon which all the phrase and accent components are superposed to form an $F0$ contour;
I: number of phrase commands;
J: number of accent commands;
A_{pi}: amplitude of the ith phrase command;
A_{aj}: amplitude of the jth accent command;
T_{0i}: time of occurrence of the ith phrase command;
T_{1j}: onset of the jth accent command;
T_{2j}: end of the jth accent command;
α_i: natural angular value of the phrase control mechanism to the ith phrase command;
β_j: natural angular value of the accent control mechanism to the jth accent command; and
θ_j: ceiling level of the accent component for the jth accent command.

Figure 14.2 illustrates this functional $F0$ generation model [FH84]. As described in [FH84], this model was proposed as a simulation of $F0$

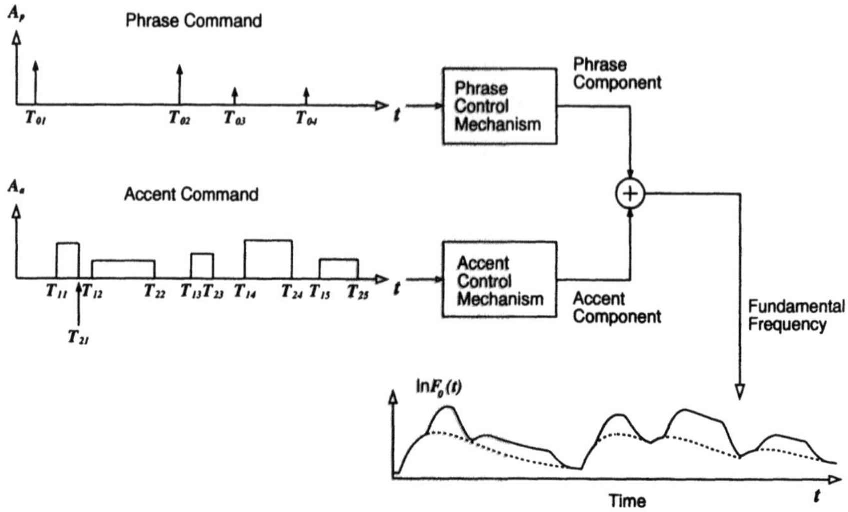

FIGURE 14.2. A functional model for the process of generating $F0$ contours of a sentence [FH84].

generation process. The portions controlled by a phrase command and an accent command are referred to with the terms major phrase and minor phrase, respectively. The $F0$ contours are decomposed into the parameters using an analysis-by-synthesis (AbS) approach in which many parameter combination sets are tested against the original $F0$ contour. In AbS, an initial value for each parameter is decided arbitrarily at first, and then optimised. $F0$ estimation error is derived from the difference between observed and estimated $F0$ contours. It is theoretically possible to reach a local minimum point by gradual changing of parameter values. When the estimation error is at a local minimum, the values are regarded as the optimal result of analysis in AbS.

14.2.3 Statistical Rule Derivation

By statistically analyzing the relationship between linguistic information and the features of a phrase command or accent command, $F0$ control characteristics can be extracted as rules which predict the value of each command feature from the linguistic information.

In the $F0$ control model, natural angular values (α_i and β_j), timing information (T_{0i}, T_{1j}, T_{2j}), and the amplitude of each command (A_{pi}, A_{aj}) are given as command features [FH84]. The natural angular values were treated as constants [FH84], and timing information was predicted with a limited number of rules [OFH84, HFK85]. The amplitude of each command is affected by the content of the utterance: the declination and the amplitude of the accent command of the $F0$ contour are changed by linguistic information [HIHS96]. Only the amplitude of each command is expected to be determined from input linguistic information. The amplitude prediction models for phrase command and accent command were generated separately.

As for a statistical computational tool, linear regression and tree regression have been used previously in prosodic modelling [KTS92, Ril92]. Each method has its own merits and disadvantages. Linear regression is based on a classification assuming one supersurface which is divided by linear formulation of factors; it cannot easily represent the dependency of multiple factors. In tree regression, no specific functional form is assumed for prediction. A binary decision tree is formed by splitting datasets by specifying the control factors, but as each split has a model parameter, the degree of freedom of the model has a tendency to be high.

MSR has been proposed [IS93] to overcome the limitations of both the linear model and the tree regression model. MSR can be regarded as a superset of the linear regression and tree regression models where the number of free parameters is improved by allowing parameter sharing within a regression tree. As with conventional tree regression models, MSR generates a binary decision tree.

In this study, we adopted MSR as the statistical computational model. The resulting binary tree can be regarded as a set of rules to predict the command amplitude, and the associated parameters are optimized to generate the appropriate amplitude of each command. By interpreting the binary tree structure, major control factors and their effect on the amplitude can be determined. For example, if a factor is frequently used in many classifications, or if the absolute value of a parameter is larger than other parameters, the factor can be considered dominant.

14.3 Experiments of $F0$ Control Rule Derivation and Their Comparison

14.3.1 Speech Data and Conditions of Parameter Extraction

To analyse the difference of $F0$ control rules, four speech databases were used. Details of these databases are shown in Table 14.1. To represent speech rate, we use mora/s. A mora (pl. morae) is a syllable-sized unit consisting of (consonant+) vowel. Speech data from three male professional narrators A, B, and C were examined for this experiment, with readings of 500 sentences extracted from newspapers and magazines [STA+90]. Because of the slow processing speed of AbS, we selected short sentences (less than 3 s) from the original speech database for the generation of $F0$ control rules. Speech databases 1, 2, and 3 were read at a normal speaking rate. Database 4 was read rapidly by the speaker of database 3. For the analysis of $F0$ control factors related to individual speaker differences, rules extracted from speech databases 1, 2, and 3 were compared, and for the analysis of differences in speaking rate, the $F0$ control rules estimated from databases 3 and 4 were compared. The initial values for AbS are given by listening. That is, the accentual phrase boundaries and accented positions were marked from listening to the speech. The $F0$ contour of each sentence was obtained using the algorithm proposed by Secrest and Doddington [SD83]. In the extraction of $F0$, the window width was 49 ms, the window shift was 5 ms, the window type was Hamming, and the LPC order was 12.

The natural angular values of the phrase control mechanism and the accent control mechanism were fixed ($\alpha = 3.0\ \mathrm{s}^{-1}$, $\beta = 20.0\ \mathrm{s}^{-1}$) [FH84]. The onset of each phrase command and the onset and offset of each accent command were based on phonetically labelled data and accent type data[1] determined according to the position of accent. The number of phrase commands and accent commands in these speech data is shown in Table 14.2.

TABLE 14.1. Parameters of the databases.

Database	1	2	3	4
Speaker	A	B	C	C
Speaking rate	norm.	norm.	norm.	rapid
Num. of sent.	237	282	187	122
Speech rate (mora/sec)	7.8	8.4	7.0	10.4

TABLE 14.2. Number of commands.

Database	1	2	3	4
Phr. cmd.	718	841	595	349
Acc. cmd.	1276	1555	999	709

[1]The *accent type* shows the location of the accent nucleus in an accentual phrase. For example, "taifuu (typhoon)" has 4 morae, and its accent type is 3 as the third mora "fu" is the accent nucleus. In standard Japanese, high tones are held from the second mora to the accent nuclear mora (only the first mora is a high tone if the accent type of a word is 1). Thus, in the case of "ta $\cdot\ \overline{\mathrm{i} \cdot \mathrm{fu}}$ $\cdot$ u", 2 morae are pronounced as high tones. If there is no accent nucleus, the accent type is zero. The $F0$ falls at the second mora of the accent word and then the level is kept constant to the end of the phrase [FS71b].

TABLE 14.3. Factors used for the control of phrase command amplitude.

Length of syntactic unit	Number of mora	Prev. maj. phr.
		Curr. maj. phr.
		Foll. maj. phr.
Lexical information	Accent type	Head of curr. maj. phr.
		Head of foll. maj. phr.
		Tail of prev. maj. phr.
		Tail of curr. maj. phr.
	Part of speech	Head of curr. maj. phr.
		Head of foll. maj. phr.
		Tail of prev. maj. phr.
		Tail of curr. maj. phr.
	Case particle	Tail of prev. maj. phr.

14.3.2 Linguistic Factors For the Control Rules

Twelve factors were selected for the control of the phrase command amplitude as shown in Table 14.3. The length of a major phrase was included because as the major phrase becomes longer, the amplitude becomes larger (preliminary experiments showed a positive correlation between the number of morae in a major phrase and the amplitude of the corresponding phrase command [HIHS96]). The depth of a phrase boundary, as it relates to the amplitude of phrase command, depends on the attributes of the accentual phrase which adjoins the phrase boundary. In Japanese, the type of case particle at the end of the previous major phrase also affects the depth of the phrase boundary [NK84]. These factors were therefore included as control factors in order to estimate the amplitude of phrase command.

The full set of 16 factors is listed in Table 14.4 which shows factors used to estimate the accent command amplitude. Previous research has reported that the amplitude of the accent command depends on whether the accent type is type zero or not [FK88, KS93, AS92]. Specifically, the amplitude of a non-zero accent command is typically larger than the amplitude of a type zero accent command. Part of speech of a word in a minor phrase and the number of minor phrases which have an accent nucleus affects the $F0$ pattern [AS92, FHT90a]. The amplitude of an accent command has a tendency to decrease when the accent command is at the end of the sentence. As the verb reflection form indicates whether the word is at the end of the sentence or not, it was included to predict the amplitude. Similar to the phrase command, the mora count of a minor phrase was used to predict the amplitude of an accent command.

TABLE 14.4. Factors used for the control of accent command amplitude.

Lexical information	Accent type	Prev. min. phr. Curr. min. phr. Foll. min. phr.
	Part of speech	Head of prev. min. phr. Tail of prev. min. phr. Head of curr. min. phr. Tail of curr. min. phr. Head of foll. min. phr. Tail of foll. min. phr.
	Reflection form	Prev. min. phr. Curr. min. phr. Foll. min. phr.
Length of syntactic unit	Number of mora	Prev. min. phr. Curr. min. phr. Foll. min. phr.
Position in syntactic unit	Number of preceding accented minor phrases	

14.4 Results

$F0$ control rules were compared between multiple speaking rates (database 3 and 4) and multiple speakers (database 1, 2 and 3). According to the result of an examination of $F0$ control rules which are derived from the different speech databases, there are common dominant factors in the $F0$ control rules. That is, the mora count of a major phrase and the accent type of an accentual phrase are important for the estimation of the amplitude of phrase command and accent command, respectively. These results are consistent across the speech data for each person [HIHS96]. However, there were significantly different effects on the amplitude of the accent command between speakers and between speaking rates with respect to certain factors.

14.4.1 The Accuracy of the $F0$ Control Rules

In order to determine the estimation accuracy by these $F0$ control rules, the correlation coefficients between observed and estimated command amplitudes were calculated for each speech database. They are shown in Table 14.5. The accuracy of estimation for the accent command was lower than for the phrase command.

TABLE 14.5. Correlation coefficients.

Database	1	2	3	4
Phrase	0.71	0.78	0.76	0.85
Accent	0.62	0.63	0.68	0.76

14.4.2 Comparison of F0 Control Rules Among Multi-Speakers

The common and different points between $F0$ control rules are shown below:

Speaker independent $F0$ control characteristics

The derived $F0$ control rules had common control factors. In $F0$ control rules which were extracted from each person's speech database, the number of morae for the phrase command was important. In the case of accent command, the accent type was important. The contribution of these factors has already been pointed out [HIHS96] as the main factor for another person's speech database. This indicates that these factors are speaker independent $F0$ control characteristics.

Relationship between factors and the amplitude of the commands

The influence of the dominant factors on the amplitude of the command is shown in Figure 14.3. Summarizing

(1) The degree of the effect caused by the difference of the accent type on the amplitude of the accent command shows a large individual difference. For example, the effect for accent type 6 of speaker C is three times of speaker A.

(2) The influence of the mora count of previous phrase for the amplitude of phrase command shows individual difference. For speakers A and C, the amplitude of the phrase command becomes small when the mora count of the phrase is small. On the contrary, for speaker B, the amplitude is lower for all non-initial phrases.

(3) The effect of the accent type on the amplitude of the accent command is larger than the effect on the amplitude of the phrase command from the difference of the number of morae of the current or previous phrase.[2]

[2]In this study, A_p and A_a were compared directly. This is reasonable since the shape of G_p and G_a are roughly equal across the first part (200 msec, about 1 mora).

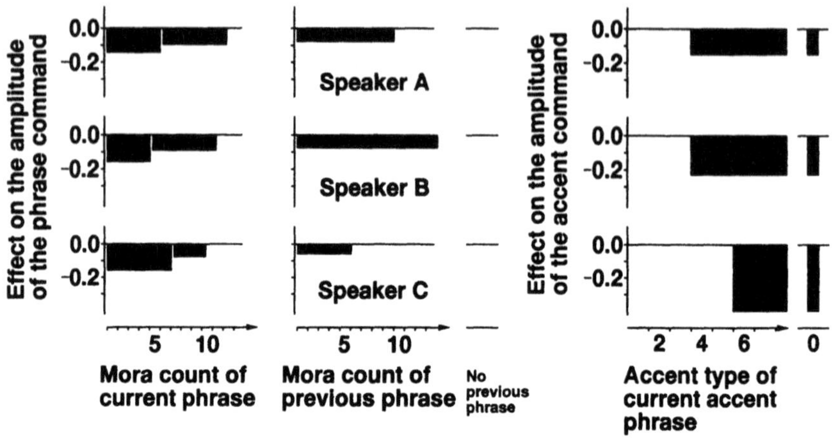

FIGURE 14.3. $F0$ control rules for phrase commands and accent commands for different speakers.

14.4.3 Differences of $F0$ Control Rules Between Different Speech Rates

The dominant factors of the $F0$ control rules which were derived from database 4 (speaker C speaking rapidly) were the mora count of the current phrase for the phrase command amplitude and the accent type of the accentual phrase for the accent command amplitude. The influence of the mora count of the current phrase and the accent type of the accentual phrase is shown in Figure 14.4. In these factors, the effect on the amplitude of the phrase command is smaller and the effect on the amplitude of the accent command is larger than the $F0$ control rules derived from the normal speaking rate (compared with Speaker C in Figure 14.3). These results can be interpreted as control by accent command is focused on instead of control by phrase command in rapid speaking style. That is, quick changing $F0$ caused by accent command is dominant in rapid speech. According to the effect for accent command in rapid speech (Figure 14.4, right), almost all their amplitudes will be small. This fact supports the results of [HHS96] statistically.

However, we have difficulty if Fujisaki's model is used for the analysis: if the duration of the accent command is short, that is, if the accentual phrase has an accent nucleus and the fall is on an early mora, the accent component starts to decrease before it saturates. As a result of this, the amplitude of accent command will be larger than the amplitude estimated intuitively by looking at $F0$ contour. In our analysis, we defined the shortest duration of on accent command to be no less than 100 ms, which is roughly equal to half the duration of an average mora. This problem is one which we intend to investigate further in the future.

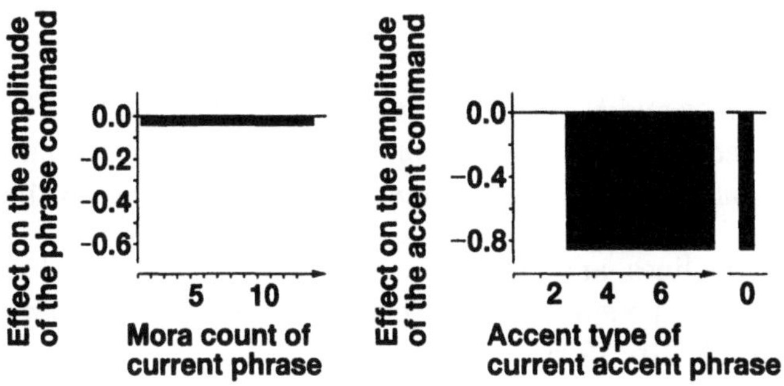

FIGURE 14.4. $F0$ control rules of the "rapid" speaking rate by speaker C.

14.5 Summary

In this paper, computational models of $F0$ were derived using four different speech corpora and their control characteristics were compared statistically to confirm the possibility of prosody rule conversion between different speakers or different speaking styles. In this modelling and comparison, the superpostitional $F0$ control model proposed by Fujisaki was employed to reduce computational complexities and the MSR method was used to extract the statistically dominant factors of $F0$ control in each speech corpus. The analyses showed the following $F0$ control characteristics :

(1) The dominant factors in $F0$ control rules are speaker independent: for the amplitude of the phrase command, the dominant factors are the numbers of morae of the current and previous phrase, and for the amplitude of the accent command, the dominant factor is the accent type.

(2) The effect on the amplitude of the accent command depending on the accent type has a large individual difference.

(3) The control of the accent command is focused on in a rapid speaking rate.

These $F0$ control characteristics in different corpora confirmed both the invariance of some dominant control parameters (e.g., length of current and previous phrases for phrase amplitude and accent type for accent amplitude) and the difference of control dominance due to speaking styles. Furthermore, most of the differences were reflected in the control of accent commands rather than of phrase commands. These results support the possibilities of computational modelling for prosody conversion and suggest the importance of $F0$ control decomposition into local and global characteristics in this conversion modelling. To establish a conversion scheme, not only further detailed analyses of control factors are needed but the correlations between control factors should also be analysed to naturally embed the control constraints existing in human's speech.

References

[AS92] M. Abe and H. Sato. Two-stage $F0$ control model using syllable based $F0$ units. *Proceedings of the International Conference on Acoustics, Speech, and Signal Processes*, pp. 53–56, 1992.

[ES95] E.Moulines and Y. Sagisaka. Voice conversion: State of the art and perspectives. Special issue of *Speech Communication*, 16:125–216, 1995.

[FH84] H. Fujisaki and K. Hirose. Analysis of voice fundamental frequency contours for declarative sentences of Japanese. *J. Acoustics Soc. J. (E)*, 5:233–242, 1984.

[FHT90a] H. Fujisaki, K. Hirose, and N. Takahashi. Manifestation of linguistic and para-linguistic information. *Proceedings of the International Conference on Acoustics, Speech, and Signal Processes*, pp. 485–488, 1990.

[FK88] H. Fujisaki and H. Kawai. Realization of linguistic information in the voice fundamental frequency contour. *Proceedings of the International Conference on Acoustics, Speech, and Signal Processes*, pp. 663–666, 1988.

[FS71b] H. Fujisaki and H. Sudo. Synthesis by rule of prosodic features of connected Japanese. *Proceedings of 7th ICA*, 3:133–136, 1971.

[HFK85] K. Hirose, H. Fujisaki, and H. Kawai. A system for synthesis of connected speech—special emphasis on the prosodic features. *Trans. of the Committee on Speech Research*, 1985. S85–43 (in Japanese.).

[HHS96] N. Higuchi, T. Hirai, and Y. Sagisaka. Effect of speaking style on parameters of fundamental frequency contour. In J. P. H. van Santen, R. Sproat, J. Olive, and J. Hirschberg, editors, *Progress in Speech Synthesis.* New York: Springer-Verlag, 1997.

[HIHS96] T. Hirai, N. Iwahashi, N. Higuchi, and Y. Sagisaka. Automatic extraction of F_0 control parameters using statistical analysis. In J. P. H. van Santen, R. Sproat, J. Olive, and J. Hirschberg, editors, *Progress in Speech Synthesis.* New York: Springer-Verlag, 1997.

[IS93] N. Iwahashi and Y. Sagisaka. Duration modelling with multiple split regression. *Proceedings of the European Conference on Speech Communication and Technology, Berlin, Germany*, pp. 329–332, 1993.

[KS92b] N. Kaiki and Y. Sagisaka. Optimization of intonation control using statistical F0 resetting characteristics. *Proceedings of the International Conference on Acoustics, Speech, and Signal Processes*, 2:49–52, 1992.

[KS93] N. Kaiki and Y. Sagisaka. Prosodic characteristics of Japanese conversational speech. *Trans. IEICE Jpn.*, E76–A:1927–1933, 1993.

[KTS92] N. Kaiki, K. Takeda, and Y. Sagisaka. Linguistic properties in the control of segmental duration for speech synthesis. In G. Bailly, C. Benoît, and T. R. Sawallis, editors, *Talking Machines: Theories, Models, and Designs*, pp. 255–263. Amsterdam: Elsevier Science, 1992.

[NK84] S. Nakajima and K. Kabeya. Relations between phrase structure and pitch contour. *Rec. Spring Meeting, Acoustics Soc. Jpn., Mar. 1984* (in Japanese), pp. 113–114, 1984.

[OFH84] E. Ohira, H. Fujisaki, and K. Hirose. Relationship between articulatory and phonatory controls in the sentence context. *Rec. Spring Meeting, Acoustics Soc. Jpn., Mar.* (in Japanese.), pp. 111–112, 1984.

[Ril92] M. Riley. Tree-based modelling of segmental durations. In C. Benoît G. Bailly and T. R. Sawallis, editors, *Talking Machines: Theories, Models, and Designs*, pp. 265–274. Amsterdam: Elsevier Science, 1992.

[SD83] B. G. Secrest and G. R. Doddington. An integrated pitch tracking algorithm for speech system. *Proceedings of the International Conference on Acoustics, Speech, and Signal Processes*, pp. 1352–1355, 1983.

[STA+90] Y. Sagisaka, K. Takeda, M. Abe, S. Katagiri, T. Umeda, and H. Kuwabara. A large-scale Japanese speech database. In *Proceedings of the International Conference on Spoken Language Processing, Kobe, Japan*, pp. 1089–1092, 1990.

15 Segmental Duration and Speech Timing

Jan P. H. van Santen

ABSTRACT Many speech technologies assume that speech can be approximated by time warping and concatenating appropriately selected "speech units", such as diphones. This paper first discusses evidence for the validity of this assumption, and then points out how modelling of speech timing can be cast in terms of time warp functions; conventional segmental duration based modelling is a special case of time warp function based modelling. Next, the paper addresses a challenge against time warp based modelling: the possible existence of long-range temporal constraints on timing, as proposed by isochrony and syllabic timing concepts. However, evidence is provided that such constraints simply do not exist in American English and Mandarin Chinese. The paper concludes with a presentation of a time warp based approach to pitch modelling.

15.1 Introduction

A major challenge in the analysis of spontaneous speech is that one has little or no control over the words and sentences being spoken. As a result, spontaneous speech data require drawing inferences under conditions of severe sparsity, by which the following is meant [vS93b, vS94b, vS94a]: Any aspect of speech, whether it is timing, pitch, or spectral parameters such as tilt, is the resultant of many factors (*prosodic* factors such as stress, word prominence, word length, and location in the phrase; and *coarticulatory* effects from neighboring segments). The combinatorics of natural language is such that the number of factorial combinations is not only very large, but that—paradoxically—one is extremely likely to encounter very rare combinations very often, the reason being that the number of distinct rare combinations is quite large.

Thus, if one were to analyse speech with the purpose of training a speech recognition system or developing acoustic-prosodic rules for text-to-speech synthesis, one cannot ignore rare events that do not occur in the speech training database, because they are certain to be encountered by the recognition or synthesis system when it is actually used. Since it is practically impossible to obtain training materials containing all conceivable combinations, in particular for spontaneous speech where one has little or no control over most prosodic factors, one needs the capability to generate accurate predictions for absent combinations (*generalization* or *interpolation*).

Generalization, in turn, requires capitalizing on knowledge of fundamental *invariances* or regularities of the speech production process. Specifically, the statistical analysis tools have to incorporate these invariances (e.g., in the form of parameter constraints, or in the mathematical structure of their equations) in order to produce accurate predictions.

In summary, the unconstrained nature of spontaneous speech puts a premium on the search for invariances. This paper focuses on one aspect of speech—its timing—and asks the following fundamental question: *How should one describe speech timing in order to maximize the odds of discovering such invariances?*

15.1.1 Modelling of Speech Timing

The conventional way in which speech timing is described is in terms of the *durations of the phonetic segments* that make up utterance. For example, one could predict durations of vowels in various contexts using a model such as

$$DUR(V; VOI, POS) = S_1\ (V) \times S_2\ (VOI) \times S_3\ (POS). \quad (15.1)$$

Here, V refers to the identity of the vowel, VOI to the voicing feature of the post-vocalic consonant, and POS to the position of the vowel in the phrase. S_1, S_2, and S_1 are functions (*scales*) that assign different numbers for different values of their arguments. For example, $S_1(/U/)$ might have the value 120, and $S_1(/I/)$ 70. Equation (15.1) is referred to as the *multiplicative model.*

There are many alternative ways of describing timing aside from segmental duration. For example, some automatic speech recognition systems [Ljo94] represent a phonetic segment as a sequence of three abstract sub-segmental "states", each with its own duration. Some approaches to text-to-speech synthesis represent the temporal pattern of an utterance in terms of the durations of its syllables; these systems compute durations of segments as an afterthought, under the assumption that segmental durations are not critical for a speech perception or production [CI91, Col92c, Cam92c].

When one represents speech in terms of underlying articulatory parameters (e.g., various geometric descriptors of tongue shape and position) or acoustic parameters (e.g., formants), further alternatives to segmental duration become visible [Her90, SB91, Col92b]. In their systems, these authors allow for the possibility—but do not require—that the parameters may change *asynchronously* in response to contextual factors. To illustrate with a hypothetical example, the onsets of anticipatory rounding of the lips and of devoicing may coincide in, say, stressed syllables, but rounding might precede devoicing in unstressed syllables.

We call these representations *multi-variate* (and possibly *asynchronous multi-variate*), and the former *template based.*

To be clear, the mere fact of using multi-variate parameter vectors as speech representation is not what distinguishes between asynchronous and template based approaches—which, in fact, commonly use parameter vectors as speech representation [OS95]. The key difference is that in template based approaches the vectors are left unchanged—only their timing is manipulated. In other words, template based approaches assume that any utterance can be approximated by concatenating a sequence of templates, with their duration as the only degree of freedom. Put differently, the trajectories of each parameter in these vectors would be *warped* with a common time warp function.[1]

The concept of asynchronous representations calls into question the existence of speech templates. If contextual factors have asynchronous effects on parameters, then context causes changes in the parameter vectors themselves, and not merely in their timing. In principle, of course, this could be handled by having specific templates for specific contexts, but this may quickly lead to an impractical number of context-specific templates.[2]

To summarize, a major question before us is whether the degree and prevalence of asynchronicity of contextual effects is such that a template based approach is impractical.

15.1.2 Goals of this Chapter

This paper summarizes collaborative research on timing performed in our laboratory. It will be argued that this research supports the following claims:

---

[1]A time warp between two sequences of entities (e.g, speech parameter vectors) is a sequence of pairs of entities from each sequence having the property that the average within-pair distance ("warp distance") is minimized [SK83]. In synthesis applications, the warp is computed from timing rules and then used to alter the timing of vectors of the template; here, the warp distance is trivially zero. In speech recognition applications, the warp is computed between input speech and candidate templates, and the template is selected that minimizes the warp distance; usually, the warp is unconstrained except for some boundary conditions, and is not based on timing rules.

[2]We computed that a diphone based system with each diphone annotated with a rather coarse coding scheme—in terms of such factors as phrasal position, stress, and word accent—would need at least 150,000 templates to cover at most 75% of randomly drawn sentences. Current synthesizers have at most a few thousand units.

(1) There is sufficiently little asynchronicity to make a template based approach practical.

(2) Modelling of timing can and should be done at the sub-segmental level, via rule based time warp functions.

(3) The duration of a small unit (e.g., a segment, a frame) is affected by its *phonological relationship* to larger units. For example, a phonetic segment is linked via the phoneme level to syllables and words; the number of syllables in that word has an effect on the duration of the segment.

(4) There are no long-range constraints on timing in the following sense:

   (a) The duration of a small unit cannot be computed solely based on the computed duration of a larger unit of which it is a constituent; one also needs to know the exact contextual constellation as well as the location of the small unit in the larger unit (contextual effects on smaller units are not *mediated* by larger units).

   (b) Timing of a larger unit is dictated by prosodic context but also *depends on* the segments it contains.

(5) Pitch contours can be modelled using (pitch) templates and using rule-generated time warps.

(6) Segment boundaries play a useful role in prediction of timing, but any salient acoustic discontinuities can play this role.

15.2 Template Based Timing: Path Equivalence

We address here the asynchronicity issue by reporting results from a study in which the effects of post-vocalic voicing on the trajectory of the preceding sonorant region were analysed [vSCR92]. Utterances were produced by a male speaker, and sentences were of the form "please say X", where "X" is a monosyllabic word. These words formed minimal pairs, differing only in the final segment (/d/ vs /t/). Figure 15.1 provides some examples of these minimal pairs. We analysed the sonorant regions of the utterances (e.g., in "meld" vs "melt" we would compare the segment sequence $m - e - l$).

Consider the segment sequence $m - e - l$ (as in "meld", "melt", etc.). At the articulatory level, this sequence is produced by a complex sequence of articulatory events involving the lips and several dimensions of tongue movement. At the acoustic level—specifically in formant space—there is a sharp increase in F_1 starting at the onset of the vowel, followed by a sharp decrease in F_2 as the /l/ approaches.

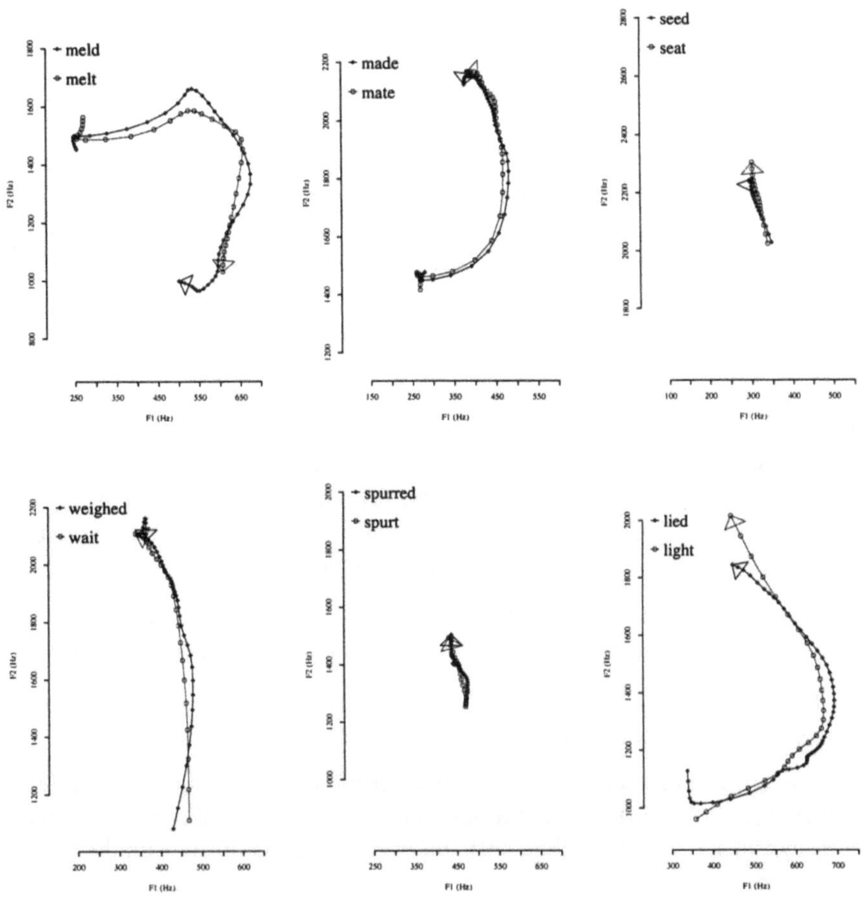

FIGURE 15.1. F_1, F_2 trajectories (centroids) for minimal word pairs. Open and closed symbols are placed at 10 ms intervals. Only sonorant portions are shown. Arrow heads indicate the end of the trajectories. See the text for computation of centroids.

When we contrast the occurrence of the sequence $m - e - l$ in the words "meld" and "melt", there is a difference in overall duration, in particular when these words occur in utterance final position. Typically, the sequence is at least 50% longer in "meld" than in "melt". Given this large temporal difference, given the complexity of the underlying articulatory events, and given the non-simultaneity of the changes in F_1 and F_2, it would appear surprising if the two occurrences could be generated by a single, common template. Instead, we would expect the changes in the formants to be *asynchronous*.

Yet, when we inspect Figure 15.1 (top left panel), we see that the F_1, F_2 trajectories are remarkably close in terms of their path.[3] Figure 15.1 shows five additional examples. Of course, this does not necessarily mean that the same would be found if we also plotted other formants, spectral tilt, and still other acoustic dimensions. Nevertheless, the degree of *path equivalence* is striking.

Now, there is an obvious link between path equivalence and templates: a set of trajectories corresponding to the same segment sequence in different contexts are all pairwise path equivalent if and only if there exists a template with which all these trajectories are path equivalent. As template, we can arbitrarily select one of the sequences. The logic behind this statement is that path equivalence is a transitive relation.

These examples suggest that a powerful coarticulatory factor—voicing of the post-sonorant obstruent—has little effect on the acoustic paths traversed in the sonorant portion.

This is not to say that other coarticulatory factors, in particular place of articulation of neighboring consonants, do not affect acoustic paths. Of course they do.[4] The point is that one can construct a set of segment sequences (templates) that have the following property:

(1) Jointly, they *span* the language (i.e., any possible sequence of phonemes in that language can be written as a sequence of templates).

(2) In the space of (appropriately restricted[5]) occurrences of a given segment sequence type, these occurrences are path equivalent.

The possibility of accurate representation of speech with such template inventories is, of course, the fundamental assumption of concatenative speech synthesis. The fact that very high levels of perceptual quality can be achieved by these systems adds credence to this assumption.

---

[3]These centroid trajectories were computed as follows. The cepstral trajectories of four of the five tokens of one word (e.g., *meld*) were time warped (without constraints on the warp slope) onto the fifth token (the "pivot"), and for each frame of the latter the median vector was computed of the five cepstral vectors mapped onto that frame. The same was done with each of the other four tokens playing the role of pivot. Subsequently, the process was repeated, with now the median vector trajectories taking the place of the original cepstral trajectories. The process was continued until convergence was reached.

[4]Also, vowel and consonant reduction (e.g., due to de-stressing) produce violations of path equivalence. We hypothesize that these phenomena might be handled by the concept of *generalized path equivalence*, where one path can be generated from a second path (but not vice versa) by *short-cutting* the latter. Mathematically, this could be described as the first path being obtained from the second by smoothing a subset of the points on the second path—leaving out, e.g., points with extreme F_2 values.

[5]For example, restricted to a particular speaker, speaking mode, or speaking rate. In addition, one may restrict certain sequences to particular coarticulatory contexts, e.g., as defined in terms of place of articulation.

15.3 Measuring Subsegmental Effects

Having stated that speech can be approximated by "appropriately warped" templates raises the issue of *how* to perform this warping. That is, both speech analysis and speech synthesis require *rules* for computing how to warp a template given a particular context in which the template is to be embedded. This section presents some preliminary results on how to *measure* warping in natural speech. Measurement is a necessary condition for developing such rules.

15.3.1 Trajectories, Time Warps, and Expansion Profiles

The striking examples of path equivalence shown in Fig.15.1 would not have been obtained if we had used individual trajectories, because of speaker variability. Van Santen *et al.* [vSCR92] used an *iterative averaging method*, and applied it to five tokens of each word. The effect of this averaging method is to reduce the effects of speaker variability.

The project on intra-segmental structure [vSCR92] was inspired by earlier work by Macchi and her colleagues at Bellcore [Mac89], showing how time warping can be used to understand intra-segmental effects of supra-segmental factors. We built on their methods by using cepstral parameters instead of formant frequencies and, more important, by using the iterative averaging procedure. The reasons for using cepstral parameters are the frequent occurrence of formant tracking errors, and the fact that formants are too sparse a representation of speech. Let me elaborate on the latter. In the formant domain, formants can be steady state in regions during which other aspects of the spectrum are changing. Time warping algorithms have great difficulty matching frames when they appear in sequences of near-identical frames, because similar warp distance values are obtained over a wide range of warps. Using the "richer" cepstral vectors counters this problem, because they capture more aspects of the spectrum than formants.

We now draw the conceptual link between time warps, expansion profiles, and speech timing. Figure 15.2 shows three hypothetical examples of trajectories, time warps, and "expansion profiles" (time derivatives of time warps). All examples involve path equivalent pairs of trajectories. Path-equivalence is not a necessary condition for time warping algorithms to work, but for the results from time warping to have a clear interpretation.

The trajectories marked "long context" are the same in the three rows of figures. This trajectory decelerates, reaches a quasi-steady-state around the peak value of F_2, and then accelerates again.

The top row displays a "short context" trajectory that is faster in all regions, but in particular in the steady-state region. The time warp shows this by accelerating around this region. The expansion profile makes it clearer by having a peak here.

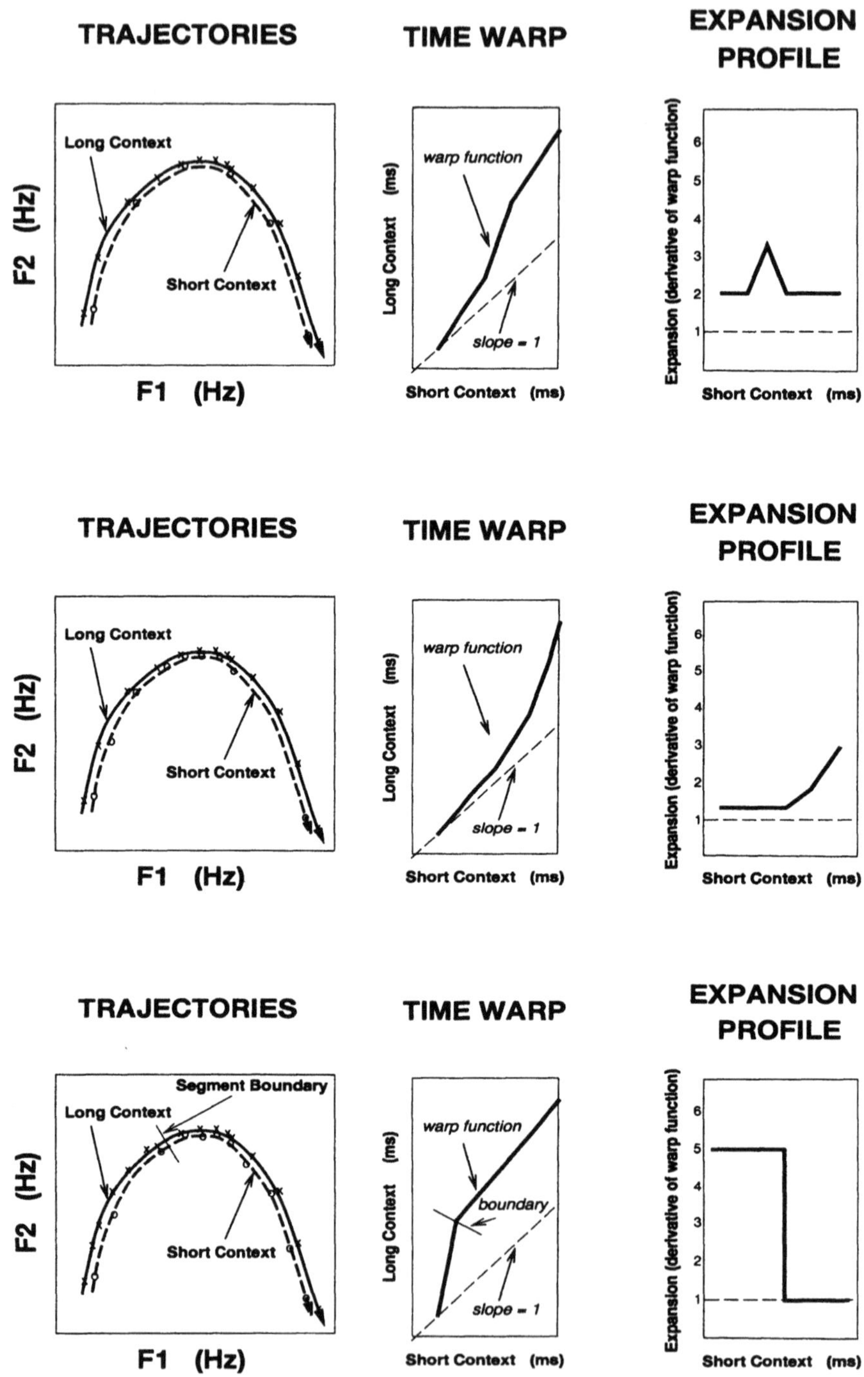

FIGURE 15.2. Trajectories, time warps, and expansion profiles for three hypothetical cases: single-peaked expansion (top panel); monotonically increasing expansion (center panel); uniform within segments: step function expansion (bottom panel).

The center row displays a case where, relative to the long context, the short context accelerates throughout the speech interval displayed.

The bottom row displays the *implicit time warping that is performed in speech synthesis based on segmental duration.* In segmental duration based synthesis, rules compute the overall duration of segments based on their identities and the context [vS94a]. During synthesis, parts of the to-be-computed speech signal are *uniformly warped*[6] so that the resulting segmental intervals match the computed durations. The result is that the expansion profile is discontinuous at the segment boundary.

These examples show that looking at expansion profiles can provide information concerning questions such as:

1. Does a given contextual factor affect some parts of a segment more than others (*intra-segmental uniformity*)?

2. If not uniform, what shapes can the expansion profiles have?

3. What happens around segment boundaries?

15.3.2 Preliminary Results

In our study [vSCR92], we found that changes in the overall duration of a segment never involve uniform warping. Instead, we found that expansion profiles were either single-peaked (as in the top panel of Fig. 15.2) or monotonically increasing (as in the center panel of Fig. 15.2). More specifically, instances of the diphthong $/a^y/$ typically produced single-peaked expansion profiles, with the expansion peak in the steady-state region [replicating earlier results [Gay68, Her90]]; while all other cases produced monotonically increasing expansion profiles. The latter was found regardless of whether the sonorant region ended on a vowel (as in "wait") or on a sonorant consonant (as in "meld").

We found no evidence for sudden changes in expansion profiles around segment boundaries. It is important to point out that this cannot be an artifact of our trajectory averaging procedure, because this procedure preserved sharp acoustic discontinuities; in fact, the time warping process is particularly keen on discontinuities, because they have a sharp impact on the cost function.

15.3.3 Modelling Time Warp Functions

The previous section suggests that time warps can be computed reliably from speech, provided that the proper speech domain is used and that

[6]Synthesizers differ in how they impose durations. Some perform linear interpolation in addition to uniform warping.

sources of speaker variability are minimized. Now the question is raised how to compute time warp functions from context by rule.

Although we analysed the effects of only one factor (voicing of the post-vocalic consonant), we conjecture that some of the findings will generalize to other factors as well. If indeed the effects of contextual factors on expansion profiles are locally single-peaked (or monotonically increasing), and *do not respect segment boundaries*, then this may require an approach toward the modelling of timing that fundamentally differs from current segmental or supra-segmental (e.g., syllabic) timing: instead of having to compute a single number for a segmental or supra-segmental chunk (its duration), one has to compute an expansion profile.

What will be particularly challenging is to model interactions between factors. For example, we know that the effects of post-vocalic voicing on vowel duration are much stronger (measured in ms or percent) for phrase-final syllables than for phrase-medial syllables [Kla73, CH88, vS92]. Modelling this interaction at the level of vowel duration is easy, and can be done with simple models such as sums-of-products models [vS93a]. These models describe the effects of two factors as the sum of the effects of the individual factors plus a term which consists of the product of their contributions.[7]

We sketch here a possible approach to what form rules for expansion profiles may take. Suppose that speech templates are *duration normalized* in the sense that the sub-intervals of all templates that contain the vowel /U/ are made of equal duration. Also assume that some speech representation is used where such sub-intervals can be thought of as consisting of sequences of N *frames*, where the i-th frame has duration $D_i(/\mathrm{U}/)$. Finally, suppose that the multiplicative model (Eq. 15.1) applies to these frames. That is, the duration of the i-th frame is given by

$$D_i(/\mathrm{U}/; VOI, POS) \;=\; S_{2,i}\,(VOI) \;\times\; S_{3,i}\,(POS) \;\times\; D_i(/\mathrm{U}/). \quad (15.3)$$

This equation specifies the expansion profile in a context defined by VOI and POS; the expansion profile is given by plotting the values of $D_i(V; VOI, POS)$ for successive values of i. The time warp can be obtained

---

[7]For N factors, the formalism is

$$\mathrm{DUR}(\mathbf{f}) = \sum_{i \in T} \prod_{j \in I_i} S_{i,j}(f_j). \quad (15.2)$$

Here, f_j is a value on the j-th factor; $S_{i,j}$ is a scale for the i-th *product term* for the j-th factor; and T and I_i are sets of integers[vS93a]. To illustrate, for the multiplicative model: $T = \{1\}$ and $I_1 = \{1, \ldots, n\}$; for the additive model: $T = \{1, \ldots, n\}$ and $I_i = \{i\}$.

by integration:

$$WARP_i(V; VOI, POS) = \sum_{j=1}^{j=i} D_j(V; VOI, POS). \qquad (15.4)$$

Finally, the total duration of the vowel is given by

$$D(V; VOI, POS) = \sum_{j=1}^{j=N} S_{2,j}\ (VOI) \times S_{3,j}\ (POS) \times D_j(/U/) \qquad (15.5)$$

Interestingly, Eq. 15.5 is the general equation for sums-of-products models [vS93a]. Thus, *modelling segmental duration via sums-of-products models is mathematically equivalent with computing sub-segmental (i.e., frame) duration with multiplicative models.*

Of course, at this point the viability of multiplicative models for computing sub-segmental duration is entirely speculative. The main point of this exercise was to show that modelling of expansion profiles may not require methods that are radically different from what is currently used for segmental duration modelling.

15.4 Syllabic Timing vs Segmental Timing

Suppose speakers roughly *keep constant* either the time interval between stressed syllable onsets (*isochrony*) or the duration of syllables in specific prosodic contexts (*syllabic timing*). Although segmental (or sub-segmental, microtiming) models are not logically incompatible with such (hypothetical) constancy they would need inelegant compensatory equations.

Focussing now on syllabic timing, syllabic timing models [CI91, Cam92c] could explain these constancies as follows:

Step 1: Compute duration of syllable from prosodic context, (largely) ignoring the segmental makeup of syllable.

Step 2: Compute segmental durations by scaling their intrinsic durations so that the scaled durations add up to the syllable duration.

The question raised in this section is whether these hypothetical constancies in fact exist in certain languages (American English and Mandarin Chinese).

To avoid any misunderstandings, it should be emphasized here that the issue of which unit to use as a *temporal unit* is orthogonal to the issue of which *phonological entities* are needed for timing prediction. To illustrate, MITalk [AHK87] predicts segmental duration. The information it uses involves the word level (e.g., location of a word in the sentence), the syllabic level (e.g., whether the syllable is stressed), and the segmental level (whether the post-vocalic consonant is voiced). Likewise, Campbell's model predicts (in its first stage) syllable durations, and uses at least some information at the segmental level (the number of segments in the syllable, and the nature of the nucleus).

There is agreement that for prediction of any temporal unit, various phonological entities are needed (e.g., phonemes, syllables, words). The issue at stake in this section exclusively concerns temporal units.

15.4.1 The Concept of Syllabic Timing

The most complete model of syllabic timing has been developed by Campbell [Cam92c]. In his model, for an n-segment syllable S occurring in context C, having duration $\Delta(S;C)$, the duration of the i-th segment in S is given by $\mu_i + k\sigma_i$, where k is chosen such that

$$\Delta(S;C) = \sum_{i=1}^{i=n} \exp(\mu_i + k\sigma_i). \tag{15.6}$$

Here, μ_i and σ_i are the intrinsic duration and "elasticity" of the i-th segment, estimated, e.g., by the mean and standard deviation of the segment in the training corpus.

Syllabic duration $\Delta(S,C)$ depends on prosodic factors (e.g., stress, word, accent). It depends on the segmental makeup of S only through the number of segments and the nature of the nucleus (short vowel, long vowel, diphthong, syllabic consonant).

Another important feature of the model is that the index k does not depend on context C. That is, given two contexts C_1 and C_2 such that for some syllable S, $\Delta(S;C_1) = \Delta(S;C_2)$, it follows that k must have the same values because all other quantities in Eq. 15.6 do. This makes a testable prediction: the model predicts that segments in a given syllable should have the same durations in any contexts that cause this syllable to have the same duration. Realizing that this prediction is obviously wrong for contexts involving phrase-final positions—in pre-boundary syllables, primarily the nucleus and coda are lengthened—Campbell added to the model a special mechanism for phrase-final lengthening, where the above equation is replaced by

$$\Delta(S;C) = \sum_{i=1}^{i=n} \exp(\mu_i + \alpha_i k\sigma_i), \tag{15.7}$$

where α_i is a constant that is equal to 1.0 for all *non*-phrase final contexts, and is 0.75 for phrase final contexts.

Another model is by Barbosa and Bailly [BB94]. Their model differs from that described in Eq. 15.6 in that it uses the *inter-perceptual center group* instead of the syllable. Inter-perceptual center groups are speech intervals between vowel onsets (except for vowels in utterance-initial syllables, in which case syllable start is used).

Regardless of the details, the essence of these models consists of two broad assumptions about the interdependence of syllable and segment duration [vSS95]:

1. **Segmental Independence:**
   The duration of a syllable is mostly independent of the identities of the segments it contains. In Campbell's model, it only depends on the number of segments and on a coarse categorization of the syllable nucleus. It should not matter, for example, whether a syllable starts with an intrinsically short consonant (e.g., a nasal) or an intrinsically long consonant (e.g., a voiceless fricative).

2. **Syllabic mediation:**
   The duration of a segment depends mostly on the (pre-computed) syllable duration and the segment's identity. In Campbell's model, when two contexts produce the same overall duration of a given syllable, then all segments should also have the same duration in the two contexts. Campbell makes an exception for contexts involving [phrase-final position, because—as is well-known—phrase-final lengthening primarily affects nucleus and coda, whereas other lengthening factors do not.

In summary, this section has argued that the concept of syllabic timing involves two broad assumptions, *segmental independence* and *syllabic mediation*, that both have to do with how the quantitative relationship between between segmental an syllabic duration is affected by contextual and other factors. [8]

The following two subsections summarize results of tests of these two assumptions; these results are reported more extensively elsewhere [vSS95].

15.4.2 Testing Segmental Independence

We analysed durations of syllables in the same prosodic context (e.g., stressed, word-initial, phase-final, accented, ...) and having the same internal structure (e.g., CVC). A prediction of segmental independence is that the variation in syllable durations should be small and random. We test this prediction by investigating if there is a systematic relationship

[8]Elsewhere, Campbell [Cam93a] has described a neural net approach to syllabic timing. The underlying concepts of this implementation are closely related, but not equivalent, to the 1992 model. For example, there is a "backstep" procedure which contradicts the segmental independence assumption.

between syllable and segmental duration. If such a relationship exists, this would obviously violate segmental independence.

There are two way of testing this prediction. First, we could simply find all examples of syllables of the given structure and in the given context, and compute the correlation between, say, the duration of the onset and the duration of the syllable. This would be invalid, however. For example, even small sentence-to-sentence speaking rate variations could cause a positive correlation.

The second way is by measuring the *average* duration of all syllables containing a particular segment in a specific location (e.g., all stressed, accented-word-initial, phrase-medial CVC syllables starting with /t/) and the average duration of that segment, for all possible segment identities that could occur in that location in these syllables. We use the notation $AV[DUR_C(/t/\cdot\cdot)]$ for the former (i.e., syllable duration), $AV[DUR_C(/t/)]$ for the latter (i.e., segmental duration); as before, C refers to a particular context. When a sufficiently large number of occurrences are analysed, correlations due to sentence-to-sentence speaking rate variations are minimized. Any correlations found must be due to a systematic association between the *intrinsic duration* of the segment and the duration of a syllable.

We note that the tests involve in-depth analyses of single speakers, leaving open the possibility that other speakers may produce different results.

We found high correlations between $AV[DUR_C(p\cdot\cdot)]$ and $AV[DUR_C(p)]$ (correlations ranging from 0.65 to 0.96), where p ranged over all possible onset consonant identities. We found this both for American English and for Mandarin Chinese, and for various contexts C and syllabic structures. Similar findings were obtained for the nucleus: $AV[DUR_C(\cdot p\cdot)]$ vs $AV[DUR_C(p)]$; and for the coda: $AV[DUR_C(\cdot\cdot p)]$ vs $AV[DUR_C(p)]$.

Not only did we find strong correlations, we also found that the slopes of syllable duration plotted against segmental duration were often close to 1.0. This means, for example, that a syllable such as "ti" (as in "timid") was on average 32 ms longer than a syllable such as "bi" (as in "bitter"), while at the same time syllable-initial /t/ is 34 ms longer than syllable-initial /b/. It was thus not only the case that there was *imperfect* compensation (i.e., compensation too weak to produce fixed-duration syllables), but that in fact there was *hardly any* compensation at all. The speaker barely (by 2 ms) compressed the /I/ in "timid" to make room for the intrinsically longer /t/.

We generally found shallower slopes in Mandarin Chinese than in American English. In particular, we found that coda consonants are shorter when the preceding vowel is intrinsically long.

In summary, it is not simply that syllabic duration only loosely depends on the intrinsic durations of its segments, it is highly dependent on it.

Accurate prediction of syllabic duration requires taking into account the full segmental makeup of a syllable.

15.4.3 Testing Syllabic Mediation

We analysed effects of stress, predicted pitch accent (American English only), tone (Mandarin Chinese only), word-initiality, and phrasal position on segmental duration, for different segment classes and for segments occurring in different within-syllable locations.

There were two main findings. First, there are numerous two-way interactions, indicating that the effects of some factors are primarily on the onset (e.g., word-initiality), others on the onset and nucleus (e.g., stress), and still others on the nucleus and coda (e.g., phrasal position).

Second, the effects also depend on an interaction between factor (e.g., stress vs phrasal position), segmental identity (e.g., alveolar stops vs. labial sonorants), and intra-syllabic position (onset vs coda).

These interactions make it obviously impossible to accurately predict segmental duration merely from knowledge of total syllable duration, even if an exception is made for phrasal position effects.

15.4.4 Syllabic Timing: Conclusions

We conclude that the constancies and independencies assumed by syllabic timing do not occur in these two languages, at least not in the corpora analysed.

It should be noted that other ideas about constancies, in particular isochrony, have not received much hard empirical evidence either [Noo91]. For example, elsewhere, I found no evidence for shortening of syllables as a function of the number of syllables between stressed syllables (stress group length) once word length effects are taken into account [vS92].

Compensatory effects—if present—rarely if ever result in any type of constancy of larger units. So, even though the larger units play critical roles in speech production, they are not a happy choice as temporal units. Speakers do not carefully control timing over long stretches of speech. For temporal units in speech production, the smaller, the better.

In fact, the term "compensatory effect" may be a misnomer, with isochrony overtones. Syllables in long words may be pronounced faster on average than syllables in short words not because a speaker attempts to be "rhythmic" but simply because such syllables contain less information. The speaker can afford making the second syllable in a long, redundant word such as "collaboration" very short without risking that the listener will not understand what was said.

15.5 Timing of Pitch Contours

The underlying assumption in the preceding sections has been the conceptual importance of time warping for speech modelling. Here, we apply this concept to timing of pitch contours.

Consider the phrases "This is strong" and "I like to walk", both produced with a contour that linearly descends until the onset of the phrase-final word, climbs to a peak, and then returns to a continuation of the linear descent just before the end of the last sonorant in the utterance. The duration of the two phrase-final words differ significantly (typical durations would be 900 ms for "strong" and 500 ms for "walk"), and there is a corresponding difference in peak location. If one imposes the pitch contour from "strong" onto "walk", or vice versa, the result clearly sounds wrong to a typical native speaker. This raises the question of what the alignment is between a pitch contour and the local segment and syllable boundaries.

Historically, alignment of this type of accent has been measured in terms of peak location. This raises two problems. First, it is well-known that the first 50–100 ms after vowel start have much higher pitch values when a vowel is preceded by a voiceless consonant than when it is preceded by a sonorant; this *perturbation* can produce a spurious peak. Second, we know that listeners hear differences in peak placement, but in the process of manipulating peak placement the locations of other points also change; there are no experiments showing the all-importance of peak timing compared to timing of other points of the contour, such as the point of steepest increase, the point where the rise appears to start, etc. Hence, it might be prudent to pay attention to these other points as well.

In this section, a method is described for modelling segmental effects on both timing and height of pitch contours that attempts to circumvent these problems.

15.5.1 Modelling Segmental Effects on Pitch Contours: Initial Approach

In collaboration with Hirschberg, a general method for modelling alignment was introduced [vSH94]. We analysed 2000 utterances collected from a female speaker, who read phrases of the type "Now I know $C_o.V.X$", where $C_o.V.X$ was a word receiving a H* accent on the first syllable; the phrases were produced as single intonational phrases with a low phrase accent and low boundary tone, as defined in [Pie80] (see Fig. 15.3). C_0 (syllable onset) consisted of 0 or more consonants, V was a stressed vowel or diphthong, and X (*post-vowel region*) was the remainder of the word following the vowel (and thus coincided with the coda—labelled C_c—in the case of monosyllabic words).

To demonstrate that prediction of alignment—even of merely the peak—requires more than a few simple rules, we measured peak placement in monosyllabic words as a function of the *phonetic class* (voiceless, non-sonorant voiced, sonorant) of C_0 and C_c. We measured peak location in four different ways: T_1: time measured from syllable onset; T_2: time measured from vowel onset; T_3: time measured from syllable onset divided by syllable duration; T_4: time measured from vowel onset divided by rhyme duration. Not surprisingly, all four measures were affected by the phonetic classes of C_0 and C_c. For example, T_1 was about 60 ms longer for sonorant codas than for voiceless codas; T_2 was 50 ms longer for sonorant onsets than for voiceless onsets; T_3 was 80% larger for voiceless codas than for sonorant codas; and T_4 was more than 150% larger for sonorant onsets than for voiceless onsets.

Thus, the dependency of peak placement on the durations of the syllable's constituents is strong but not simple, and depends on phonetic class. One way to analyse this dependency involves the following linear model:

$$\begin{aligned} &T_{\text{peak}}(D_{C_o}, D_{s-\text{rhyme}}; C_o, v, C_c) \\ &\quad = \alpha_{C_o,C_c} \times D_{C_o} + \beta_{C_o,C_c} \times D_{s-\text{rhyme}} + \mu_{C_o,C_c}. \end{aligned} \tag{15.8}$$

According to this model, peak time for a syllable whose onset duration is D_{C_o} and s-rhyme[9] duration is $D_{s-\text{rhyme}}$ is a weighted combination of these two durations plus a constant, which, like the weights, may depend on C_o and C_c.

The α, β, and μ parameters can be estimated with ordinary multiple regression. We call the α and β parameters *alignment parameters.* Across the nine possible combinations of onset and coda phonetic classes, correlations between observed and predicted peak locations ranged between 0.61 and 0.87, with a median of 0.77.[10]

Of course, this analysis still has the problem of being confined to the peak. To extend this linear model to other points on a contour, two problems had to be addressed. First, a definition of "point" must be provided. Second, we have to take into account the perturbations on pitch caused by obstruent onsets.

We noted that our recordings were singularly consistent in that one could invariably draw a straight line through some frame preceding the syllable onset by about 50 ms (in the center of the /O/ of "know") and the last sonorant frame in the phrase. This line had the further property that the

[9]The *s-rhyme* [vSH94] of a syllable consists of any non-syllable-initial sonorants in the syllable onset, the vowel or diphthong, and any sonorants in the coda. Thus, the s-rhymes of "pink" and "pin" are the same (/In/), while the s-rhymes of "pit", "strict", "blend", and "lend", are /I/, /rI/, /len/, and /en/, respectively.

[10]Similar results were obtained recently for Mexican Spanish [PSH95].

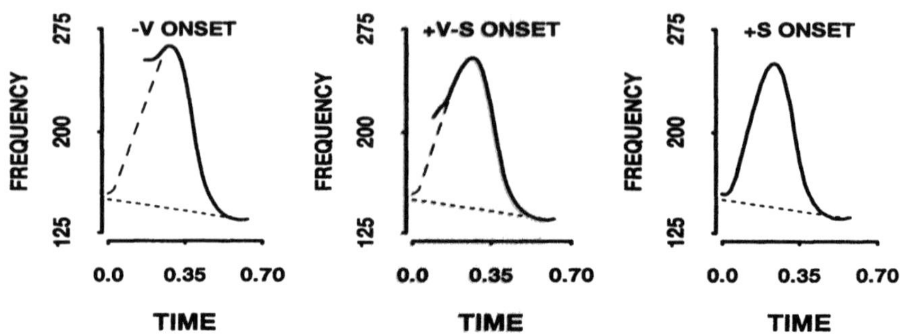

FIGURE 15.3. Solid lines: averaged contours for syllables with − V (voiceless), + V − S (voiced obstruent), and + S (sonorant) onsets; all have + S codas. Dotted line: local phrase curve; dashed line: estimated "underlying" contour. From [vSH94], reprinted with permission from the Acoustical Society of Japan.

pitch curve between these two points was positioned strictly above it, so that subtraction of the line from the pitch curve would produce a curve that would both start and end at a value of 0 (Fig 15.3). We called the curve resulting from this subtraction the *deviation curve.*

For syllables with sonorant onsets, the definition of "points" seemed straightforward: we defined the *pre-peak P%* point as the point where the deviation curve reached P% of the peak value of the deviation curve; similarly, we defined the *post-peak P%* point. We call the deviation curve divided by the peak value of the deviation curve the *relative deviation curve.* By performing regression analyses for sufficiently many percentage points, it would be possible to predict a smooth relative deviation curve from the durations of the onset and s-rhyme for any sonorant-initial syllable. Thus, the alignment for pre/post-peak P% point would be given by

$$\begin{aligned} T_P(D_{C_o},\ D_{s-\text{rhyme}};\ C_o,\ v,\ C_c) \\ = \alpha_{P,C_o,C_c} \times D_{C_o} + \beta_{P,C_o,C_c} \times D_{s-\text{rhyme}} + \mu_{P,C_o,C_c}. \quad (15.9) \end{aligned}$$

These alignment parameters are not sufficient, however, for predicting the deviation curve; they only predict the relative deviation curve, and do not pay attention to F_0 excursion. Moreover, for syllables with non-sonorant onsets, onset perturbations make it impossible to measure these percentage points directly. This required a more complicated approach in which, rather than directly measuring percentage points and performing regression, we use a model that incorporates the effects of onset perturbation, intrinsic F_0, and other factors; estimating the parameters of this model involves non-linear optimization.

15.5.2 *Alignment Parameters and Time Warps*

The goal of the current section is to show that the time warping concept can be applied to timing of pitch contours. Here, we draw the link between alignment parameters and time warps.

Suppose that we would compute some type of centroid of H* contours for a set of all-sonorant syllables (right panel in Fig. 15.3), compute its relative deviation curve, sample the result at N equal spaced points, thereby generating a sequence $(t_1, F_{0[1]}), \cdots, (t_N, F_{0[N]})$. This sequence can be considered as a template for a particular pitch accent class. For a given syllable with given durations, the alignment parameters allow us to compute via Eq. 15.9 a set of times $T(1), \ldots, T(N)$. Then the pairs $< t_1, T(1) >, \ldots, < t_N, T(N) >$ form a time warp function that can be used to warp the relative deviation values in the template to the time scale of the syllable.

In other words, the alignment parameters allow us to compute a time warp; this time warp function varies depending on the segmental makeup of the syllable and the durations of these segments.

15.5.3 *Modelling Segmental Effects on Pitch Contours: A Complete Model*

The key assumptions of the complete model are the following:

(1) The effects of obstruent onset perturbation can be described as a rapidly decreasing *perturbation curve* (e.g., an exponential decay curve, reaching a value of less than 0.01 of the value at $t = 0$ at $t = 75$ ms.) This curve is independent of the post-vowel region X, the nucleus V, their durations, and the pitch accent type of the target syllable. The curve only depends on the phonetic class of the onset.

(2) As in the Fujisaki model [Fuj88], we assume that a complete pitch contour is a combination[11] of multiple curves (Fujisaki assumes an underlying phrase curve and accent curves; we also include the perturbation curve).

(3) A key role is played by the concept of a stress group[12] [MPH93]. It is assumed that there is a one-to-one relation between stress groups and accent curves.

[11]For example, the linear sum, or the sum in the logarithmic domain, or whatever generalized addition operator.

[12]We define a (left-headed) stress group, or foot, as a sequence of one or more syllables where the first syllable is accented and the remaining syllables—if any—are not.

(4) The accent curves are generated by multiplying the relative deviation curve with a constant that depends on the overall duration of the sonorant interval (to produce larger pitch excursions in slow speech) and on vowel height (to produce higher pitch for high vowels). In contrast to the effects of obstruent onset perturbation, these constants are strongly dependent on the pitch accent type of the target syllable. In fact, for deaccented syllables, there is no effect of vowel height.

A model developed along these lines was applied to several pitch contour types, including the single-peaked contours discussed earlier, continuation rise contours, and yes/no question contours. In total, we analysed 2052 single-peaked contours, 42 continuation rise contours, and 1219 yes/no question contours—all from a single speaker. The following results were obtained consistently across these pitch contour classes:[13]

(1) The effect of the phonetic class of the onset is primarily that of the onset perturbation. Differences in pitch alignment due to the onset are entirely due to the fact that voiceless onsets have longer durations than voiced obstruents, and the latter longer than sonorants.

(2) Alignment is systematically different for different coda classes, and cannot be reduced to differences in duration between these classes or the durational effects they have on the preceding vowel.

(3) The start of the stressed syllable is an important landmark in the specific sense that we found that the value of the μ_{C_0,C_c} parameters could be set equal to zero without any effect on the goodness-of-fit.

(4) The alignment parameters α_{i,C_0,C_c} and β_{i,C_o,C_c} increase with i, reflecting the necessary fact that points later in the template occur later in the observed pitch contour. More interestingly, the α parameters (reflecting the duration effects of the onset) quickly reach a value of 1.0 (for H* contours, around the point corresponding to the peak), while the β points slowly increase, and often do not reach a value of 1.0.

(5) For polysyllabic words, syllable boundaries other than the onset of the stressed syllable do not play a special role. For example, peak locations vary in a 100 ms interval surrounding the boundary between the first and second syllables; their exact location depend on the particular segmental and durational constellation of the word, and do not appear to be associated with perceptually significant differences.

---

[13] Analyses were performed using multiple regression with as dependent variables locations and heights of anchor points, and as independent variables the durations of subintervals of the syllable. Standard tests for additional variance explained were used to test each of the results reported.

This call into question a literalist interpretation of the "H*" notation, according to which the peak should always occur inside the stressed syllable.

15.5.4 Summary

In summary, these results show both striking invariances and changes of pitch contours as a result of segmental factors. The high degree of predictability of these pitch contours across a wide range of segmental constellations show that there is a close coordination between pitch and segmentals in speech production.

Conclusions

This paper asserted that time warps should play a conceptually central role in segmental timing. The basis for this claim is the belief that most contextual factors other than outright coarticulation have fairly mild (path-preserving, or generalized-path preserving, non-asynchronous) effects on the local spectrum—effects that hence can be captured largely through temporal distortions. Under this assumption, we can discuss speech timing purely in terms of time warps on templates.

We found that context-induced time warps in natural speech are smooth, but are not uniform within phones. Hence, we need rules that allow us to go beyond segmental duration, and compute these non-uniform warps (or expansion profiles—their derivatives) for any context in which a template may occur. We suggested how one might construct these rules, basically by applying segmental duration models to individual template frames.

Although timing via time warping appears to focus on microscopic timing, there is no mathematical reason that one could not incorporate long-range invariances such as dictated by isochrony and syllabic timing concepts; but this incorporation would be inelegant. However, we found strong evidence against syllabic timing in two corpora, American English and Mandarin Chinese. It is also becoming clear that, except for unusual speaking conditions (e.g., certain types of poetry readings), there is no evidence for isochrony to hold in the corpora studied in these languages [Noo91]. Thus, for now long-range *invariances* need not terribly concern us.

This is not to say that there are no long-range *effects*, because there are several phonological constituency relations that are known to affect timing. Research on many languages has indicated effects on segmental duration (and hence on time warps) of factors such as position in the utterance, phrase, word, and syllable. We currently do not know whether these effects are truly compensatory—suggesting that the speaker desperately, but in vain, attempts to keep the duration of the larger unit constant; or whether these effects are the result of the need to acoustically emphasize syntactic boundaries, or are a matter of communicational redundancy.

Of course, our data on Mandarin Chinese indicated that codas are shorter when they are preceded by intrinsically long tautosyllabic vowels. These tentative results qualify more readily for being called "compensatory" than the results on phonological constituency relations.

The work on segmental effects on pitch contours complements the work on segmental timing, and shows that rule based time warping can also be applied here. We showed how, once segmental durations (or any acoustically salient features) have been computed, one can accurately predict the alignment of pitch contours; in fact, we were also able to model segmental effects in the frequency domain (i.e., effects on the height of pitch contour, such as intrinsic pitch).

In summary, we started this paper by explaining why the unconstrained nature of spontaneous speech puts a premium on the search for invariances, or, equivalently, for accurate mathematical models. In the area of speech timing—both segmental and pitch timing—there is much controversy concerning even the most basic issue, which is how to describe speech timing. This paper proposed approaches where *rule based time warps* play an important role. We believe that progress in the analysis or synthesis of spontaneous speech requires addressing issues at this fundamental level.

Acknowledgments

The work on acoustic trajectories was done in collaboration with John Coleman and Mark Randolph. Syllabic timing is a joint project with Chilin Shih. Template based pitch modelling involves collaboration with Julia Hirschberg and Bernd Möbius. I am very grateful for the thought, energy, and time that these colleagues have devoted to these projects. Any farfetched, or erroneous, conclusions drawn in this paper based on this joint work are entirely my responsibility. I also want to thank Joseph Olive, Richard Sproat, and Pilar Prieto for many helpful discussions. Finally, challenging reviews by Nick Campbell and Robert Port, who have—and continue to have—fundamentally different views on many of the issues discussed, have contributed to this paper in significant ways.

References

[AHK87] J. Allen, S. Hunnicut, and D. H. Klatt. *From Text to Speech: The MITalk System.* Cambridge, UK: Cambridge University Press, 1987.

[BB94] P. Barbosa and G. Bailly. Characterization of rhythmic patterns for text-to-speech synthesis. *Speech Communication* 15:127–137, 1994.

[Cam92c] W. N. Campbell. Syllable-based segmental duration. In G. Bailly, C. Benoît, and T. R. Sawallis, editors, *Talking Machines: Theories, Models, and Designs*, pp. 211–224. Asterdam: Elsevier Science, 1992.

[Cam93a] W. N. Campbell. Automatic detection of prosodic boundaries in speech. *Speech Communication*, 13:343–354, 1993.

[CH88] T. H. Crystal and A. S. House. Segmental durations in connected-speech signals: Current results. *J. Acoust. Soc. Am.*, 83:1553–1573, 1988.

[CI91] W. N. Campbell and S. D. Isard. Segment durations in a syllabic frame. *Journal of Phonetics*, 47:19:37, 1991.

[Col92b] J. S. Coleman. 'Synthesis-by-rule' without segments of rewrite-rules. In G. Bailly, C. Benoît, and T. R. Sawallis, editors, *Talking Machines: Theories, Models, and Designs*, pp. 43–60. Amsterdam: Elsevier Science, 1992.

[Col92c] R. Collier. A comment on the prediction of prosody. In G. Bailly, C. Benoît, and T. R. Sawallis, editors, *Talking Machines: Theories, Models, and Designs*, pp. 205–208. Amsterdam: Elsevier Science, 1992.

[Fuj88] H. Fujisaki. A note on the physiological and physical basis for the phrase and accent components in the voice fundamental frequency contour. In O. Fujimura, editor, *Vocal Fold Physiology: Voice Production, Mechanisms and Functions.* New York: Raven, 1988.

[Gay68] Th. Gay. Effect of speaking rate on diphthong formant movements. *J. Acoust. Soc. Am.*, 44:1570–1573, 1968.

[Her90] S. R. Hertz. The delta programming language: An integrated approach to nonlinear phonology, phonetics, and speech synthesis. In J. Kingston and M. E. Beckman, editors, *Papers in Laboratory Phonology I: Between the Grammar and Physics of Speech*, pp. 215–257. Cambridge, UK: Cambridge University Press, 1990.

[Kla73] D. H. Klatt. Interaction between two factors that influence vowel duration. *J. Acoust. Soc. Am.*, 54:1102–1104, 1973.

[Ljo94] A. Ljolje. High accuracy phone recognition using context clustering and quasi-triphonic models. *Computer Speech and Language*, 8:129–151, 1994.

[Mac89] M. J. Macchi. Using dynamic time warping to formulate duration rules for speech synthesis. *J. Acoust. Soc. Am.*, 85:S1(U49), 1989.

[MPH93] B. Möbius, M. Pätzold, and W. Hess. Analysis and synthesis of F_0 contours by means of Fujisaki's model. *Speech Communication* 13, pp. 53–61, 1993.

[Noo91] S. G. Nooteboom. Some observations on the temporal organisation and rhythm of speech. In *Proceedings of the XIIème International Congress of Phonetic Sciences, Aix-en-Provence, France*, 1991.

[OS95] J. P. Olive and R. W. Sproat. Principles of speech synthesis. In W. B. Kleijn and K. K. Paliwal, editors, *Speech Coding and Synthesis*. Amsterdam: Elsevier, 1995.

[Pie80] J. B. Pierrehumbert. The Phonology and Phonetics of English Intonation. PhD thesis, Massachusetts Institute of Technology, Distributed by the Indiana University Linguistics Club, 1980.

[PSH95] P. Prieto, J. P. H. van Santen, and J Hirschberg. Tonal alignment patterns in Spanish. *Journal of Phonetics*, 23: 1995.

[SB91] K. N. Stevens and C. A. Bickley. Constraints among parameters simplify control of Klatt formant synthesizer. *Journal of Phonetics*, 19:161–174, 1991.

[SK83] D. Sankoff and J. B. Krusal. *Time Warps, String Edits, and Macromolecules: The Theory and Practice of Sequence Comparison.* London: Addison-Wesley, 1983.

[vS92] J. P. H. van Santen. Contextual effects on vowel duration. *Speech Communication*, 11:513–546, 1992.

[vS93a] J. P. H. van Santen. Analyzing N-way tables with sums-of-products models. *Journal of Mathematical Psychology*, 37:327–371, 1993.

[vS93b] J. P. H. van Santen. Timing in text-to-speech systems. *Proceedings of the European Conference on Speech Communication and Technology, Berlin, Germany*, pp. 1397–1404, 1993.

[vS94a] J. P. H. van Santen. Assignment of segmental duration in text-to-speech synthesis. *Computer Speech and Language*, 8:95–128, 1994.

[vS94b] J. P. H. van Santen. Using statistics in text-to-speech system construction. *Proceedings of the ESCA/IEEE Workshop on Speech Synthesis, Mohonk, NY*, pp. 240–243, 1994.

[vSCR92] J. P. H. van Santen, J. C. Coleman, and M. A. Randolph. Effects of post-vocalic voicing on the time course of vowels and diphthongs. *J. Acoust. Soc. Am.*, 4.2:2444–2447, 1992.

[vSH94] J. P. H. van Santen and J. Hirschberg. Segmental effects on timing and height of pitch contours. In *Proceedings of the International Conference on Spoken Language Processing, Yokohama, Japan*, pp. 719–722, 1994.

[vSS95] J. P. H. van Santen and C. Shih. Syllabic and segmental timing in Mandarin Chinese and American English (in preparation).

16

Measuring temporal compensation effect in speech perception

Hiroaki Kato
Minoru Tsuzaki
Yoshinori Sagisaka

ABSTRACT The perceptual compensation effect between neighboring speech segments is measured in various word contexts to explore the following two problems: (1) whether temporal modifications of multiple segments perceptually affect each other, and (2) which aspect of the stimulus correlates with the perceptually salient temporal markers. Experiment 1 utilizes an acceptability rating of temporal unnaturalness for words with temporal modifications. It shows that a vowel (V) duration and its adjacent consonant (C) duration can perceptually compensate each other. This finding demonstrates the presence of a time perception range wider than a single segment (V or C). The results of the first experiment also show that rating scores for compensatory modification between C and V do not depend on the temporal order of modified pairs (C-to-V or V-to-C) but rather on the loudness difference between V and C; acceptability decreases when the loudness difference between V and C becomes high. This suggests that perceptually salient markers locate around major loudness jumps. Experiment 2 further investigates the influence of the temporal order of V and C by utilizing a detection task instead of the acceptability rating.

16.1 Introduction

To achieve natural sounding synthesized speech by rule-based synthesis techniques, a number of specific rules to assign duration have been proposed to replicate the segmental durations found in natural speech [Cam92a, FK89, HF80, KS92a, ST84]. Each of the segmental durations produced by such duration-setting rules generally has a certain amount of error compared to the corresponding naturally spoken duration. The effectiveness of a durational rule should ideally be evaluated by how much these errors would be accepted by human listeners, who are the final recipients of synthesized speech. However, in almost all previous researches, the average of absolute error of each segmental duration from its standard has been adopted as the measure for objective evaluation of such durational rules. Although we will not deny the effectiveness of this traditional approach, i.e., the effort to minimize the mean acoustic error, we also find it crucial to investigate the "perceptual" basis to the evaluation of durational modification and to test the validity of the implicit premise of the traditional approach. An implicit premise of this approach is that the perceptual distortion for the entire speech is equal to the sum of perceptual distortions for each of its segments. A possible problem with this premise is that it neglects interrelationships among such acoustic errors. Relation factors between errors in adjacent segments, such as difference in relative direction of deviations (the same or opposite), may affect the total impression of perceived distortions, even when the total amount of error remains the same. If such a contextual effect on perceptual evaluation could be specified quantitatively, we could obtain a more valid (closer to human evaluation) measure than the simple mean of acoustic errors for evaluating durational rules. For the purpose of developping better evaluation criteria for durational rules, the current study examined the following two contextual effects that may affect perceptual evaluation for temporal modification of speech segments: (1) processing range in time perception and (2) perceptual salience of temporal markers. Both problems were probed by measuring the perceptual compensation effect between two neighboring speech segments.

16.1.1 Processing Range in Time Perception of Speech

The first purpose of the current study was to explore whether there is a processing range wider than a single segment, i.e., a phoneme, in the time perception of speech. A considerable number of acoustic studies have indicated that a segmental duration may depend on the surrounding contexts at various levels [Cam92a, FK89, HF80, HKO67, KTS92, TSK89, vS92, Sat77b]. The results of these studies suggest that there are processing ranges wider than a single segment in the domain of speech production. However, none of these works provided direct evidence for the presence of such a wide processing range in the domain of speech perception because their studies were limited to the description of naturally produced speech.

In psychophysical "non-speech" studies, it has been demonstrated that successive intervals in a sequence of beats perceptually interact with each other. Schulze (1978) conducted a perceptual test in which subjects had to detect one of the following three types of displacement within an isochronous rhythmic pattern: (1) lengthening of one interval, (2) lengthening of one interval and shortening of its adjacent interval, and (3) lengthening of two or more successive intervals. If each interval in the test sequence was independently compared to its counterpart in the standard sequence (an isochronous one), then the number of modified intervals would be a crucial factor affecting displacement detection. Thus, the highest detectability should be observed for (3), the second highest for (2), and the lowest for (1). However, it turned out that the detectability for (1) was equivalent to or higher than that for (2). This suggests the presence of a global process ranging over two or more intervals in perceiving the regular rhythmic pattern.

In "speech" research, on the other hand, several studies have looked at the perception of temporal modifications for speech segments [CG75, FNI75, Hug72a, Hug72b, Kla76], although only a few have addressed the perceptual phenomena caused by interactions among multiple modifications. It has been reported that speech stimuli with multiple durational modifications in opposite directions between consonants (C) and vowels (V) tend to be heard as more natural than those with multiple durational modifications in the same direction [ST84, HF83]. Sato (1977a) moreover found that a lengthening of a consonant duration may perceptually cancel out the same amount of shortening of the adjacent vowel. These observations imply a perceptual compensation phenomenon between C durations and their adjacent V durations. However, one should be prudent in concluding that such perceptual compensation would be commonly observed because each of these studies employed a fairly small number of speech samples: two sentences in Sagisaka et al.'s study., three non-sense words in Hoshino et al.'s, and the first to third syllables of one word "sakanayasan" (a fish dealer) in Sato's.

In the current study, therefore, we tried to directly test the hypothesis that there is a wider processing range than a single segment in the time perception of speech. To do this, we tried to collect a sufficient number of subjective responses using a sufficient number of stimulus samples by measuring perceptual compensation effects with the following procedure. First, we chose thirty C and V pairs from fifteen four-mora Japanese words. Each of the chosen words were temporally modified in four ways: (1) single V, (2) single C, (3) V and C in opposite directions, and (4) V and C in the same direction (*Fig. 16.1*). Temporal distortion was rated for each of the modified words by human listeners. The obtained rating scores were mapped using psychological scaling to assure an interval scale and then pooled for each of the four modification conditions.

If the traditional premise, i.e., adopting the mean acoustic error as an evaluation measure of durational rules, was valid, then the subjective distortion for multiple modifications should be the same as the sum of subjective distortions for each of the single modifications constituting the whole of the multiple modifications. Thus, the estimation scores for both "double modified" conditions (3) and (4) would each be expected to become equal to the sum of the scores for the "single modified" conditions (1) and (2). Otherwise, the results would suggest that the interaction between adjacent modifications had affected the perceptual evaluation; this supports the presence of a wider processing range than a single segment in the time perception of speech. In particular, if the mean estimation score for condition (3) was significantly lower than the sum of those for conditions (1)

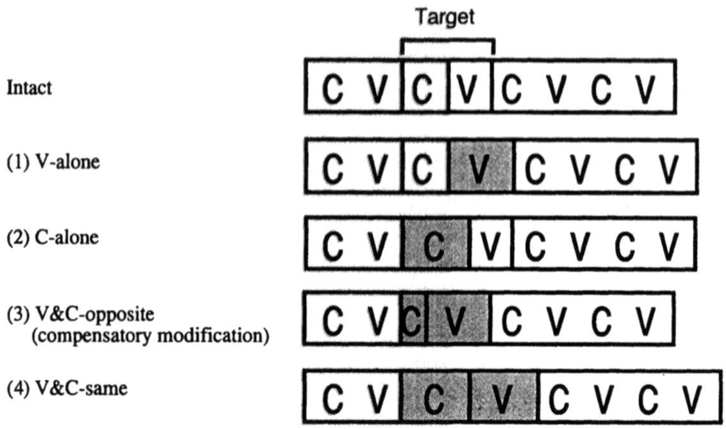

FIGURE 16.1. Schematic diagrams showing the four manners of temporal modification performed on each of the chosen word samples. Highlighted segments were temporally modified.

and (2), this would imply a general tendency of perceptual compensation effect (Hoshino et al., 1983; Sagisaka et al., 1984).

16.1.2 Contextual Effect on Perceptual Salience of Temporal Markers

Since temporal structures such as rhythm or tempo can be perceived in speech, there should be markers that give us such temporal information about speech. Therefore, one could deal with issues of time perception of speech by considering these temporal markers. Problems that arise from this approach are that one cannot explicitly specify the locations of such temporal markers and that the markers do not necessarily have the same perceptual salience. In the current study, therefore, we assumed the location of temporal markers to be at phoneme boundaries. We then tried to explore the stimulus context that correlates with the perceptual salience of temporal markers.

As shown in Figure 16.1, the compensatory modification of two neighboring segments does not destroy the temporal structure outside the modified pair. Therefore, if each modification were made on stable portions, such as vowel plateaus, fricatives, nasals, and pre-burst closures, then generally speaking only the boundary between the two modified segments would be temporally displaced. In such a case, if the boundary portion contains some perceptually salient temporal marker, this modification would have a strong effect on perceptual evaluation. In this way, the change in perceptual evaluation, such as an acceptability rating, caused by the compensatory modification can measure the perceptual salience of the temporal markers located between the two modified segments. The current study utilized

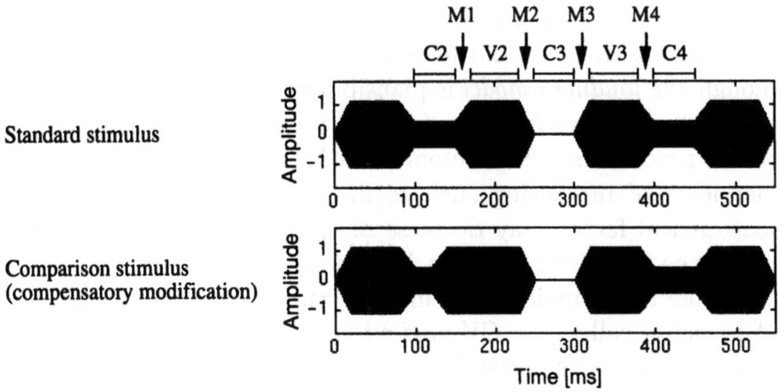

FIGURE 16.2. Time waveforms of the examples of non-speech stimuli used in Kato and Tsuzaki (1994). Each V or C indicates a target portion to be modified. Each M indicates the location of a temporal marker considered. The level of V is 73 dB SPL (= 9.85 sone), the level of louder C is 64 dB SPL (= 5.28 sone), and that of softer C is "silence", which each models the average level of vowels, nasals, or pre-plosive closures. All signals are 1 kHz pure tones.

this measure to test the following two possible models for predicting the perceptual salience of temporal markers in speech.

The first model is called *the loudness model.* This model assumes that the perceptual salience of a temporal marker would correlate with the amount of change in perceived intensity, i.e., loudness, around the marker in question. This model is based on the idea that spoken language perception is governed by the same psychoacoustic laws that determine the perception of non-speech stimuli. In the current study, we chose the magnitude of the loudness difference or jump between two modified segments from among various psychophysical variables. This is because a previous non-speech study suggested that the perceptual salience of a temporal marker correlates with this sort of loudness jump.

Kato and Tsuzaki (1994) measured detectability of the marker displacement in the pure tone stimuli that modelled the overall loudness contours of four-mora words. Their subjects listened to the pair of standard and comparison stimuli and were asked to rate the difference between them (an example of the stimulus pair is shown in Figure 16.2). Each of the comparison stimuli had the compensatory temporal modification in two (e.g., C2 and V2) of five consecutive steady-state portions (C2 to C4), i.e., the boundary or temporal marker between the modified portions (e.g., M1) was solely displaced relative to its standard counterpart. The results showed that the displacements of the markers with large loudness jumps (M2, M3) were more easily detected than those of the markers with small loudness jumps (M1, M4). *The loudness model* of the current study assumed the same perceptual effect of loudness jump to be also valid for the speech stimuli. If the temporal modification of the segment boundary having a larger loudness jump made a larger effect on perceptual evaluation, then *the loudness model* would be supported.

Although *the loudness model* is plausible in terms of the psychophysical measurement, perceptual effects for spoken language do not necessarily depend on psychoacoustical factors alone but also on factors relating to a higher level of processing such as linguistic segmentation rules. Any psychoacoustical factor may decrease or lose its effect on the perceptual outcome when linguistic contexts have a considerably strong effect. We therefore take a linguistic factor into account for the second model. The second model is called *the CV model* ("CV" stands for a pair consisting of a consonant and its succeeding vowel). This model assumes that the CV is a predominant unit in the time perception of speech and the unit boundary, i.e., consonant onset, is perceptually the most salient. The second model is likely to be supported from linguistic considerations because a CV unit usually coincides with a mora, a phonological segmentation unit in Japanese.

Several studies have repeatedly cited the evidence of CV units in the domain of speech production (e.g., [CS91, ST84]); the compensatory relationship between a C duration and its succeeding V duration was observed in both Sagisaka et al.'s and Campbell et al.'s studies that performed acoustical analyses on a large database of spoken Japanese. These previous studies seem to support *the CV model* which predicts that the displacement of a consonant onset probably has the largest effect on perceptual evaluation. However, these studies can not directly support the dominance of the CV unit in speech perception because they are based on the observation of physical characteristics of "naturally spoken" speechand do not make an empirical assessment of the subjective evaluation of these stimuli.

In the domain of speech perception, on the other hand, several pioneering studies by Hoshino and Fujisaki (1983) and Sato (1977a) have looked into the temporal compensation between C and its adjacent V. Although both studies suggest the presence of a compensation effect, they seem to disagree in results for the compensation unit. Hoshino and Fujisaki reported the advantage of CV-unit compensation over VC-unit compensation, while Sato's results supported the advantage of VC-unit compensation. Discrepancies between these studies can be partly attributed to differences in the speech utterances and the psychophysical procedures and, as has been pointed out, to the shortage in the number of samples employed. Thus, it is still an open question whether CV is a more significant unit for perceptual compensation than VC; *the CV model* therefore needs to be tested. Using a relatively large number of speech samples, the current study compared the perceptual evaluations of compensatory modifications for CV and for VC. If the compensatory modification for CV had a smaller perceptual effect than that for VC, then *the CV model* would be supported.

The second aim of the experiments described in this chapter was to provide a direct comparison of the two models (*loudness* and *CV*) which each refer to either psychoacoustic or linguistic consideration. One should be, however, aware that these two models are not mutually exclusive. It could be the case that both models are supported, or both rejected together. The models are assumed to reflect two different processing levels. Even when only one model was supported, it would not mean that the other model should be rejected but just mean that the supported one had a stronger effect than the other one had.

16.2 Experiment 1—Acceptability Rating

In Experiment 1, acceptability for temporal modification of V, C, or both V and C within a four-mora word was measured to find evidence of a perceptual compensation effect between multiple segments and to test the two models (*loudness* and *CV*) for the temporal compensation effect.

16.2.1 Method

Subjects

Six adults with normal hearing participated in Experiment 1. All were native speakers of Japanese.

Stimuli

Fifteen four-mora Japanese words were chosen from a speech database of commonly used words [STA+90] as the original material (see *Table 16.1*). The underlined CVC sequences were the targets of the modifications; the temporal position of the target vowels were chosen from the first three out of four morae.

TABLE 16.1. Speech tokens chosen in Experiment 1. The underlined CVC sequences are the target portions. The left column indicates the temporal positions of targets in a word; i.e., C_i or V_i is the ith consonant or ith vowel in a word.

Target	Roman transcription				
$C_1V_1C_2$	bakugeki	gakureki	hanareru	nagedasu	sakasama
$C_2V_2C_3$	hanahada	imasara	kasanaru	katameru	mikakeru
$C_3V_3C_4$	hanahada	korogasu	rokugatsu	tachimachi	tamatama

Each of the paired target segments (CV or VC) was temporally modified in four ways: (1) V-alone, (2) C-alone, (3) V and C in opposite directions, and (4) V and C in the same direction, as to be seen in Figure 16.1. Each modification was either to lengthen or to shorten the segment(s). The size of a modification was either 15 ms or 30 ms. When two segments were modified, i.e., (3) or (4), the absolute modification size of one segment was equal to the other. The modifications were made by a cepstral analysis and resynthesis technique with the Log Magnitude Approximation (LMA) filter [IK78], and were carried out at a 2.5-ms frame interval. The duration change was achieved by deleting or doubling the synthesis parameters frame by frame. The target portions were carefully trimmed out so as to exclude the transient portions at both ends of the vowels and the burst and release portions of the plosives or affricates. That is, since the temporal markers were assumed to be at the VC or CV boundaries, the editing procedures modified durations at locations remote from these boundaries. In addition to the above modified stimuli, we prepared unmodified stimuli for reference. In total, 435 word stimuli were prepared. [1]

Procedure

The stimuli were fed diotically to the subjects through a D/A converter (MD-8000 mkII, PAVEC), a low-pass filter (FV-665, NF Electronic Instruments, fc = 5,700 Hz, -96 dB/octave), and headphones (SR-Λ Professional, driven by SRM-1 MkII, STAX) in a sound-treated room. The average presentation level was 73 dB (A-weighted) which was measured with a sound level meter (Type 2231, Brüel & Kjær) mounted on an artificial ear (Type 4153, Brüel & Kjær). The subjects were told that each stimulus word was possibly subjected to a temporal modification. They listened to each of the randomly presented word stimuli and were asked to rate each stimulus regarding how acceptable the temporal modification was, if perceived at all, as an exemplar of that token using seven subjective categories ranging from "quite acceptable" to "unacceptable". [2] Each subject rated each stimulus eight times in total. The obtained responses were

---

[1] 15 CVCs × 29 variations of modification; i.e., 2 absolute modifications (= 15 ms, 30 ms) × 2 modification directions (= lengthening, shortening) × 7 modification manners (= V alone, pre-C alone, post-C alone, V and pre-C in the same direction, V and post-C in the same direction, V and pre-C in opposite directions, V and post-C in opposite directions) + 1 (= intact for reference).

[2] If listeners were asked to estimate "naturalness", they would tend to use such a strict criterion that the range of temporal modifications having informative estimation results would be very restricted. To obtain information for a reasonably wide range of modifications, we chose the "rating of acceptability" over the "rating of naturalness".

pooled over all subjects for each category, and then each stimulus was mapped on a unidimensional psychometric scale in accordance with Torgerson's Law of Categorical Judgment [Tor58]. [3] The scaled value of each "modified" stimulus was then adjusted by subtracting the scaled value of its corresponding "intact" stimulus. Thus, the measure obtained for each stimulus corresponded to the amount of loss of acceptability from the intact reference stimulus.

16.2.2 Results and Discussion

Figure 16.3 shows the loss of acceptability pooled over the fifteen stimulus words for each of the four manners of temporal modification, i.e., V alone, C alone, V and C in opposite directions (V&C-opposite), and V and C in the same direction (V&C-same). For comparison, the right-most bar indicates the sum of the V-alone and C-alone values.

Each obtained value for loss of acceptability was assured to be on an interval scale by the Torgerson's model. Accordingly, if the two modifications of "double modified" conditions had evaluations of the

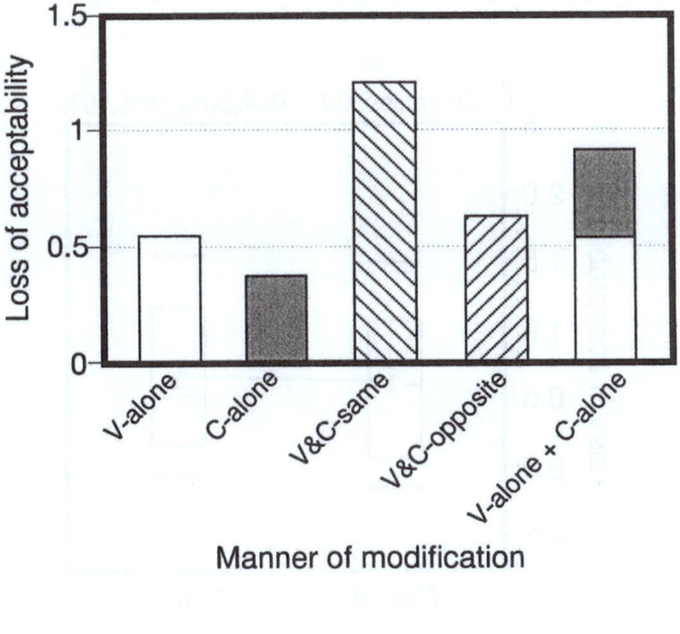

FIGURE 16.3. Scaled loss of acceptability pooled over fifteen word stimuli for each manner of temporal modification.

---

[3] A method of psychological scaling using outputs of a rating scale method. Each of the categorical boundaries and the stimuli used in the rating is mapped on a uni-dimensional interval scale.

acceptability independently from one another, then the expected loss of acceptability in these conditions should be a simple sum of those expected in the corresponding "single modification" conditions. That is, the loss of acceptability in both the "V&C-same" and "V&C-opposite" cases should approach the sum of those in the "V alone" and "C alone" cases (i.e., V-alone + C-alone). As shown in Figure 16.3, however, this was not the case. The value for "V&C-opposite" was smaller than that for "V-alone + C-alone", and the value for "V&C-same" was larger than that for "V-alone + C-alone". Multiple comparisons using Tukey-Kramer's HSD indicated that both differences were significant ($p < 0.05$).

These results imply that simultaneous modifications with a V duration and its adjacent C duration are not independent in terms of the acceptability evaluation. Either they perceptually compensate each other when in opposite directions or they perceptually enhance each other when in the same direction. This suggests that a process having a time span wider than a single segment (C or V) is involved in the time perception of speech.

To interpret the results in terms of perceptual markers, the following analyses focused on the size of the effect obtained when displacing a segment boundary within a pair of fixed duration. If a salient marker occurs at the boundary, its displacement should have a large effect.

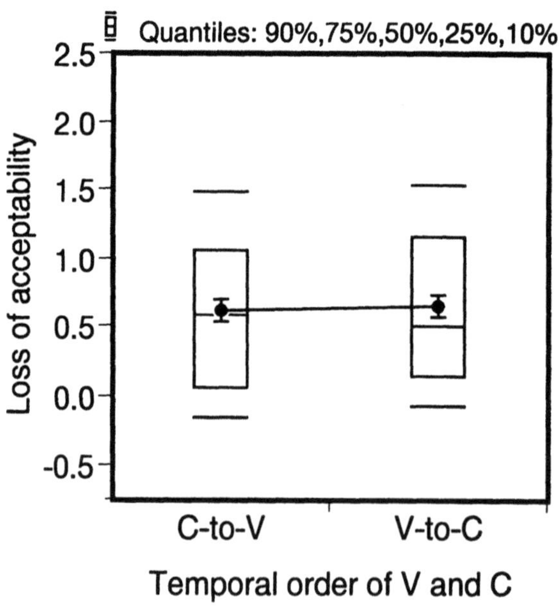

FIGURE 16.4. Loss of acceptability caused by compensatory modification as a function of the temporal order of V and C. The dots and error bars show the group averages and the standard errors, respectively. Quantile boxes are also shown.

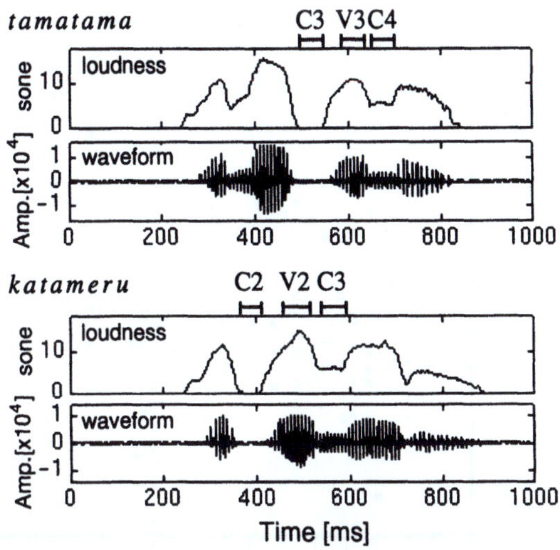

FIGURE 16.5. Examples of loudness contours and time waveforms of the word stimuli used in Experiment 1. The horizontal bars at the top of each figure indicate the target portions to be modified. Upper: the word /tamatama/. Lower: the word /katameru/.

First, *the CV model* was evaluated. This model assumes that the CV (mora) is a predominant unit in the time perception of speech and that the consonant onset, i.e., the hypothesized unit boundary, is perceptually the most salient. Therefore, this model predicts a larger loss of acceptability for the displacement of V-to-C boundaries, i.e., consonant onsets, than for the displacement of C-to-V boundaries. As shown in Figure16.4, only a small difference in the loss of acceptability could be observed due to the temporal order of V and C. A *t*-test did not indicate this difference to be significant [$t(118) = 0.188$, $p = 0.851$]. Consequently, *the CV model* was not supported here. This result suggests that perceptually salient markers are not generally located around V-to-C boundaries.

Next, *the loudness model* was evaluated. Figure16.5 shows waveforms of two stimuli used in Experiment 1 and their corresponding loudness contours, which were calculated in accordance with ISO 532B [ISO75, ZFW$^+$91][4] every 2.5 ms. As predicted from the examples of Figure16.5, every V target in this experiment was louder than its adjacent C portions; that is, each of the boundaries between two modified segments always had some change in loudness. In light of this fact, we defined "loudness jump", calculated by subtracting the median loudness of C from that of

[4]Although ISO-532B does not always provide excellent approximations for non-steady-state signals like speech, we adopted this method due to the advantage of its psychophysical basis instead of adopting power or intensity which incorporates no psychophysical consideration.

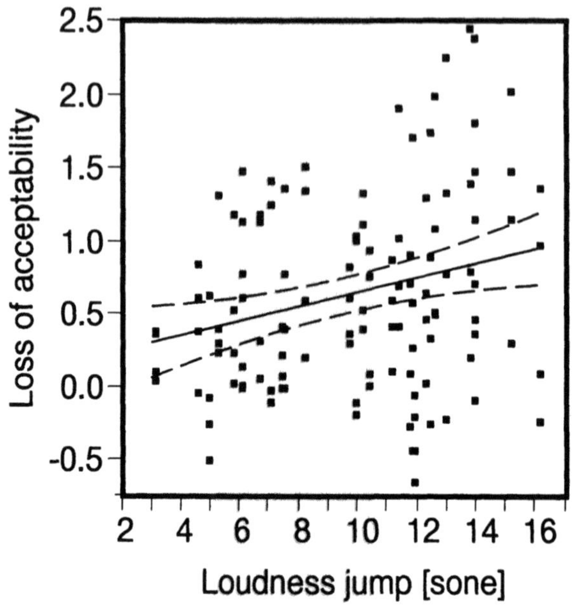

FIGURE 16.6. Loss of acceptability caused by compensatory modification as a function of the loudness jump between V and C. The solid line and dashed lines show the regression line and its 95-% confidence curves.

V, as our explanatory variable in *the loudness model.* Thus, the employed loudness jumps were always positive. This model predicts a larger loss of acceptability for the displacement of segmental boundaries having large loudness jumps than for those having small loudness jumps. In addition to the factor of loudness jump, we included three factors capable of affecting acceptability: the temporal order of V and C (V-to-C or C-to-V), the temporal position of the modified vowel (1, 2, or 3), and the amount of each single modification (15 ms or 30 ms).

The effects of the above four factors and their interactions on the amount of loss of acceptability were tested by a four-way factorial General Linear Model (GLM[5]) (SAS Institute, 1990). The main effect of the loudness jump was significant [$F(1, 96) = 10.51$, $p < 0.005$]. The loss of acceptability increased with increasing loudness jump (Figure 16.6). The interaction between the loudness jump and the amount of modification was significant [$F(1, 96) = 4.15$, $p < 0.05$], that is the effect of loudness jump was higher for the longer (30 ms) modification condition. The temporal order of V and C was again not significant [$F(1, 96) = 0.087$, $p = 0.769$]. No other main effect nor interaction was significant.

[5]An extended version of the analysis of variance or ANOVA. GLM copes with continuous values as explanatory variables as well as nominal values.

Summarizing, there was no evidence for *the CV model* within the scope of Experiment 1. On the contrary, the results of the GLM analysis supported *the loudness model*; a large amount of loudness jump between modified segments generally caused a considerable loss of acceptability. This suggests that perceptually salient temporal markers tend to locate around major loudness jumps. The observed loudness effect is of the same tendency as one observed in the previous non-speech study [KT94a] as mentioned in Introduction.

However, the results of Experiment 1 did not agree, in every aspect, with those of the previous study. Kato and Tsuzaki (1994) reported that the direction of the marker slope (rising or falling) affected the listeners' ability to detect the temporal modifications as well as the loudness jump did. That was, the detectability of temporal displacement for a rising marker (e.g., M1 or M3 in Figure 16.2) was significantly higher than that for a falling marker (e.g., M2 or M4 in Figure 16.2). Such tendency that a rising marker is more perceptually salient than a falling marker has been also reported by Kato and Tsuzaki (1995); they measured the discrimination thresholds for pure tone durations marked by rising or falling slopes and found that the rising slopes more accurately marked the auditory durations than the falling slopes did. Therefore, we thought that by applying these previous observations directly to Experiment 1, the displacements of the C-to-V transition (always a rising slope) would have a greater effect on the perception than the displacements of the V-to-C transition (always falling one). This was, however, not the case. What could have brought about such an inconsistency between the factors of slope direction in the previous studies and temporal order of V and C in the current study?

Two major differences existed between the previous experiments and the current experiment (Experiment 1). The first one was a physical difference between the pure tone stimuli and the speech stimuli. While the rising and falling slopes compared in the previous experiments were the exact mirror images of each other in the time axis, Experiment 1 used 30 different slopes (V-to-C or C-to-V transitions). Such a wide stimulus variation in Experiment 1 possibly obscured the potential effect of slope direction.

The second difference was in the experimental procedure; Experiment 1 employed the acceptability rating of single stimuli while the previous experiments used a detection or discrimination task. The task in Experiment 1 could be broken down, from an analytical viewpoint, into the following two stages: 1) a detection stage—each subject had to detect the difference between the temporal structure of the presented stimulus and that of his/her internal exemplar of that token even though a single stimulus was presented in each trial, and 2) a rating stage—the degree of acceptability was rated. That is, Experiment 1 required the subjects to do a rather central or higher level process in addition to a simple detection task similar to the ones used in the previous experiments. Therefore, even though the displacements of the C-to-V transition were detected more easily than those of the V-to-C transition, the rated score possibly showed no difference with regard to the temporal order of V and C if the subjects were more tolerant of the displacement of C-to-V transition than that of V-to-C transition. This would most likely occur if the mora (CV) functioned as a perceptual unit at a higher cognitive level in the acceptability rating task. In other words, the greater salience of weak-strong (C-to-V) boundaries revealed by the previous psychoacoustic experiment was compensated by the greater linguistic importance of boundaries between CV units. This is indirect support for *the CV model.*

Experiment 2 was therefore designed to test the second possibility: whether the task of Experiment 1, which possibly involved a higher cognitive process, functioned to cancel out the potential effect of slope direction or temporal order of V and C. This experiment adopted a detection task similar to those in the previous non-speech studies and employed stimuli similar to Experiment 1's, i.e., we tried to separate out the influence of the higher level processes possibly functioning at the rating stage. If the temporal displacements of the C-to-V transition were detected more easily than those of the V-to-C transition, the hypothesis that the inconsistency between the results of Experiment 1 and those of the previous experiments was due to the difference in task would be supported. This would suggest the possibility that the CV unit (mora) functioned at the stage of the acceptability rating in Experiment 1.

16.3 Experiment 2—Detection Test

The purpose of Experiment 2 was to test the possibility that the task of Experiment 1, the acceptability rating which possibly involved a higher cognitive process, functioned to cancel out the potential effect of temporal order of V and C.

16.3.1 Method

Subject Six adults with normal hearing participated in Experiment 2. They were the same subjects as in Experiment 1.

Design and stimuli The factors of loudness jump, temporal order of V and C, and temporal position of the modified vowel in a word were included in a factorial design. The stimuli were a reduced set of those used in Experiment 1; the modification manner was compensatory and the amount of each modification was 30 ms. In total, 75 word stimuli were employed. [6]

Procedure The detectability index (d') was measured for the difference between the intact unmodified tokens and each of the modified tokens by the method of constant stimuli. The experimental apparatus was the same as in Experiment 1. The subjects listened to the standard (intact) and the comparison (possibly modified) stimuli with a 0.7-s inter-stimulus interval and were asked to rate the difference between them. The subjects were allowed to use numerical categories 1 to 7 when they perceived any difference (the larger number corresponding to a larger subjective difference) or 0 when they perceived no difference. Twenty percent of the trials were control trials in which each comparison stimulus was the same as the standard stimulus. In total, twelve judgments were collected from each subject for each comparison stimulus. The obtained responses were pooled over all subjects for each category, then the detectability index, d', for each comparison stimulus was estimated in accordance with the Theory of Signal Detection [GS66].

16.3.2 Results and Discussion

The effects of loudness jump, temporal order of C and V, temporal position of the modified vowel, and the interactions among these three factors on the obtained detectability d' were tested by a three-way factorial General Linear Model (GLM). The main effect of loudness jump was significant [$F(1, 48) = 33.1$, $p < 0.0001$]. On the other hand, the main effect of temporal order of V and C was not significant [$F(1, 48) = 0.65$, $p = 0.422$]. No other main effect nor interaction was significant. Figures 16.7 and 16.8 show the obtained d' as a function of loudness jump and temporal order of V and C.

These results are in good agreement with those obtained in Experiment 1. Even though the detection task was adopted in Experiment 2, there was no significant effect for the temporal order of V and C, i.e., the direction of marker slope. Thus, we can safely state that the inconsistency between the results of Experiment 1 and those obtained in the previous non-speech studies was not due to the difference in experimental task but to the difference in stimuli. This finding, therefore, does not support the hypothesis that the mora unit functioned as a factor cancelling the effect of slope direction at the acceptability rating stage. Yet, we cannot exclude the possibility that the mora unit functioned in Experiment 2 even though the task was of the detection type. We are, however, willing to say in a practical

---

[6] 15 CVCs $\times$ 5 variations of modification; i.e., 2 temporal orders of V and C in target segments (= V-to-C or C-to-V) $\times$ 2 modification directions of vowel (= lengthening, shortening) + 1 (= intact for reference).

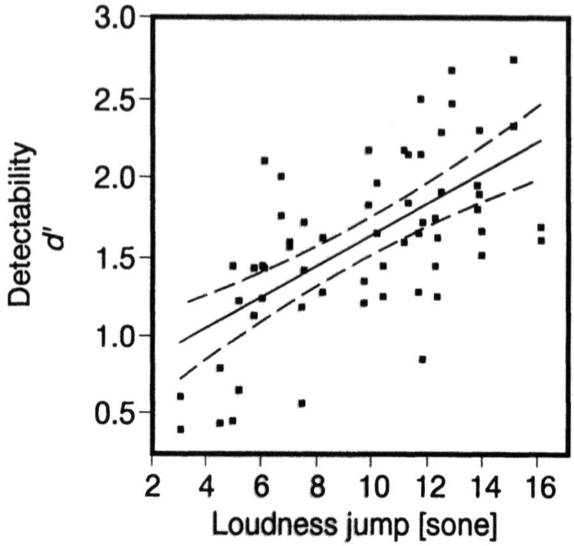

FIGURE 16.7. Detectability index *d'* for 30-ms compensatory modification as a function of the loudness jump between V and C. The solid line and dashed lines show the regression line and its 95-% confidence curves.

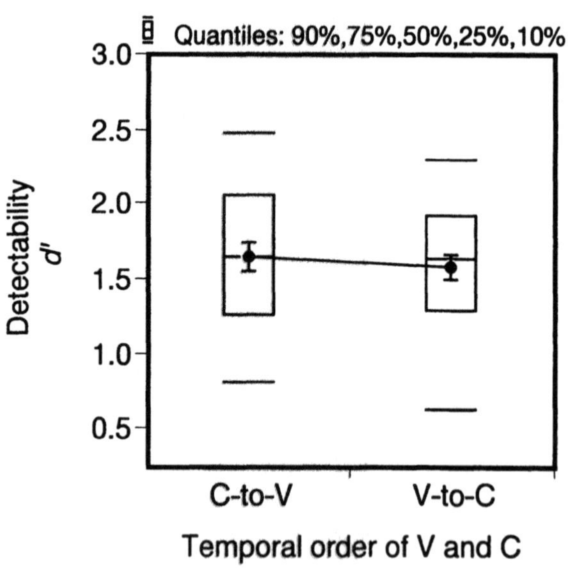

FIGURE 16.8. Detectability index *d'* for 30-ms compensatory modification as a function of the temporal order of V and C. The dots and error bars show the group averages and the standard errors, respectively. Quantile boxes are also shown.

sense that such influence of the mora unit should be taken as a secondary effect preceded by more general processes based on the loudness jump. Note that we adopted the loudness jump as a representative of the psychophysical auditory basis in contrast with more central or speech-specific ones. We should do further investigation to explore whether the loudness jump has an advantage over other psychoacoustical indexes, e.g., the change in an auditory spectrum.

Conclusion

The experimental results of the current study showed that a perceptual compensation effect was generally observed between V durations and their adjacent C durations. This suggests that a range having a wider time span, corresponding to a moraic range or wider, than a single segment (C or V) functions in the time perception of speech. Furthermore, the results supported that the acoustic-based psychophysical feature (loudness jump) is a more essential variable than the phonological or phonetical feature (CV or VC) to explain the perceptual compensation effect at such a wider range. Large jumps in loudness were found to function as salient temporal markers. Such large jumps generally coincide with the C-to-V and V-to-C transitions. This is probably one reason why previous studies have been successful, to some extent, in explaining perceptual phenomena by assuming a unit comprising CV or VC. However, the results of the current experiments indicated that the perceptual estimation is more closely related to loudness jumps per se than to their role of boundaries between linguistic units, be they CV or VC units. The practical conclusion of this study is that duration compensation may occur between adjacent C and V segments, particularly when the loudness jump between them is small. Thus the traditional evaluation measure of durational rules, based on the sum of absolute deviations of the duration of each segment from its standard, is not optimum from the perceptual viewpoint. We can expect to obtain a more valid (closer to human evaluation) measure than a traditional mean acoustic error by taking into account the perceptual effects described above.

References

[Cam92a] W. N. Campbell. *Multi-level timing in speech.* PhD thesis, University of Sussex, Department of Experimental Psychology, 1992. Available as ATR Technical Report TR-IT-0035.

[CG75] R. Carlson and B. Granström. Perception of segmental duration. In A. Cohen and S. G. Nooteboom, editors, *Structure and process in speech perception*, pp. 90–106. Heidelberg: Springer-Verlag, 1975.

[CS91] W. N. Campbell and Y. Sagisaka. Moraic and syllable-level effects on speech timing. Technical Report SP 91-107, IEICE, 1991.

[FK89] G. Fant and A. Kruckenberg. Preliminaries to the study of Swedish prose reading and reading style. Technical Report 2, Royal Institute of Technology, 1989.

[FNI75] H. Fujisaki, K. Nakamura, and T. Imoto. Auditory perception of duration of speech and non-speech stimuli. In G. Fant and M. A. A. Tatham, editors, *Auditory Analysis and Perception of Speech*, pp. 197–219. London: Academic Press, 1975.

[GS66] D. M. Green and J. A. Swerts. *Signal Detection Theory and Psychophysics.* New York: John Wiley, 1966.

[HF80] N. Higuchi and H. Fujisaki. Durational control of segmental features in connected speech. Technical Report S80-40, Acoust. Soc. Jpn., 1980. in Japanese with English abstract.

[HF83] M. Hoshino and H. Fujisaki. A study on perception of changes in segmental durations. *Technical Report H83-8/S82-75*, 1983.

[HKO67] S Hiki, Y. Kanamori, and J. Oizumi. On the duration of phonemes in running speech. *Journal of the Institute of Electrical Communication Engineers of Japan*, 50:849–856, 1967. in Japanese.

[Hug72a] A. W. F. Huggins. Just noticeable differences for segment duration in natural speech. *J. Acoust. Soc. Am.*, 51(4):1270–1278, 1972.

[Hug72b] A. W. F. Huggins. On the perception of temporal phenomena in speech. *J. Acoust. Soc. Am.*, 51(4):1279–1290, 1972.

[IK78] S. Imai and T. Kitamura. Speech analysis synthesis system using the log magnitude approximation filter. *Trans. Institute of Electronics and Communication Engineers*, J61-A:527–534, 1978. in Japanese with English figure captions.

[ISO75] ISO. Acoustics - method for calculating loudness level. *International Organization for Standardization, ISO 532-1975(E)*, 1975.

[Kla76] D. H. Klatt. Linguistic uses of segmental duration in English: acoustic and perceptual evidence. *J. Acoust. Soc. Am.*, 59:1208–1221, 1976.

[KS92a] N. Kaiki and Y. Sagisaka. The control of segmental duration in speech synthesis using statistical methods. In E Vatikotis-Bateson, Y Tohkura, and Y Sagisaka, editors, *Speech Perception, Production and Linguistic Structure*, pp. 391–402. Ohmsha (Tokyo)/ IOS Press (Amsterdam), 1992.

[KT94a] H. Kato and M. Tsuzaki. Intensity effect on discrimination of auditory duration flanked by preceding and succeeding tones. *J. Acoust. Soc. Japan (E)*, 15(5):349–351, 1994.

[KT94b] H. Kato and M. Tsuzaki. Temporal discrimination of part of tone marked by two amplitude changes — comparison among on-marker and off-marker, and their combinations. *Proceedings of the Fall meeting of Acoustics Society Japan*, pp. 555–556, 1994.

[KTS92] N. Kaiki, K. Takeda, and Y. Sagisaka. Linguistic properties in the control of segmental duration for speech synthesis. In G. Bailly, C. Benoît, and T. R. Sawallis, editors, *Talking Machines: Theories, Models, and Designs*, pp. 255–263. Amsterdam: Elsevier Science, 1992.

[SAS90] SAS Institute Inc. *The GLM procedure*, SAS/STAT User's Guide edition, 1990.

[Sat77a] H. Sato. Segmental duration and timing location in speech. *Technical Report S77-31*, 1977. in Japanese with English abstract and English figure captions.

[Sat77b] H. Sato. Some properties of phoneme duration in Japanese nonsense words. *Proceedings of the Fall Meeting of Acoustics Society Japan*, pp. 43–44, 1977. in Japanese with English figure captions.

[Sch78] H. H. Schulze. The detectability of local and global displacements in regular rhythmic patterns. *Psychological Research*, 40:173–181, 1978.

[ST84] Y. Sagisaka and Y. Tohkura. Phoneme duration control for speech synthesis by rule. *Transactions of the Institute of Electronics, Information and Communication Engineers of Japan*, J67-A(7):629–636, 1984.

[STA+90] Y. Sagisaka, K. Takeda, M. Abe, S. Katagiri, T. Umeda, and H. Kuwabara. A large-scale Japanese speech database. In *Proceedings of the International Conference on Spoken Language Processing, Kobe, Japan*, pp. 1089–1092, 1990.

[Tor58] W. S. Torgerson. *Theory and Methods of Scaling*. New York: John Wiley, 1958.

[TSK89] K. Takeda, Y. Sagisaka, and H. Kuwabara. On sentence-level factors governing segmental duration in Japanese. *J. Acoust. Soc. Am.*, 89:2081–2087, 1989.

[vS92] J. P. H. van Santen. Contextual effects on vowel duration. *Speech Communication*, 11:513–546, 1992.

[ZFW+91] E. Zwicker, H. Fastl, U. Widmann, K Kurakata, S. Kuwano, and S. Namba. Program for calculating loudness according to DIN 45631 (ISO 532b). *J. Acoust. Soc. Japan (E)*, 12(1):39–42, 1991.

17

Prediction of Major Phrase Boundary Location and Pause Insertion Using a Stochastic Context-free Grammar

Shigeru Fujio
Yoshinori Sagisaka
Norio Higuchi

ABSTRACT

In this paper, we present models for predicting major phrase boundary location and pause insertion using a stochastic context-free grammar (SCFG) from an input part of speech (POS) sequence. These prediction models were made with similar ideas as both major phrase boundary location and pause insertion have similar characteristics. In these models, word attributes and left/right-branching probability parameters representing stochastic phrasing characteristics are used as input parameters of a feed-forward neural network for the prediction. To obtain the probabilities, first, major phrase characteristics and pause characteristics are learned through the SCFG training using the inside–outside algorithm. Then, the probabilities of each bracketing structure are computed using the SCFG. Experiments were carried out to confirm the effectiveness of these stochastic models for the prediction of major phrase boundary locations and pause locations. In a test predicting major phrase boundaries with unseen data, 92.9% of the major phrase boundaries were correctly predicted with a 16.9% false insertion rate. For pause prediction with unseen data, 85.2% of the pause boundaries were correctly predicted with a 9.1% false insertion rate.

17.1 Introduction

Appropriate F_0 control is needed for the generation of synthetic speech with natural prosody. The F_0 pattern of a Japanese sentence can be described by partial ups and downs grouped over one or two *bunsetsu* (accent phrase) and superimposed on a gentle downslope. At most boundaries between accent phrases, the downslope is maintained; such accent phrases being in the same prosodic group, but at major phrase boundaries, the underlying F_0 declination is reset.

Prediction of such major phrase boundaries and the insertion of pauses is indispensable if we are to synthesize speech with natural prosody. In Japanese speech synthesis, various heuristic rules have been employed for phrasing and pause allocation[HFKY90][HS80]. Though these rules are based on the analysis of prosodic characteristics in relation to the complexity of phrase dependency structure, there have been very few attempts to directly create a computational model for pause or boundary allocation. Statistical modelling has been proposed only recently[SK92][KS92c][SS95].

As allocation of phrase boundaries and pauses is not unique, statistical analyses were performed using about 500 sentence utterances with hand-tagged phrase dependency structure[SK92][KS92c]. These analyses produced computational models using linear regression, but required phrase dependency structure which is quite difficult to automatically extract from plain text, and this prevented a fully computational treatment. In [SS95], part of speech (POS) information of neighboring phrases were employed for the allocation of phrase boundaries and pauses using an explicit determination of phrase dependency structure. All combinations of modifying and modified phrases are included and their reliabilities are calculated based on the number of possible combinations. The phrase dependency structure is predicted based on their reliabilities.

In this paper, a computational model is proposed for the allocation of phrase boundaries and pauses in which phrase dependency structure is computationally parameterized through the training of a stochastic context-free grammar (SCFG) with a hand-tagged spoken corpus. Though the proposed model has not yet been fully optimized in combination with SCFG training, a feed-forward type neural net is employed to optimize using input phrase dependency structure parameters obtained from a trained SCFG. The modelling details and experimental results are shown in the following sections.

17.2 Models for the Prediction of Major Phrase Boundary Locations and Pause Locations

The previous analyses showed that major phrase boundary location and pause insertion are closely related to phrase dependency structure and have similar characteristics[SK92][KS92c]. However, it is quite difficult to automatically derive a phrase dependency structure for a sentence because phrase dependency structure is determined not only by syntax but also by semantics.

The prediction models for major phrase boundary location and pause insertion presented in this chapter both have the same design and training procedure, as illustrated in Figure 17.1. The structural constraints described by hand transcribed phrasal dependencies are approximately

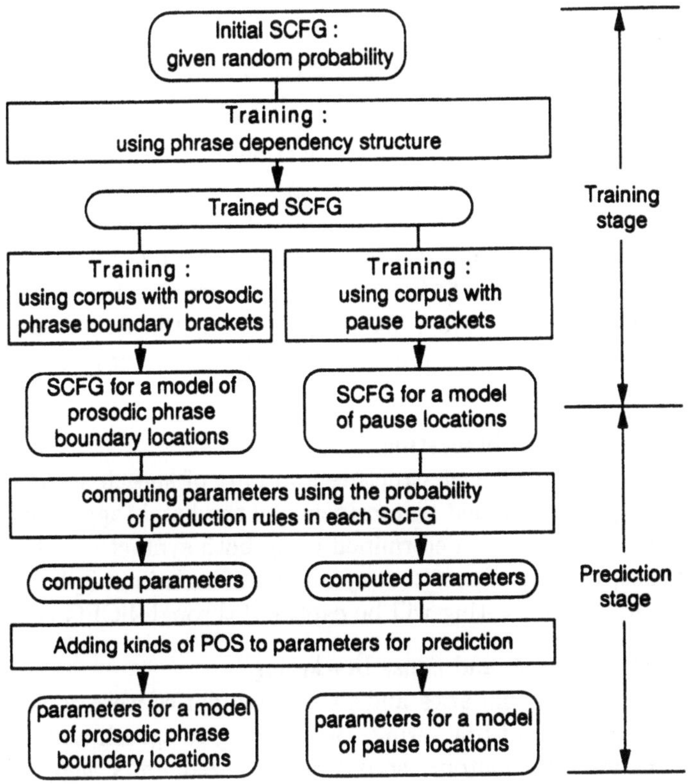

FIGURE 17.1. Flow of generation of the parameters for the model.

captured by SCFGs. The parameters reflecting structural constraints are computed using the probability of production rules in each SCFG, and are then used in models for the prediction of major phrase boundary locations and pause locations. As POS before and after boundaries have commonly been used for predicting major phrase boundary locations and pause locations, they were included as parameters to obtain models. Each model predicts a location using these parameters in conjunction with a feed-forward type neural network, which will be described in Sect. 17.2.4.

17.2.1 Speech Data

503 sentences of the speech data[STA+90] were used for training the SCFG and a neural network. The sentences were read by ten Japanese professional announcers or narrators, giving a total of 5030 sentences which include 70020 boundaries as training and test data. The sentences of speech data were parsed morphemically, bracketed, and labelled for part-of-speech.

17.2.2 Learning Major Phrase Boundary Locations and Pause Locations Using a SCFG

Parameters representing phrase dependency structures are needed for the prediction of major phrase boundary locations and pause locations. We expect that SCFG to capture these structures through training. For the training of the SCFG, an efficient inside–outside algorithm has already been proposed[LY89] and an extension of this has been applied to partially bracketed text corpora[PS92]. We extend this algorithm to learn prosodic phrase dependency structure[SP94]. For training the SCFG using this method, the sentence data were labelled with two types of bracketing information as follows:

(1) Phrase dependency bracketing
The phrase dependency structure and part of speech sequence were hand-tagged by trained transcribers. As the hand-tagged phrase dependency structure is determined using both syntactic and semantic relations, we only expect that syntactic information implicitly manifested in this bracketing will be captured through SCFG training.

(2) Prosodic bracketing and pause bracketing
Accent phrase boundaries and pause boundaries were considered in this second level of bracketing. By listening to speech and observing the analysed F_0 contour, accent-phrase sized units were manually bracketed. For a model of major phrase boundary locations, prosodic phrase bracketing was automatically carried out by finding F_0 resets. Reset is defined as an increase in the F_0 averaged across two successive accent phrases. And, for a model of pause insertion, pause bracketing was obtained by grouping all the constituents segmented by pauses.

Corpora with phrase dependency brackets were first used to train the SCFG from scratch. Next, this SCFG was retrained using the same corpora with prosodic brackets and then with pause brackets.

Determining a set of effective terminal symbols and an appropriate number of non-terminal symbols is important for obtaining a SCFG. Considering the limitations of data size and computational cost, POS and post-positional particles were selected as terminal symbols. Inclusion of post-positional particles was effective for increasing the accuracy of a SCFG, because these particles indicate the syntactic attributes of a phrase. Postpositional particles that occurred more than 50 times in the speech data were used as terminal symbols. As a result, the following four sets of terminal symbols were selected:

(1) POS tags alone (n=23):
13 kinds of content words (e.g., adjective, adverb, verb, auxiliary verb, nominal noun, verbal noun, proper noun, adjectival noun, quantifier and pronoun) and ten kinds of function words (e.g., auxiliary verb,

case particle, conjunctive particle, modal particle, adverbial particle, final particle, adnominal particle, and coordinate particle) are used. Three of the content words and one of the function words are inflectional.

(2) POS tags (n=22) plus the following tags:

   (a) Seven classes of case particles:
   "*ga*", "*no*", "*ni*", "*wo*", "*de*", "*to*" and others.

   (b) Two classes of conjunctive particles:
   "*te*" and others.

   (c) Two classes of modal particles:
   "*wa*" and others.

17.2.3 Computation of Parameters for the Prediction Using a SCFG

We propose two parameters P_m and Q_n to represent phrasal dependencies captured by the SCFG. Figure 17.2 illustrates these two parameters. As shown in Figure 17.2, the left-branching probability P_m represents the probability that a word is part of a left-branching structure which includes the previous m words. Similarly, the right-branching probability Q_n represents the probability that the word is part of a right-branching structure which includes the next n words. These probabilities represent phrase dependency structures, and can be calculated using the inner/outer probabilities which are defined in the inside–outside algorithm in the following fashion.

Let $a[i, j, k]$ be the probability that the non-terminal symbol i will generate the pair of non-terminal symbols j and k. Let $b[i, m]$ be the

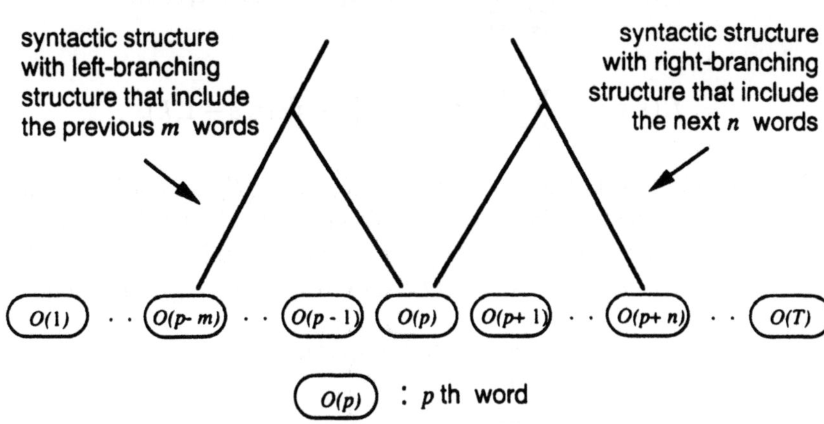

FIGURE 17.2. Syntactic structure with left/right-branching structure.

probability that the non-terminal symbol i will generate a single terminal symbol m. In the inside–outside algorithm[LY89], the inner probability $e(s,t,i)$ is defined as the probability of the non-terminal symbol i generating the observation $O(s),\ldots,O(t)$ and can be expressed as follows:

CASE 1: $s=t$: $\quad e(s,s,i)=b[i,O(s)];$

CASE 2: $s\neq t$:

$$e(s,t,i)=\sum_{j,k}\sum_{r=s}^{t-1}a[i,j,k]e(s,r,j)e(r+1,t,k).$$

The outer probability $f(s,t,i)$ is the probability that, in the rewrite process, i is generated and that the strings not dominated by it are $O(1),\ldots,$
$O(s-1)$ to the left and $O(t+1),\ldots,O(T)$ to the right. Hence:

$$\begin{aligned} f(s,t,i) &= \sum_{j,k}\Big[\sum_{r=1}^{s-1} f(r,t,j)a[j,k,i]e(r,s-1,k) \\ &\quad + \sum_{r=t+1}^{T} f(s,r,j)a[j,i,k]e(t+1,r,k)\Big] \end{aligned}$$

$$and \quad f(1,T,i)=\begin{cases}1 & \text{if } i=S(\text{start symbol})\\ 0 & \text{otherwise}\end{cases}.$$

The non-terminal symbol i can have two possible rules $j\rightarrow ik$ or $j\rightarrow ki$. $f_l(s,t,i)$ is the probability when only the rules $j\rightarrow ik$ are considered and $f_r(s,t,i)$ is the probability when only the rules $j\rightarrow ki$ are considered. These are expressed as follows:

$$f_l(s,t,i)=\sum_{j,k}\sum_{r=t+1}^{T} f(s,r,j)a[j,i,k]e(t+1,r,k),$$

$$f_r(s,t,i)=\sum_{j,k}\sum_{r=1}^{s-1} f(r,t,j)a[j,k,i]e(r,s-1,k).$$

The probability that the observation $O(1),\ldots,O(T)$ has a left-branching structure which includes the observation $O(s),\ldots,O(t)$, and the probability that the observation $O(1),\ldots,O(T)$ has a right-branching structure which includes the observation $O(s),\ldots,O(t)$ are, respectively, given as follows:

$$\sum_i e(s,t,i)f_l(s,t,i),$$

$$\sum_i e(s,t,i)f_r(s,t,i).$$

The probability generated for the entire observation $O(1), \ldots, O(s), \ldots, O(t), \ldots, O(T)$ is $e(1, T, S)$.

Therefore, P_m and Q_n at the pth word are given as follows:

$$P_m = \frac{\sum_i e(p-m, p, i) f_l(p-m, p, i)}{e(1, T, S)},$$

$$Q_n = \frac{\sum_i e(p, p+n, i) f_r(p, p+n, i)}{e(1, T, S)}.$$

17.2.4 Prediction Model Using a Neural Network

A feed-forward neural network was employed to predict major phrase boundary locations and pause locations. For training this neural network, fast back-propagation learning methods[HSWS89] were used. This neural network has four layers: an input layer with 50 units, two hidden layers with 25 units, and an output layer with 2 units.

The input parameters are as follows:

(1) P_m and Q_n at the following words (where m,n=1,2,3,4, and 5 or over):

- **(a)** the content word preceding the word before the boundary;
- **(b)** the word before the boundary;
- **(c)** the word after the boundary; and
- **(d)** the content word following the word after the boundary.

(2) The class of the terminal symbols of the following:

- **(a)** the five words preceding the boundary and
- **(b)** the five words following the boundary.

The output parameters are set to 1 or 0 to mark the presence or absence of a major phrase boundary or pause boundary.

17.3 Experiments

We carried out several experiments to investigate the effect of the numbers of terminal and non-terminal symbols on prediction accuracy and to evaluate the effectiveness of the proposed models.

17.3.1 Learning the SCFG

17.3.1.1 Influence of the Number of Terminal Symbols

To compare the effectiveness of each set of terminal symbols defined in Sect. 17.2.2, the following experiments were carried out. SCFGs were trained separately with the four sets of terminal symbols. In these tests, 15 non-terminal symbols were used in each case. As an index for the evaluation of these SCFGs, the percentage of compatible predicted bracketings[1] for the corpora with bracketings based on F_0 resetting characteristics were computed. Table 17.1 shows the scores of SCFGs. The score of experiments for held-out data in Table 17.1 were obtained as follows. The corpora were divided into ten parts, and nine parts were used for training, while the remaining one was used as a test corpus. Ten experiments were carried out using each part as test corpora, and the average results were computed.

The results show that the inclusion of case particles improved the accuracy of the SCFG. Case particles occur more often than other particles and the precise classification was possible only for case particles. This is thought to be the reason why the case particles improved the accuracy.

17.3.1.2 Influence of the Number of Non-terminal Symbols

As is well known, the training speed of a SCFG using inside–outside algorithm is proportional to $O(N^3)$ where N stands for the number of non-terminal symbols. Though smaller N is desirable from the viewpoint of calculation cost, too small an N causes a decrease of the description ability for the corresponding grammar. To determine N appropriately, several SCFGs with a different number of non-terminal symbols were trained, and the compatibility scores of the SCFGs were evaluated. Each SCFG used the same terminal symbols: 22 POS with seven classes of case particles.

Table 17.2 shows the compatibility scores for the SCFGs using different numbers of non-terminal symbols. The results show that the accuracy of SCFG was gradually improved by increasing the number of non-terminal symbols for the training data. For the test data, the compatibility scores were almost saturated at 15.

17.3.2 Accuracy of the Prediction

17.3.2.1 Prediction of Major Phrase Boundary Locations

The prediction model was trained using 7002 samples in 503 sentences uttered by one speaker. As the major phrase boundary locations differ

---

[1] Compatibility is defined as the ratio of the number of appropriate brackets to the sum of the numbers of appropriate and inappropriate brackets. If there is an overlap between brackets given manually and that predicted by the model like $(a\ b)\ c$ and $a\ (b\ c)$, it is an inappropriate bracket.

TABLE 17.1. Comparison among compatibility scores of SCFGs using different terminal symbols.

Corpus	Terminal symbols	Compatibility score(%)
Training data	23 POS	88.4
	22 POS + 7 classes of case particles	90.5
	22 POS + 2 classes of conjunctive particles	88.7
	22 POS + 2 classes of modal particles	89.5
Test data	23 POS	87.7
	22 POS + 7 classes of case particles	88.3
	22 POS + 2 classes of conjunctive particles	87.6
	22 POS + 2 classes of modal particles	87.5

among speakers, four speaker's utterances were analysed. As major phrase boundaries are found where all speakers reset F_0 and boundaries are not found where no speaker resets F_0, the evaluation of prediction results was carried out for the 680 boundaries where all speakers reset F_0 and the 5261 boundaries where no speaker resets F_0.

The results are shown in Table 17.3, in which experimental results of the prediction using POS only are also shown to confirm the validity of the use of P_m and Q_n.

As the results in column ALL-RESET in Table 17.3 show, major phrase boundaries were predicted successfully using the model which was controlled by the parameters obtained by a SCFG, and this model will be effective for the prediction of major phrase boundary locations. However, as the results in column NO-RESET show that insertion error rate is not so low, and it is necessary to reduce this error rate. The results in column "other boundaries" of NO-RESET show that some of the insertion errors

TABLE 17.2. Comparison of the compatibility scores for SCFGs with different numbers of non-terminal symbols.

Corpus	Number of non-terminal symbols	Compatibility score(%)
Training data	10	86.2
	15	90.5
	20	90.4
	25	91.2
Test data	10	85.3
	15	88.3
	20	88.4
	25	89.1

TABLE 17.3. Prediction accuracy for major phrase boundary locations. ALL-RESET: boundaries where all speakers reset F_0; NO-RESET : boundaries where no speaker resets F_0.

Parameter used	Data	Percentage of correct prediction			
		ALL-RESET	NO-RESET		Total
			Accent phrase boundaries	Other boundaries	
P_m, Q_n, POS	training	99.4 (676/680)	79.2 (1248/1575)	98.1 (4061/4140)	93.6 (5985/6395)
	test	92.9 (632/680)	51.7 (814/1575)	95.1 (3937/4140)	84.2 (5383/6395)
POS only	training	93.4 (635/680)	53.3 (839/1575)	91.4 (3784/4140)	82.2 (5258/6395)
	test	85.3 (580/680)	50.3 (792/1575)	85.8 (3553/4140)	76.0 (4862/6395)

occurred at non-accent phrase boundaries. Though these insertion errors generate unnatural speech synthesis, most of these insertion errors occurred at boundaries within compound nouns. These errors can be reduced by appending compound nouns to the dictionary.

TABLE 17.4. Prediction accuracy for pause locations. All-Pause: boundaries where all speakers insert a pause; No-Pause: boundaries where no speaker inserts a pause.

Parameter used	Data	Percentage of correct prediction			
		ALL-PAUSE	NO-PAUSE		Total
			Accent phrase boundaries	Other boundaries	
P_m, Q_n, POS	training	99.7 (371/372)	80.7 (917/1136)	98.5 (4063/4125)	95.0 (5351/5633)
	test	85.2 (317/372)	67.6 (768/1136)	96.4 (3976/4125)	89.8 (5061/5633)
POS only	training	93.3 (346/372)	53.3 (606/1136)	89.5 (3693/4125)	82.5 (4645/5633)
	test	86.8 (322/372)	52.1 (592/1136)	89.1 (3676/4125)	81.5 (4590/5633)

17.3.2.2 Prediction of Pause Locations

The pause prediction model was also trained using the same 7002 samples in 503 sentences uttered by one speaker. As the pause locations also differ among speakers, ten speaker's utterances were analysed. As with the prediction of major phrase boundaries, the evaluation of prediction results was carried out for the 371 boundaries where all speakers insert a pause and the 5261 boundaries where none of them insert a pause. The results are shown in Table 17.4.

The results in column ALL-PAUSE show that pause locations were predicted successfully using the model which was controlled by the parameters obtained by a SCFG. Though the insertion error rate is quite large, as shown in the column NO-PAUSE, it does not necessarily mean that this frequent pause insertion needs to be reduced. Pausing is obligatory at some phrase boundaries and the lack of a pause would be quite problematic in many cases. On the contrary, extra-pausing at boundaries where subjects do not pause may not be perceived as unnatural. Perceptual characteristics should be analysed and be reflected in further modelling.

Additionally, we checked the similarity between major phrase boundary location and pause location in speech data by substituting the prediction models for an arbitrarily chosen speaker. The prediction model trained by using major phrasing characteristics was used for the prediction of pauses. After training the phrase prediction model, it was tested to predict pauses using the same training data. In this experiment, a high prediction accuracy of 98.1% of the pause boundaries was obtained. This score is higher than the accuracies in open experiments using a pause prediction model trained by different sentences but pausing characteristics. These results suggest high correlation of these two characteristics.

Conclusion

We have presented a computational model for predicting major phrase boundary locations and pause locations without any information of syntactic or semantic bracketings based on phrase dependency structure. These models were designed using neural networks that were given as input parameters a part-of-speech sequence and probability parameters P_m [left-branching probability] and Q_n [right-branching probability], which represent the stochastic phrasing characteristics obtained by SCFGs trained using phrase dependency bracketings and bracketings based on major phrase boundary locations and pause locations.

In a test predicting major phrase boundaries with unseen data, the proposed model correctly predicted 92.9% of the major phrase boundaries with a 16.9% false insertion rate, and 85.2% of the pause boundaries with a 9.1% false insertion rate with unseen data. These results show that the proposed models are effective. Future work should consider a prediction model which includes perceptual characteristics.

Acknowledgments

We would like to thank Dr. Y. Schabes and Dr. F. Pereira for providing the program for inside–outside training.

References

[HFKY90] K. Hirose, H. Fujisaki, H. Kawai, and M. Yamaguchi. Manifestation of linguistic and para-linguistic information in the voice fundamental frequency contours of spoken Japanese. In *Proc. ICSLP*, pp. 485–488, 1990.

[HS80] K. Hakota and H. Sato. Prosodic rules in connected speech synthesis. *Trans. IECE Japan*, J63-D:715–722, 1980 (in Japanese).

[HSWS89] P. Haffner, H. Sawai, A. Waibel, and K. Shikano. Fast back-propagation learning methods for large phonemic neural networks. In *Rec. Spring Meeting, Acoust. Soc. Jpn.*, pp. 27–28, Mar. 1989.

[KS92c] N. Kaiki and Y. Sagisaka. Pause characteristics and local phrase dependency structure in Japanese. In *Proc. ICSLP*, pp. 357–360, 1992.

[LY89] K. Lari and S. J. Young. The estimation of stochastic context-free grammars using the inside–outside algorithm. *Computer Speech and Language*, 4:35–56, 1989.

[PS92] F. Pereira and Y. Schabes. Inside–outside reestimation from partially bracketed corpora. In *Proc. ACL*, pp. 128–135, 1992.

[SK92] Y. Sagisaka and N. Kaiki. Optimization of intonation control using statistical F_0 resetting characteristics. In *Proceedings of the International Conference on Acoustics, Speech, and Signal Processes*, pp. 49–52, 1992.

[SP94] Y. Sagisaka and F. Pereira. Inductive learning of prosodic phrasing characteristics using stochastic context-free grammar. In *Rec. Spring Meeting, Acoust. Soc. Jpn.*, pp. 225–226, Mar. 1994.

[SS95] K. Suzuki and T. Saito. N-phrase parsing method for Japanese text-to-speech conversion and assignment of prosodic features based on N-phrase structures. *Trans. IEICE Japan*, J78-D-II:177–187, Feb. 1995 (in Japanese).

[STA+90] Y. Sagisaka, K. Takeda, M. Abe, S. Katagiri, T. Umeda, and H. Kuwabara. A large-scale Japanese speech database. In *Proceedings of the International Conference on Spoken Language Processing, Kobe, Japan*, pp. 1089–1092, 1990.

Part IV

Prosody in Speech Recognition

18
Introduction to Part IV

Sadaoki Furui

18.1 The Beginnings of Understanding

This section consists of five papers on how to use prosodic information (prosodic features of speech), such as pitch, energy, and duration cues, in automatic speech recognition. As earlier chapters have shown, prosodic information plays an important role in human speech communication. In the last few years, speech recognition systems have dramatically improved, and automatic speech understanding is now a realistic goal. With these developments, the potential role of recognizing prosodic features has become greater, since a transcription of the spoken word sequence alone may not provide enough information for accurate speech understanding; the same word sequence can have different meanings associated with different prosody. Meaning is affected by phrase boundaries, pitch accents, and tone (intonation). For example, phrase boundary placement (detection) is useful in syntactic disambiguation, and tone is useful in determining whether or not an utterance is a yes–no question. In English, there are many noun–verb or noun–adjective pairs in which a change in the word accent indicates a change in the word meaning. Phrase boundary placement is also useful for reducing the search space, that is, reducing the number of calculations in continuous speech recognition.

A variety of algorithms have been proposed for analyzing and recognizing prosodic features. However, the suprasegmental nature of these features poses a challenge to computational modelling, and there still exist various difficulties and problems in using these features. First, we do not have reliable methods to automatically extract prosodic features, such as fundamental frequency contours, from speech waves. Second, we do not have computational (quantitative) models which can precisely model prosodic features and are automatically trainable without using hand-labelled data. Third, researchers regularly disagree on prosodic transcriptions even in carefully articulated speech, partly because prosodic parses can be ambiguous. Fourth, we do not know the best way to combine suprasegmental prosodic measures with segmental phonetic measures extracted from speech in recognition decision.

The first paper, "A multi-level model for recognition of intonation labels," by Mari Ostendorf and Ken Ross describes a new computational model of prosody aimed at recognizing detailed intonation patterns, both pitch accent and phrase boundary location, and their specific tonal markers. This model uses a multi-level hierarchical representation to capture acoustic feature dependence on different time scales. It assumes that each phrase in an utterance is composed of a sequence of syllable-level tone labels which are represented as a sequence of acoustic feature vectors (fundamental frequency and energy) partly depending on the segmental composition of the syllable. The variable lengths are explicitly modelled in a probabilistic representation of the complete sequence, using a dynamical system model at the syllable level that builds on existing models of intonation. Recognition and training algorithms are described, and experimental results for prosodic labelling of carefully read speech are reported. The performance is compared to the consistency among human labellers at this task.

The second paper "training prosody-syntax recognition models without prosodic labels" by Andrew Hunt presents two prosodic recognition models: the canonical correlation analysis (CCA) model and the linear discriminant analysis (LDA) model. They are based on multi-variate statistical techniques that identify a linear relationship between sets of acoustic and syntactic features, and are capable of resolving syntactic ambiguities using acoustic features measured from the speech signal. These models can learn an intermediate representation which is appropriate for both the acoustic and syntactic feature sets. This obviates the need for hand-labelled prosodic data in training. Despite the unsupervised prosodic training, the recognition models achieve accuracies of up to 75% when resolving the syntactic ambiguities of professionally read speech corpora.

The third paper, "Disambiguating recognition results by prosodic features" by Keikichi Hirose proposes a method to use pitch contours to check the feasibility of hypotheses in Japanese continuous speech recognition. In this method a pitch contour is generated top-down for each recognition hypothesis and compared with the observed contour. The pitch contour is generated based on prosodic rules formerly developed for text-to-speech conversion. The comparison is performed only for ambiguous hypotheses, and the hypothesis with the contour that best matches the observed one is selected. This method can detect recognition errors accompanied by wrong accent types and/or syntactic boundaries. The method is also evaluated in terms of its performance for the detection of phrase boundaries.

The fourth paper, "Accent phrase segmentation by $F0$ clustering using superpositional modelling," by Mitsuru Nakai *et al.* proposes an automatic method for detecting accent phrase boundaries in Japanese continuous speech by using pitch information. In the training phase, hand-labelled accent patterns are parameterized using the superpositional model proposed by Fujisaki and clustered to make accent templates which are represented by the centroid of each cluster. In the segmentation phase, N-best boundaries are automatically detected by a one-stage DP matching procedure between a sequence of reference templates and the pitch contour. Bigram probabilities of accent phrases are successfully used as additional information, since there is a strong correlation between adjacent templates. It is reported that 90% of accent phrase boundaries are correctly detected in speaker independent experiments using a continuous speech database.

The last paper, "Prosodic modules for speech recognition and understanding in VERBMOBIL" by Wolfgang Hess *et al.* describes the computational modules that were developed in the framework of the German spoken language project VERBMOBIL. The project deals with automatic speech-to-speech translation of appointment scheduling dialogs. The prosodic modules detect phrase boundaries, sentence modality, and accents using pitch and energy extracted from speech waves. They are designed to work in both bottom-up and top-down methods, and are trained using prosodically labelled corpora. An accuracy of 82.5% was obtained for unaccented vs accented syllables and an accuracy of 91.7% was obtained for phrase boundary detection in spontaneous dialogs. Although these results are considerably lower than those obtained for read speech, the presence of the modules improves the performance of speech understanding systems. Also the incorporation of prosodic information considerably reduces the number of parse trees in the syntactic and semantic modules and thus decreases the overall search complexity.

Although quite a large number of papers, including these five, have reported the usefulness of prosodic features in speech recognition, there is no advanced large-vocabulary continuous speech recognition system that really uses these features. No method in this section has yet been implemented in continuous speech recognition systems. The major reason for this is that the reliability of these features is still not high enough. Even in the basic process of pitch extraction, we always encounter the serious problem of half and double pitch periods. It is crucial to establish reliable methods for extracting and stochastically modelling these features. It is also important to evaluate the methods using a large database of speech corpora, since speech intrinsically has a wide variability. The methods proposed in this section also need extended evaluation using large corpora.

The development of sophisticated, statistically well-formed models should improve the handling of speaker and context variability in prosody. Such models require large amounts of training data. It is unlikely that such large corpora can be labelled by hand. Therefore, how to automatically train (learn) the model parameters using an unlabelled speech database is one of the key issues. This is analogous to the use of Viterbi training of hidden Markov models for phones used in large-vocabulary continuous speech recognition. If automatic labelling of prosodic markers is possible, it can facilitate corpus collection, and a large corpus of prosodically labelled speech can facilitate further research on the mapping between meaning and prosodic labels, which is still not fully understood and is needed for improved speech understanding systems.

19
A Multi-level Model for Recognition of Intonation Labels

M. Ostendorf
K. Ross

ABSTRACT Prosodic patterns can be an important source of information for interpreting an utterance, but because the suprasegmental nature poses a challenge to computational modelling, prosody has seen limited use in automatic speech understanding. This work describes a new computational model of prosody aimed at recognizing detailed intonation patterns, both pitch accent and phrase boundary location and their specific tonal markers, using a multi-level representation to capture acoustic feature dependence at different time scales. The model assumes that an utterance is a sequence of phrases, each of which is composed of a sequence of syllable-level tone labels, which are in turn realized as a sequence of acoustic feature vectors (fundamental frequency and energy) depending in part on the segmental composition of the syllable. The variable lengths are explicitly modelled in a probabilistic representation of the complete sequence, using a dynamical system model at the syllable level that builds on existing models of intonation. Recognition and training algorithms are described, and initial experimental results are reported for prosodic labelling of radio news speech.

19.1 Introduction

In the last few years, speech recognition systems have improved dramatically, so that automatic speech understanding is now a realistic goal and an active area of research. With these developments, recognition of prosodic patterns has become more important, since a transcription of the spoken word sequence alone may not provide enough information for accurate speech understanding: the same word sequence can have different meanings associated with different tunes. Meaning is affected by the *placement* of intonation markers, i.e., phrase boundaries and pitch accents, but also by the specific *tone* associated with these markers. For example, phrase boundary placement is useful in syntactic disambiguation, while the phrase boundary tone is useful in determining whether or not an utterance is a yes–no question. Therefore, the goal of this work is to automatically recognize specific tone labels, as well as accent and phrase boundary placement.

A variety of algorithms have been proposed for analyzing prosodic contours and recognizing/labelling abstract prosodic features. Excluding the rule-based methods, since we take as a premise that a model for intonation recognition should be automatically trainable, the different approaches can be grouped into two classes: those that model complete fundamental frequency (F_0) contours and those that use a transformation of local F_0 patterns and other cues given an utterance segmentation. The complete F_0 models include hidden Markov models (HMMs) [LF87, BOPSH90, CW92, JMDL93], and minimum mean-squared error techniques such as template matching [NSSS95] and analysis-by-synthesis [Geo93, HHS95]. These models work well for representing F_0 contours, but have the disadvantage that duration cues, known to be very important, are effectively ignored. This disadvantage can be overcome by assuming knowledge of syllable or other segmentation times and using a transformation of the variable-length observations in the syllable to obtain the feature set used in prosody recognition, as in [WO94, tB93b, KBK$^+$94, Cam94a]. Although the feature transformation approach yields good performance for automatic labelling given a known word sequence, it has the disadvantage that it is difficult to make direct comparisons of hypotheses with different segmentations (see [ODK96] for a discussion of this problem), which is a critical part of the recognition problem for unknown word sequences.

Our approach, like those based on hidden Markov models, uses a stochastic model that represents the complete sequence of acoustic observations (energy and F_0) given a sequence of abstract prosodic labels. However, like the transformation models, it is able to take advantage of duration cues explicitly by using a stochastic segment model [ODK96, OR89] to represent observation sequences in a syllable. Thus, it combines advantages of both approaches. In addition, the model represents acoustic feature dependence with a hierarchy of levels (segment, syllable, phrase) to account for interactions among factors operating at different levels. In principle, the multi-level model enables more accurate prosody recognition, albeit at a higher computational cost. The model uses the notions of accent filtering and superposition, building on the well-known Fujisaki model [HF82, FK88], but to facilitate use in recognition it includes additive Gaussian terms so that it is a probabilistic model. In addition, the model includes more free parameters than the Fujisaki model to accommodate the phonological theory of English intonation described by Beckman and Pierrehumbert [BP86].

The remainder of the paper is organized as follows. Sec. 19.2 describes the model structure and parameter estimation algorithm. The recognition search algorithm, which involves two levels of dynamic programming, is outlined in Sec. 19.3. Experimental results are presented in Sec. 19.4 for a prosodic labelling task using a speaker-dependent radio news corpus.

Finally, in Sec. 19.5, we discuss areas for further improvement of the intonation model, and its potential uses in automatic speech understanding.

19.2 Tone Label Model

The general approach to intonation modelling described here reflects to some extent the hierarchical representation of intonation in many linguistic theories. That is, we represent both phrase-level phenomena and the linear sequence of tones that occurs on syllables within the phrase, as well as finer-grained segmental effects. This multi-level model is described below, followed by discussions of the two main types of component models: acoustic and phonotactic.

19.2.1 Multi-level Model

The intonation model assumes that an utterance is a sequence of phrases, each of which is composed of a sequence of syllable-level tone labels, which are in turn realized as a sequence of acoustic feature vectors. Taking a probabilistic approach similar to that typically used in speech recognition, our goal is to find the jointly most likely combination of accent and phrase tones given the acoustic observations and optionally the word sequence $\mathcal{W}$. To be more explicit, let us denote $y_1^T = \{y_1, \ldots, y_T\}$ as the frame-based observation sequence of F_0 and energy parameters and optionally their derivatives, $s_1^N = \{s_1, \ldots, s_N\}$ as the sequence of syllable duration information that is available from a recognizer given an hypothesized word sequence, and $\gamma_1^N = \{\gamma_1, \ldots, \gamma_N\}$ as the corresponding sequence of segmental characteristics of each syllable. The segmental characteristics γ_i of the i-th syllable are encoded in a variable-length string of phone class identifiers, e.g. (voiced obstruent, vowel, sonorant consonant) or (vowel, voiceless obstruent). Given this information, we would like to recognize the tone label sequence $\alpha_1^N = \{\alpha_1, \ldots, \alpha_N\}$, i.e., one label per syllable, and the phrase boundary positions indicated by $\beta_1^M = \{\beta_1, \ldots, \beta_M\}$. The maximum likelihood recognition rule is then

$$\begin{aligned}(\hat{\alpha}_1^N, \hat{\beta}_1^M) &= \operatorname*{argmax}_{\alpha_1^N, \beta_1^M} \mathrm{P}(\alpha_1^N, \beta_1^M | y_1^T, s_1^N, \gamma_1^N, \mathcal{W}) \\ &= \operatorname*{argmax}_{\alpha_1^N, \beta_1^M} \mathrm{P}(y_1^T, s_1^N, \alpha_1^N, \beta_1^M | \gamma_1^N, \mathcal{W}), \qquad (19.1)\end{aligned}$$

(Note that β_1^M is redundant given α_1^N, since phrase tones are included in the set of values for α_i, but the discussion is simplified if we explicitly indicate phrases.) In Sec. 19.3 we discuss the solution of this maximization equation; here we present the details of the models that it is based on.

To simplify the general model for practical implementation purposes, it is necessary to make some Markov and conditional independence assumptions. To balance this requirement with the observation that prosodic cues interact over long time scales (i.e., are "suprasegmental"), the Markov assumptions are introduced at multiple levels of a hierarchy. For example, we model dependence of pitch range across phrases to capture discourse-related downtrends in F_0, but assume that there is no dependence across phrases for tone labels as is consistent with most phonological theories of intonation. In addition, we assume that the symbolic phrase and tone variables β_1^M and α_1^N are conditionally independent of the segmental composition of syllables γ_1^N, given the word sequence $\mathcal{W}$.

To simplify the discussion, let the underbar notation denote a subsequence, e.g., $\underline{\alpha}(\beta_i)$ is the sequence of labels α_j that comprise phrase β_i. The first step involves Markov assumptions at the phrase level:

$$\begin{aligned} \mathrm{P}(y_1^T, s_1^N, \alpha_1^N, \beta_1^M | \gamma_1^N, \mathcal{W}) &= \mathrm{P}(\beta_1^M | \gamma_1^N, \mathcal{W})\, \mathrm{P}(y_1^T, s_1^N, \alpha_1^N | \beta_1^M, \gamma_1^N, \mathcal{W}) \\ &= \prod_{i=1}^{M} \mathrm{P}(\beta_i | \beta_{i-1}, \mathcal{W})\, \mathrm{P}(\underline{y}(\beta_i), \underline{s}(\beta_i), \underline{\alpha}(\beta_i) | \\ &\qquad \beta_i, \beta_{i-1}, \underline{y}(\beta_{i-1}), \underline{\gamma}(\beta_i), \mathcal{W}), \end{aligned} \tag{19.2}$$

where in addition we assume that phrase-level durations $\underline{s}(\beta_i)$ and tone sequences $\underline{\alpha}(\beta_i)$ are conditionally independent of the subsequences in the previous phrases. Next, decompose the frame-based F_0 and energy observations into a normalized vector contour and a pitch range variable, $\underline{y}(\beta_i) = (\underline{\tilde{y}}(\beta_i), y_i^p)$, where range y_i^p is measured as the peak F_0 value for the phrase and

$$\tilde{y}_t = \frac{y_t - y^b}{y_i^p - y^b} \quad \text{for all } y_t \in \underline{y}(\beta_i).$$

The pitch baseline y^b is speaker-dependent, or constant over a time period that is much longer than a phrase, so it can be treated as an adaptable model constant. Assuming that the only dependence of variables across phrases is in the pitch range term and that pitch range depends on the word sequence (e.g., discourse factors) but not segmental characteristics, then

$$\begin{aligned} &\mathrm{P}(\underline{y}(\beta_i), \underline{s}(\beta_i), \underline{\alpha}(\beta_i) | \beta_i, \beta_{i-1}, \underline{y}(\beta_{i-1}), \underline{\gamma}(\beta_i), \mathcal{W}) \\ &\quad = \mathrm{P}(y_i^p | \beta_i, \beta_{i-1}, y_{i-1}^p, \mathcal{W})\, \mathrm{P}(\underline{\tilde{y}}(\beta_i), \underline{s}(\beta_i), \underline{\alpha}(\beta_i) | \beta_i, \underline{\gamma}(\beta_i), \mathcal{W}). \end{aligned} \tag{19.3}$$

Next, we assume that tone labels α_j are conditionally Markov given the phrase structure, and that the syllable-level observations $\underline{\tilde{y}}(\alpha_j)$ and durations s_j are conditionally independent given the tone information, i.e.,

$$\mathrm{P}(\underline{\tilde{y}}(\beta_i), \underline{s}(\beta_i), \underline{\alpha}(\beta_i) | \beta_i, \underline{\gamma}(\beta_i), \mathcal{W})$$

$$= \prod_{j:\alpha_j \in \underline{\alpha}(\beta_i)} \mathrm{P}(\alpha_j|\alpha_{j-1},\beta_i,\mathcal{W})\,\mathrm{P}(\underline{\tilde{y}}(\alpha_j)|s_j,\alpha_j,\gamma_j)\,\mathrm{P}(s_j|\alpha_j,\gamma_j). \tag{19.4}$$

Combining equations 19.2-19.4, gives the complete model

$$\mathrm{P}(y_1^T, s_1^N, \alpha_1^N, \beta_1^M|\gamma_1^N, \mathcal{W}) = \prod_{i=1}^{M} \mathrm{P}(\beta_i|\beta_{i-1},\mathcal{W})\,\mathrm{P}(y_i^p|\beta_i,\beta_{i-1},y_{i-1}^p,\mathcal{W})$$
$$\cdot \prod_{j:\alpha_j \in \underline{\alpha}(\beta_i)} \mathrm{P}(\alpha_j|\alpha_{j-1},\beta_i,\mathcal{W})\,\mathrm{P}(\underline{\tilde{y}}(\alpha_j)|s_j,\alpha_j,\gamma_j)\,\mathrm{P}(s_j|\alpha_j,\gamma_j). \tag{19.5}$$

Equation (19.5) shows that the model includes two distributions to represent prosodic label phonotactics $\{\mathrm{P}(\beta_i|\beta_{i-1},\mathcal{W}), \mathrm{P}(\alpha_j|\alpha_{j-1},\beta_j,\mathcal{W})\}$ and three distributions to describe the acoustic characteristics of prosodic events $\{\mathrm{P}(y_i^p|\beta_i,\beta_{i-1},y_{i-1}^p,\mathcal{W}), \mathrm{P}(\underline{\tilde{y}}(\alpha_j)|s_j,\alpha_j,\gamma_j), \mathrm{P}(s_j|\alpha_j,\gamma_j)\}$. The introduction of the phonotactic model is analogous to the use of a language model in speech recognition and improves recognition performance relative to a model based on the acoustics alone.

Note that both acoustic and phonotactic models include terms at both the phrase and syllable level, and the acoustic model also includes conditioning terms from the segment level. It is particularly important for the acoustic model to have all these conditioning terms in order to capture the interactions of factors at different time scales. Individually, the types of simplifying assumptions made here are similar to those used in speech recognition with HMMs and n-gram language models. However, since these assumptions are made at multiple time scales and are *not* made at the frame level, the model is much more powerful than an HMM. Of course, such complexity is only practical because the feature dimensionality and intonation symbol set is relatively small. The optional word sequence conditioning, first proposed in [OVHM94], also complicates the component models by introducing a large number of conditioning events to account for. However, the dimensionality of the conditioning space can be reduced using decision tree distribution clustering as in [RO96]. Thus, the complexity of the full model is much less than most continuous word recognition systems, and it would therefore be a small additional cost in a speech understanding system. Moreover, since the phonotactic models have complexity similar to existing accent and phrase prediction algorithms in synthesis [Hir93a, WH92], the model described by Eq. (19.5) is only slightly more complex than existing prosody synthesis systems.

19.2.2 Acoustic Models

As shown above, the acoustic realization of a sequence of tones is represented with three component models: phrase-level pitch range, normalized F_0 contour over a syllable, and syllable duration. Each component is described in further detail below.

19.2.2.1 Pitch Range

As shown in Eq. (19.3), pitch range y_i^p for the i-th phrase is separated out to remove this large source of variability from the contour model. The variability of pitch range is then represented in the term $\text{p}(y_i^p|\beta_i, \beta_{i-1}, y_{i-1}^p, \mathcal{W})$. Conditioning on the previous phrase type and pitch peak is included since many researchers have observed trends of decreasing pitch range within paragraphs and even sentences. In addition, as researchers develop a better understanding of how pitch range behaves as a function of the discourse structure and as simple algorithms for extracting discourse segmentation from text become available, we would want to incorporate text dependencies to allow for variations other than simple downward trends. In order to condition the distribution on a variety of factors, we prefer a parametric distribution model. Since $y_i^p > 0$, it might be reasonably described with a Gamma distribution, or $\log y_i^p$ could be described by a Gaussian distribution. We choose the Gaussian approach, since it simplifies the problem of conditioning on the previous phrase peak y_{i-1}^p. In the experiments reported here, however, we have not incorporated this probability term since the phrase boundaries are given and pitch range is mainly important for evaluating hypothesized phrase boundaries.

19.2.2.2 Contour Model

For the syllable-level model of the F_0/energy vector contour, $\text{p}(\underline{\tilde{y}}(\alpha_j)|s_j, \alpha_j, \gamma_j)$, we use a stochastic segment model [ODK96, OR89] to represent syllable-sized sequences of vector observations $y_1^l = [y_1, \ldots, y_l]$, where l is the (random) length of the syllable. Here, an observation y_t is a two-dimensional vector of F_0 and energy at time t. In a segment model, the probability of the observation sequence given label α and length l, $p(y_1, \ldots, y_l|l, \alpha)$, is characterized by a family of Gaussian distributions that vary over the course of the segment. Two components must be specified in order to define the family for all possible lengths l: the vector observation process model and the intra-segmental timing function. An alternative to hidden Markov models, the stochastic segment model has the advantage of more explicitly representing the time trajectory of a sequence of vectors and can be thought of as a sort of probabilistic version of template matching.

The particular version of the segment model used here is a state-space dynamical system, which was originally proposed for phone modelling [DRO93] and is used here for intonation modelling to represent syllable-level units. The dynamical system model represents a sequence of observation vectors y_t in terms of a hidden trajectory x_t, i.e.,

$$x_{k+1} = F_t x_t + u_t + w_t, \tag{19.6}$$

$$y_t = H_t x_t + b_t + v_t. \tag{19.7}$$

The hidden trajectory is characterized by a sequence of Gaussian observations associated with target values (u_t) and smoothed by a first-order filter (F_t). The Gaussian vectors w_t, which allow for modelling error, have zero mean and covariance matrix Q_t. This trajectory is then scaled (H_t) and combined with deterministic and random additive terms (b_t and v_t, respectively), where the random sequence v_t can be thought of as observation noise (e.g., pitch tracking error) and is associated with a covariance matrix R_t. Looking at the model from the perspective of Eqs. (19.6) and (19.7), i.e., as a generative process, it is similar to many source- or target-filtering intonation models used in synthesis [HF82, FK88, APL84, Sil87] except that it includes the random terms w_t and v_t. By making the model explicitly stochastic, we can benefit from existing maximum likelihood training and recognition search algorithms.

To use this model for intonation recognition, each prosodic label type α is associated with a set of parameters $\Theta_\alpha = \{F_r, u_r, Q_r, H_r, b_r, R_r; r \in \mathcal{R}_\alpha\}$ that vary over different regions r of the syllable. The set of regions $\mathcal{R}_\alpha$ provide for changes in the distribution characteristics over time, depending on the particular label α. The number of regions per syllable should be at least three to handle accent-phrase tone combinations (e.g., H-L-H targets); here we used six. The mapping of observations to these regions are specified by a timing function, which must capture the change in target shape over the course of the syllable, or alternatively the relative location of the target maximum or minimum. For phone modelling in speech recognition, using a small number of regions with a linear time mapping works well. However, for syllable modelling, where timing depends on the number of phones in the syllable as well as their identity, the linear warping is not a reasonable approximation. Here, we use a mapping that is linear outward from the vowel center, which seems to work reasonably well. Further study of intra-syllable timing, as in [vS96], would be invaluable for improving this component of the model.

Traditional techniques for finding the dynamical system parameters are difficult to use here, particularly since there are missing observations in the unvoiced regions. For this work, we used a new iterative method for maximum likelihood (ML) parameter estimation that relies on an algorithm developed by Digalakis *et al.* [DRO93] for speech recognition. This approach uses the two-step, iterative expectation-maximization algorithm [DLR77] for ML parameter estimation of processes that have unobserved components, which in this case are the state vectors and unvoiced data. During the first, or expectation step, the expected first- and second-order sufficient statistics for the unobserved data are estimated conditioned on the observed data and the most recent parameter estimates. These quantities are used to find the next set of parameter estimates in the second step, which maximizes the expected log-likelihood of hidden and observed data. These two steps are repeated until the relative change in likelihood of the model is small.

Because of training problems with limited data and local optima, as well as recognition search complexity, it is not practical to make the full set of parameters $\{F_r, u_r, Q_r, H_r, b_r, R_r\}$ dependent on all levels of the model (segment, syllable, and phrase), and so parameter tying is used. Here, parameter dependence is specified according to linguistic insights, but tying might also be determined by automatic clustering. To improve the accuracy of the model and capture contextual timing and segmental effects, each syllable-level model is represented by a sequence of six regions that have different model parameters, and models are conditioned on the prosodic context (analogous to triphones in speech recognition). Segmental phonetic effects, which can be incorporated if a recognition hypothesis is available, are included as tone-label-independent terms to avoid a significant increase in the number of parameters. For example, H_r and b_r are conditioned on the broad phonetic class of phones in the region of the syllable to capture effects of vowel intrinsic pitch and F_0 movements due to consonant context. The effect of phrase position is incorporated by conditioning the target values and timing on the position of the syllable in the phrase (beginning, middle or end). Further details on the parameter dependencies are described in [RO94].

19.2.2.3 Syllable Duration Model

Previous work has shown that patterns of duration lengthening within a syllable can be an important cue to phrase boundaries and pitch accents [Cam93a, Cam94a, WO94]. Syllable duration varies as a function of a number of factors, including the syllable-level tone label as well as segmental composition. Several different theories have been proposed to account for the interaction of these factors, from syllable-level [CI91, Cam92c] to segment-level [vS94a] to sub-phonetic timing control [vS96]. Any of these theories can be accommodated in the duration model $p(s_j|\alpha_j, \gamma_j)$. For example, consider a syllable comprised of K phones, in which case $s_j = [d_{j,1}, \ldots, d_{j,K}]$ is a vector of phone durations $d_{j,k}$ and $\gamma_j = [\gamma_{j,1}, \ldots, \gamma_{j,K}]$ is a vector of phone (or phone class) labels. Two models can be defined using Gamma distributions either for each segment duration or for the total syllable duration. Let $p(x) \sim G(\rho, \lambda)$ denote that $p(x) = \frac{\lambda^{\rho+1}}{\Gamma(\rho+1)} x^{\rho} e^{-\lambda x}$, where $\Gamma(\cdot)$ is the gamma function. For purposes of parameter tying, note that scaling the first parameter $\rho' = C\rho$ results in a scaling of both the mean and the variance of the distribution. A phone-level model that assumes conditional independence of durations given the segmental and prosodic contexts would be

$$p(s_j|\alpha_j, \gamma_j) = \prod_{k=1}^{K} p(d_{j,k}|\alpha_j, \gamma_{j,k})$$

$$\text{where } p(d_{j,k}|\alpha_j, \gamma_{j,k}) \sim G(C_{\alpha_j}\rho(\gamma_{j,k}), \lambda(\gamma_{j,k})) \qquad (19.8)$$

assuming that the effect of prosodic context corresponds to a simple scaling of an inherent duration associated with segmental (or triphone) identity. More sophisticated models can be constructed that do not have too many additional free parameters by using results from synthesis, e.g., [vS94a]. A syllable-level model might be

$$p(s_j|\alpha_j, \gamma_j) = p(l_j|\alpha_j, \gamma_j) \sim G(\mu_{\gamma_j}^{-1} \rho(\alpha_j), \lambda(\alpha_j)), \qquad (19.9)$$

where $l_j = \sum_k d_{j,k}$, $\mu_{\gamma_j} = \sum_k \mu_{\gamma_{j,k}}$, and we assume that the inherent duration due to prosodic context is scaled according the segmental composition of the syllable. Again, the model can be made more sophisticated by using results from synthesis research. Clearly, there are several alternatives for the duration model, depending on the theory of timing that one adheres to. With the theory-neutral goal of minimum recognition error rate, we plan to test both classes of models, though the results reported here use the syllable-level model described by Eq. (19.9).

19.2.3 Phonotactic Models

The phonotactic models can be easily represented by an n-gram model or a probabilistic finite state grammar, also the most common approaches used in speech recognition. For example, in the experiments here we define a grammar of allowable tone sequences using Beckman and Pierrehumbert's theory of intonation [BP86], and estimate the grammar transition probabilities using smoothed maximum likelihood estimates. This is essentially a bigram grammar

$$p(\underline{\alpha}(\beta_i), \beta_i) = \prod_{j:\alpha_j \in \underline{\alpha}(\beta_i)} p(\alpha_j|\alpha_{j-1}, \beta_i),$$

with the grammatical constraint that a phrase accent must precede a boundary tone. The dependence on β_i also provides the grammatical constraint that the type of boundary tone depends on whether the phrase boundary corresponds to an intermediate or full intonational phrase. For problems where the word sequence is known, more accurate results can be obtained by conditioning the tone label on the word sequence $\mathcal{W}$, e.g., using

$$p(\underline{\alpha}(\beta_i), \beta_i|\mathcal{W}) = \prod_{j:\alpha_j \in \underline{\alpha}(\beta_i)} p(\alpha_j|f(\alpha_{j-1}, \beta_i, \mathcal{W})),$$

where $f(\alpha_{j-1}, \beta_i, \mathcal{W})$ represents an equivalence class of conditioning events, which can be learned, e.g., by the decision tree accent prediction model described in [RO96]. The phrase phonotactic model can be similarly defined, again using decision tree distribution clustering if the optional word sequence conditioning is used. The phrase phonotactic model describes the likelihood of full vs intermediate phrase boundaries in sequence.

19.3 Recognition Search

Recognition with stochastic models generally involves finding the most likely label sequence $\hat{\alpha}_1^n$ as outlined in Eq. (19.1). Using the framework given by Eq. (19.5), the maximization becomes

$$\max_{\alpha_1^N, \beta_1^M} p(y_1^T, s_1^N, \alpha_1^N, \beta_1^M | \gamma_1^N, \mathcal{W})$$

$$= \max_{M, \beta_1^M} \prod_{i=1}^{M} p(\beta_i | \beta_{i-1}, \mathcal{W}) p(y_i^p | \beta_i, \beta_{i-1}, y_{i-1}^p, \mathcal{W})$$

$$\cdot \left[\max_{N_i, \underline{\alpha}(\beta_i)} \prod_{j:\alpha_j \in \underline{\alpha}(\beta_i)} p(\tilde{\underline{y}}(\alpha_j) | s_j, \alpha_j, \gamma_j) p(s_j | \alpha_j, \gamma_j) \right.$$

$$\left. p(\alpha_j | \alpha_{j-1}, \beta_j, \mathcal{W}) \right], \tag{19.10}$$

where N_i is the number of syllables in phrase β_i. Again, the word sequence conditioning factor $\mathcal{W}$ is optional. With Markov assumptions, the maximization can be implemented efficiently using dynamic programming. However, the double maximization in the above equation illustrates that when phrase-level parameters are used a two-level dynamic programming algorithm is needed: one level for hypothesized phrase boundaries, and a second level embedded within the phrase scoring routine to find the most likely tone sequence. At the tone level, the probability of a syllable is computed as the product of the probability of the innovation sequence [DRO93], based on Kalman filtering techniques that are standard in statistical signal processing. In unvoiced regions, the F_0 values are treated as missing data and the innovations are computed from the energy observations.

The two-level dynamic programming algorithm is potentially very expensive, if both syllable and phrase boundaries are hypothesized at every time frame. However, there are several options for reducing the search space by restricting the set of candidate syllable and phrase boundaries. First, consider that there are two types of recognition problems: automatic labelling, where the word sequence is known, and recognition of intonation markers for an unknown word sequence. In automatic labelling, the search cost is much less because the syllable boundaries are given. In recognition applications, the set of possible syllable boundaries can be restricted by a preliminary pass of phone recognition or via N-best word sequence hypothesis rescoring [OKA+91], both of which also provide segmental context for the models. Given a set of possible syllable boundaries, the subset of possible phrase boundaries can be restricted by using detected pauses and local F_0 peaks. For example, one could hypothesize new phrase boundaries when a local F_0 peak has low probability in the current phrase, i.e., an implausible amount of upstep, taking an approach analogous to the search space reduction technique proposed by Geoffrois [Geo93] for the Fujisaki model. A more sophisticated approach would be to use a first-pass phrase boundary detection algorithm, e.g. [NSSS95, Hub89, WO94].

19.4 Experiments

Prosodic labelling experiments were conducted using data from a single speaker (F2B) from the Boston University radio news corpus [OPSH95b], a task chosen to facilitate comparison with other results. Approximately 48 and 11 minutes of data were available for training and testing, respectively. Energy and F_0 contours were computed using Waves+ 5.0 software, and the model was trained with 10 iterations of the EM algorithm.

The corpus was prosodically labelled using the ToBI system [SBP+92], but because of sparse data for infrequently observed tone types, the ToBI tone labels were grouped into four types of accent labels ("unaccented", "high", "downstepped high", and "low"), two intonational phrase boundary tone combinations (L–L% and a few H–L% grouped as "falling", and L–H% or "rising"), and the standard three intermediate phrase accents (L–, H–, and !H–). Since a single syllable can have a combination of accent and boundary tones, the total number of possible syllable labels α is 24, though a larger set of models (roughly 600) is used here by conditioning on stress level and neighboring prosodic label context. The available training data seemed sufficient for robust training of these models, based on comparison of training and test F_0 prediction errors, although additional data would be useful to model a larger number of tone types.

In the results reported below, we compare performance to the consistency among human labelers at this task, to provide some insight into the difficulty of this task. Unlike orthographic transcription, where human disagreement of word transcriptions is rare even in noisy and casual speech, disagreements in prosodic transcriptions occur regularly even in carefully articulated speech, in part because prosodic "parses" can be ambiguous just as syntactic parses can be [Bec96a].

Since the task here was prosodic labelling, a good estimate of word and phone boundaries can be obtained using automatic speech recognition constrained to the known word sequence, and this information is used in controlling the model parameters and reducing the search space. In these preliminary experiments, the problem is also simplified by using hand-labelled intermediate phrase boundary placement rather than hypothesized phrase boundaries, so the results give a somewhat optimistic estimate of performance. However, the only word sequence information used so far is lexical stress in that pitch accents are not recognized on reduced syllables, and the duration model is rather simplistic.

Testing the model with the independent test set but known intermediate phrase boundaries results in recognition accuracy of 85% for the four classes of syllables, which corresponds to 89% accuracy (or, 84% correct vs 9% false detection) for accent location irrespective of specific tone label. These figures are close to the consistency among human labelers for this data, which is 81% accuracy for tone labels that distinguish more low tone categories and 91% for accent placement [OPSH95b]. A confusion matrix is given in Table 19.1. Not surprisingly, the down-stepped accents are frequently confused with both high accents and unaccented syllables. Low tones are rarely recognized because of their low prior probability. Although the results are not directly comparable to previous work [WO94] because of the additional side information used here and differences in the test sets, it is gratifying to see that improved accent detection accuracy is obtained in our study.

Phrase tone recognition results, for the case where intermediate phrase boundaries is known, are summarized in Table 19.2. The overall 5-class recognition accuracy is 63%, with the main difficulty being the distinction between intermediate vs intonational phrase boundaries (79% accuracy). Since the use of a relatively simple duration model led to a reduction of error rate of over 20% (from 73% accuracy), it is likely that further improvements can be obtained with a more sophisticated model. However, even with more reliable recognition of phrase size, there is room for improvement in tone recognition, since human labelers label L% vs H% with consistency of 93% [OPSH95b] (vs. 85% for the automatic labelling). It may be that human labelers are less sensitive than the automatic algorithm to phrase-final glottalization (or creak), which we know is frequent in this corpus. Or, it may simply be a matter of improving the timing function, which currently does not distinguish phrase-final syllables as different from phrase-internal syllables. The phrase tone !H– is rarely recognized correctly, but the human labelers were much less consistent in marking this tone as well.

TABLE 19.1. Confusion table of hand-labelled vs recognized pitch accents for a test set of 3366 syllables.

	Hand-labelled			
Recognized	Unaccented	High	Downstepped	Low
Unaccented	91% (2120)	7% (52)	25% (57)	63% (52)
High	7% (157)	89% (644)	39% (89)	17% (14)
Downstepped	2% (50)	3% (23)	35% (80)	15% (12)
Low	0% (5)	1% (5)	1% (2)	5% (4)

TABLE 19.2. Confusion table of hand-labelled vs recognized phrase tones, given intermediate phrase boundaries for a test set of 596 syllables. "I" indicates an intonational phrase boundary and "i" indicates an intermediate phrase boundary.

	Hand-labelled				
Recognized	I: falling	I: rising	i: L–	i: H–	i: !H–
I: falling	88% (230)	24% (38)	53% (55)	7% (3)	4% (1)
I: rising	7% (19)	62% (98)	16% (17)	22% (10)	15% (4)
i: L–	2% (6)	4% (7)	19% (20)	7% (3)	11% (3)
i: H–	1% (2)	10% (16)	9% (9)	61% (28)	63% (17)
i: !H–	1% (3)	0% (0)	3% (3)	4% (2)	7% (2)

19.5 Discussion

In summary, we have described a new stochastic model for recognition of intonation patterns, featuring a multi-level representation and using a parametric structure motivated by linguistic theory and successful intonation synthesis models. The formulation of the model incorporates two key advances over previous work. First, it uses a stochastic segment model to combine the advantages of feature transformation and frame-based approaches to intonation pattern recognition. Like the transformation approaches, F_0, energy and duration cues are used together in the model, but these observations are modelled directly as with the frame-based approaches. Second, its use of a hierarchical structure facilitates separation of the effects of segmental context, accent, and phrase position to improve recognition reliability. Mechanisms for search space reduction are proposed to counter the higher cost of using multiple levels.

Preliminary experimental results are presented for prosodic labelling based on known intermediate phrase boundaries, where good results are achieved relative to those reported in other studies. Further work is needed to assess the performance/computational cost trade-offs of the different possible search space reduction techniques for hypothesized phrase boundaries. Although we expect a small loss in accuracy due to use of hypothesized phrase boundary locations, we also expect a gain due to the use of other components of the model not yet evaluated. In particular, we have not taken advantage of word sequence conditioning, which has been beneficial in other work on prosodic labelling of spontaneous speech using decision trees where error reductions of 20–34% were obtained [Mac94].

Initial development of this model has been on a naturally occurring, but rather controlled and careful style of speaking, speaker-dependent radio news speech, primarily because of the availability of prosodically labelled data. However, we can make some observations about the expected performance for the task of labelling speaker-independent spontaneous speech based on our other work. Using the decision tree approach to prosodic labelling [WO94] in experiments on the ATIS corpus of human-computer dialogs, error rates increased by 20–50% in moving from speaker-dependent radio news to speaker-independent spontaneous speech (e.g., from 88% accuracy to 82% accuracy with text conditioning) [Mac94]. Given the error rate increases experienced in the first word recognition experiments with the ATIS corpus and the low word error rates now reported, we are optimistic that prosodic labelling accuracy can similarly be improved. In addition, we note that human transcription consistency is also slightly lower on spontaneous speech than on radio news speech. Of course, it remains to be seen whether our new algorithm is robust to speaker differences, but we believe the phrase peak normalization will be an important advantage of the approach described here.

There are several applications that might take advantage of this work. In speech understanding, prosodic labels could be recognized explicitly, for use in subsequent natural language understanding and dialog state updating processes. Alternatively, the acoustic model could be used jointly with a probabilistic model of the prosodic labels given the word sequence (e.g., a prosody/syntax mapping), as in [VO93a], to obtain a score of the prosodic likelihood of an utterance. Automatic labelling of intonation markers can facilitate corpus collection, and large corpora of prosodically labelled speech can facilitate further research on the mapping between meaning and tone labels, which is still not fully understood and is needed for improved speech synthesis as well as speech understanding. Finally, further refinements to intonation label recognition algorithms can lead to better synthesis models, and in fact, the model proposed here has also been successfully used for generating F_0 and energy contours for text-to-speech synthesis [RO94].

References

[APL84] M. Anderson, J. Pierrehumbert, and M. Liberman. Synthesis by rule of English intonation patterns. In *Proceedings of the International Conference on Acoustics, Speech, and Signal Processing*, pp. 2.8.1–2.8.4, 1984.

[Bec96a] M. Beckman. The parsing of prosody. *Language and Cognitive Processes*, 1996.

[BOPSH90] J. Butzberger, M. Ostendorf, P. Price, and S. Shattuck-Hufnagel. Isolated word intonation recognition using hidden

Markov models. In *Proceedings of the International Conference on Acoustics, Speech, and Signal Processing*, pp. 773–776, 1990.

[BP86] M. Beckman and J. Pierrehumbert. Intonational structure in Japanese and English. In J. Ohala, editor, *Phonology Yearbook 3*, pp. 255–309. New York: Academic, 1986.

[Cam92c] W. N. Campbell. Syllable-based segmental duration. In G. Bailly, C. Benoît, and T. R. Sawallis, editors, *Talking Machines: Theories, Models, and Designs*, pp. 211–224. Amsterdam: Elsevier Science, 1992.

[Cam93a] W. N. Campbell. Automatic detection of prosodic boundaries in speech. *Speech Communication*, 13:343–354, 1993.

[Cam94a] W. N. Campbell. Combining the use of duration and F_0 in an automatic analysis of dialogue prosody. In *Proceedings of the International Conference on Spoken Language Processing, Yokohama, Japan*, Vol. 3, pp. 1111–1114, 1994.

[CI91] W. N. Campbell and S. D. Isard. Segment durations in a syllabic frame. *Journal of Phonetics*, 47:19:37, 1991.

[CW92] F. Chen and M. Withgott. The use of emphasis to automatically summarize a spoken discourse. In *Proceedings of the International Conference on Acoustics, Speech, and Signal Processing*, Vol. 1, pp. 229–232, 1992.

[DLR77] A. P. Dempster, N. M. Laird, and D. B. Rubin. Maximum likelihood from incomplete data via the EM algorithm. *Journal of the Royal Statistical Society*, 37:1–38, 1977.

[DRO93] V. Digalakis, J. R. Rohlicek, and M. Ostendorf. ML estimation of a stochastic linear system with the EM algorithm and its application to speech recognition. *IEEE Trans. on Speech and Audio Processing*, 1:431–442, 1993.

[FK88] H. Fujisaki and H. Kawai. Realization of linguistic information in the voice fundamental frequency contour. *Proceedings of the International Conference on Acoustics, Speech, and Signal Processing*, pp. 663–666, 1988.

[Geo93] E. Geoffrois. A pitch contour analysis guided by prosodic event detection. *Proceedings of the European Conference on Speech Communication and Technology, Berlin, Germany*, pp. 793–796, 1993.

[HF82] K. Hirose and H. Fujisaki. Analysis and synthesis of voice fundamental frequency contours of spoken sentences. In *Proceedings of the International Conference on Acoustics, Speech, and Signal Processing*, pp. 950–953, 1982.

[HHS95] T. Hirai, N. Higuchi, and Y. Sagisaka. A study of a scale for automatic prediction of prosodic phrase boundary based on the distribution of parameters from a critical damping model. *Proceedings Spring Meeting, Acoustics Soc. Jpn*, I:315–316, 1995 (in Japanese).

[Hir93a] J. Hirschberg. Pitch accent in context: Predicting prominence from text. *Artificial Intelligence*, 63:305–340, 1993.

[Hub89] D. Huber. A statistical approach to the segmentation and broad classification of continuous speech into phrase-sized information units. In *Proceedings of the International Conference on Acoustics, Speech, and Signal Processing*, pp. 600–603, 1989.

[JMDL93] U. Jensen, R. Moore, P. Dalsgaard, and B. Lindberg. Modelling of intonation contours at the sentence level using CHMMs and the 1961 O'Connor and Arnold scheme. *Proceedings of the European Conference on Speech Communication and Technology, Berlin, Germany*, pp. 785–788, 1993.

[KBK+94] R. Kompe, A. Batliner, A. Kießling, U. Kilian, H. Niemann, E. Nöth, and P. Regel-Brietzmann. Automatic classification of prosodically marked boundaries in German. *Proceedings of the International Conference on Acoustics, Speech, and Signal Processing*, 2:173–176, 1994.

[LF87] A. Ljolje and F. Fallside. Recognition of isolated prosodic patterns using hidden Markov models. *Computer Speech and Language*, 2:27–33, 1987.

[Mac94] D. Macanucco. Automatic recognition of prosodic patterns. unpublished Boston University course report, 1994.

[NSSS95] M. Nakai, H. Singer, Y. Sagisaka, and H. Shimodaira. Automatic prosodic segmentation by f_0 clustering using superposition modelling. In *Proceedings of the International Conference on Acoustics, Speech, and Signal Processing*, 1995.

[ODK96] M. Ostendorf, V. Digalakis, and O. Kimball. From HMMs to segment models: A unified view of stochastic modelling for speech recognition. *IEEE Trans. on Acoustics Speech and Signal Processing*, 1990.

[OKA+91] M. Ostendorf, A. Kannan, S. Austin, O. Kimball, R. Schwartz, and J. R. Rohlicek. Integration of diverse recognition methodologies through reevaluation of N-best sentence hypotheses. *Proceedings of the DARPA Workshop on Speech and Natural Language*, pp. 83–87, 1991.

[OPSH95b] M. Ostendorf, P. J. Price, and S. Shattuck-Hufnagel. The Boston University Radio News Corpus. Technical Report ECS-95-001, Boston University ECS Dept., 1995.

[OR89] M. Ostendorf and S. Roukos. A stochastic segment model for phoneme-based continuous speech recognition. *IEEE Trans. on Acoustics, Speech, and Signal Processing*, 37:1857–1869, 1989.

[OVHM94] M. Ostendorf, N. Veilleux, M. Hendrix, and D. Macannuco. Linking speech and language processing through prosody. *J. Acoustics Soc. Am.*, 95:2947, 1994.

[RO94] K. Ross and M. Ostendorf. A dynamical system model for generating F_0 for synthesis. *Proceedings of the ESCA/IEEE Workshop on Speech Synthesis, Mohonk, NY*, pp. 131–134, 1994.

[RO96] K. Ross and M. Ostendorf. Prediction of abstract prosodic labels for speech synthesis. *Computer, Speech and Language*, 1996.

[SBP+92] K. Silverman, M. Beckman, J. Pitrelli, M. Ostendorf, C. Wightman, P. Price, J. Pierrehumbert, and J. Hirschberg. ToBI: A standard for labelling English prosody. In *Proceedings of the International Conference on Spoken Language Processing, Banff, Canada*, Vol. 2, pp. 867–870, 1992.

[Sil87] K. E. A. Silverman. The structure and processing of fundamental frequency contours. Ph.D. thesis, University of Cambridge, 1987.

[tB93b] L. ten Bosch. On the automatic classification of pitch movements. *Proceedings of the European Conference on Speech Communication and Technology, Berlin, Germany*, pp. 781–784, 1993.

[VO93a] N. Veilleux and M. Ostendorf. Probabilistic parse scoring with prosodic information. In *Proceedings of the International Conference on Acoustics, Speech, and Signal Processing*, Vol. II, pp. 51–54, 1993.

[vS94a] J. P. H. van Santen. Assignment of segmental duration in text-to-speech synthesis. *Computer Speech and Language*, 8:95–128, 1994.

[vS96] J. P. H. van Santen. Segmental duration and speech timing. In *Computing Prosody: Approaches to a Computational Analysis of the Prosody of Spontaneous Speech.* New York: Springer-Verlag, 1997. This volume.

[WH92] M. Wang and J. Hirschberg. Automatic classification of intonational phrase boundaries. *Computer Speech and Language* 6:175–196, 1992.

[WO94] C. W. Wightman and M. Ostendorf. Automatic labelling of prosodic patterns. *IEEE Trans. on Speech and Audio Processing*, 2:469–481, 1994.

20
Training Prosody-Syntax Recognition Models without Prosodic Labels

Andrew J. Hunt

ABSTRACT [1]
This chapter presents three prosodic recognition models which are capable of resolving syntactic ambiguities using acoustic features measured from the speech signal. The models are based on multi-variate statistical techniques that identify a linear relationship between sets of acoustic and syntactic features. One of the models requires hand-labelled break indices for training and achieves up to 76% accuracy in resolving syntactic ambiguities on a standard corpus. The other two prosodic recognition models can be trained without any prosodic labels. These prosodically unsupervised models achieve recognition accuracy of up to 74%. This result suggests that it may be possible to train prosodic recognition models for very large speech corpora without requiring any prosodic labels.

20.1 Introduction

As speech technology continues to improve, prosodic processing should have a greater potential role in spoken language systems for interpreting a speaker's intended meaning. Prosodic features of utterances can aid the processing of higher linguistic levels such as syntax and semantics, and can be used to detect a number of dialogue characteristics such as turn-taking and topic shift. However, the implementation of prosodic processing is often limited by a lack of appropriate prosodically labelled data.

The ToBI prosody transcription system [SBP$^+$92] is one initiative to increase the availability of prosodically labelled data for English. A number of speech corpora are currently available with hand-labelled break index and intonation labels. However, it is unlikely that hand-labelled prosodic data will ever be available for some of the very large speech corpora now

[1]Research was carried out while affiliated with the Speech Technology Research Group, University of Sydney and ATR Interpreting Telecommunications Research Labs.

used in the training and evaluation of speech recognition systems. For example, the NAB and Switchboard corpora contain tens to hundreds of hours of speech but are available with only text transcriptions.

Therefore, it is important to consider how prosodic models could be trained on these large corpora. One approach, which was used by Veilleux and Ostendorf [VO93b] on the ATIS speech understanding task, is to hand label a section of the corpus with prosodic labels and to use that data for training. Another approach might be to bootstrap a prosodic model by training on a prosodically labelled corpus and then re-estimating its parameters on the larger corpus. A third approach might be the combination of these two approaches by hand-labelling some bootstrap training data and then re-estimating on the complete model.

The current work investigates yet another approach; the design of a model which requires no prosodic labels for training. Two prosody-syntax recognition models which can be trained without any prosodic labelling are presented; the canonical correlation analysis (CCA) model and the linear discriminant analysis (LDA) model. Both models analyse the relationship between prosodic phrasing and syntax and are capable of resolving a range of syntactic ambiguities. Training of the models requires the sampled speech signal, phoneme labels, and syntactic analyses. Typically, phoneme labels and syntactic analyses are not available for large corpora. However, given the text transcription for an utterance, reliable phoneme labels can be obtained by forced recognition using most speech recognition systems. Reliable automatic syntactic analysis is more difficult to obtain. For the current work, hand-corrected parse diagrams from an automated parser were used.

A third prosodic recognition model is presented that uses the same acoustic and syntactic features and the same mathematical framework as the CCA and LDA models, but which uses hand-labelled break indices in training; the break index linear regression (BILR) model. This model is presented as a benchmark model from which we can determine the effect of the unsupervised prosodic training on recognition accuracy.

Section 20.2 describes the speech data and the acoustic and syntactic analysis used as input to the prosody-syntax models. The three prosody-syntax recognition models are presented in Sec. 20.3. Section 20.4 presents results and analysis from the application of the models to two professionally read speech corpora. Section 20.5 compares the models and discusses their application to automatic speech understanding.

20.2 Speech Data and Analysis

20.2.1 Speech Data

Two professionally read speech corpora were used for training and testing the prosody-syntax models; the *radio news corpus* [OPSH95b] and the *ambiguous sentence corpus* [POSHF91]. Both corpora have been used previously in the training and testing of prosody-syntax recognition models.

The radio news corpus consists of a series of news stories on topical issues read by four professional news readers from Boston. In the current work, the first five news stories from a single female speaker (speaker f2b) were used, providing data on 1491 sentence-internal word boundaries. The corpus was available with phonetic labels obtained by forced recognition using the Boston University SSM recognizer.

The ambiguous sentence corpus was originally developed to test the ability of human subjects to resolve syntactic ambiguities using prosodic cues. It has since been used to evaluate the capability of prosody-syntax recognition models to resolve syntactic ambiguities. It contains 35 pairs of structurally ambiguous sentences each read by 4 professional news readers (3 female, 1 male, including the female speaker of the radio news data described above)—in total, 280 utterances. The phonetic labels provided with the corpus were obtained by forced recognition using the SRI recognizer. The disambiguation task is to determine the correct syntactic representation for an utterance from the speech signal. Due to parsing problems six sentence pairs were unavailable for the recognition tests. Therefore, the recognition results presented here are based on 232 utterances which have in total 1708 sentence-internal word boundaries.

There are substantial differences in the styles of the two corpora. These differences are important because in some of the recognition tests, the radio news data is used for training and the ambiguous sentence data is used for testing. The average length of sentences in the radio news corpus is considerably longer than that of the ambiguous sentence corpus; 19 words per sentence compared with 7.6 words. The types of syntactic forms observed in the two corpora are also substantially different.

20.2.2 Acoustic Feature Set

Ten acoustic features were used in the current work. All ten features were automatically extracted at each sentence-internal word boundary in both speech corpora using the phonetic and word labels provided with the corpora. The feature set included:

(1) Duration of a pause at the word boundary (or zero if there is no pause);

(2) number of syllables in the word preceding the boundary;

(3) stress label on any phone in the word preceding the boundary (as marked by the recognition system which labelled the database).

The remaining seven features were measurements of the pre-boundary syllable (that is, the last syllable of the word preceding the boundary):

(1) Number of phonemes in the rhyme;

(2) duration of the rhyme;

(3) duration of the syllabic nucleus;

(4) phonetic identity of the syllabic nucleus (a categorical feature);

(5) energy in the syllabic nucleus (log form);

(6) average power of the syllabic nucleus (log form);

(7) stress label on the syllabic nucleus (as marked by the recognition system which labelled the database).

The three durational features were included because previous work has shown that the primary acoustic correlates of syntactic and prosodic boundaries are durational (e.g., [Kla75]). Segmental-normalized rhyme and syllabic nucleus durations and pause length are also correlated with break indices [WSOP92] and have been successfully used in previous automatic prosodic recognition models (e.g., [VO93a]). The remaining features were selected to compensate for non-syntactic effects upon the three durational features. In brief, phonetic identity can compensate for the inherent duration [Kla75], the two stress features, energy and power can compensate for stress-induced segment lengthening [Cam93a], and the number of phonemes in the rhyme can compensate for the reduction in phone duration that typically accompanies an increase in the syllable size [CH90].

20.2.3 Syntactic Feature Set

The *link grammar* [ST91] provided the syntactic framework for modelling prosodic phrasing. The link grammar identifies links between syntactically related words. These links are presented in a *link diagram.* Each link

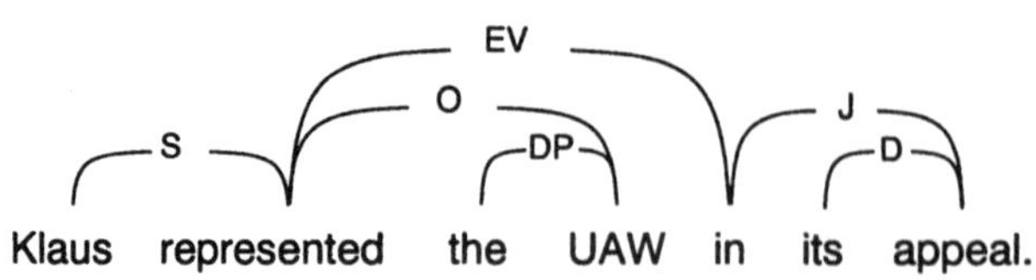

FIGURE 20.1. Example link diagram.

has an associated *link label* which indicates a *surface-syntactic relation* (in Melcuk's terminology of a dependency grammar [Mel88]). Figure 20.1 shows the link diagram for a sentence from the radio news corpus. The links in the diagram have the following syntactic functions: *S* links the subject to a verb, *O* links an object to the verb, *D* and *DP* link determiners to nouns, *EV* links a preposition to its head verb, and *J* links a preposition to the head noun within a prepositional phrase. Around 45 different link labels were encountered in parsing the two corpora described above, however, less than 20 links account for more than 90% of the links in the corpora.

A theory of syntactic influence on prosodic phrasing using the syntactic framework of the link grammar has been presented previously [Hun94b, Hun95a, Hun95b]. The following is a brief summary of the results for that work which are relevant to the CCA and LDA models. Three hypotheses regarding the influence of syntax on prosodic phrasing were proposed and tested:

(1) Each surface-syntactic relation (link label) has an *intrinsic prosodic coupling strength.*

(2) Longer links will tend to have weakened prosodic coupling strength.

(3) Increasing the syntactic coupling of a word to its left will tend to decrease its prosodic coupling to its right.

A set of eight syntactic features was extracted from the syntactic analysis of utterances. The first and most important of the eight syntactic features, the link label, represents the surface syntactic relation of the most immediate link crossing a word boundary and was selected to model the first hypothesis. The remaining seven features represent the syntactic structure around the word boundary and reflect hypotheses (2) and (3). They are:

(1) Distance from the current word boundary to the left end of the most immediate link crossing the word boundary,

(2) Distance from the current word boundary to the right end of the most immediate link crossing the word boundary,

(3) Number of links covering the most immediate link,

(4) Number of links connected to the left of the preceding word,

(5) Number of links connected to the right of the preceding word,

(6) Number of links connected to the left of the following word,

(7) Number of links connected to the right of the following word.

All eight features can be extracted from the output of the link parser (which implements the link grammar). The previous work showed that linear models using these eight features can reliably predict break indices. Moreover, the roles of the eight features in the models were in agreement with the theoretical predictions.

20.3 Prosody-Syntax Models

20.3.1 Background

Veilleux and colleagues have developed and evaluated a series of prosody-syntax recognition models using the architecture shown in Figure 20.2(A) [VOW92, OWV93, VO93a, VO93b]. Two separate decision trees are used to predict an intermediate representation of prosodic labels, or a stochastic distribution of the labels, from acoustic and syntactic feature sets. The probability of a syntactic parse given the acoustic features can be calculated by comparison of the predictions of prosodic labels from the two domains. Their acoustic feature set contained segmental normalized duration, pause information, and in some cases pitch and energy measurements. Their syntactic feature set was derived from Treebank syntactic bracketings.

The series of models developed by Veilleux and colleagues reliably resolved syntactic ambiguities. For the ambiguous sentence corpus 69% accuracy was achieved using break indices as an intermediate representation [VOW92, OWV93, VO93a], and 73% accuracy was achieved using a combination of break indices and prominence [VO93a]. This approach has also been applied to the ATIS speech understanding task [VO93b]. The decision tree model using both break indices and prominence was retrained on a section of the ATIS corpus which had been hand-labelled with prominence and break indices with new notation to indicate disfluencies. The addition of a prosodic score to other recognition scores provided significant improvements in the detection of the user's intended meaning.

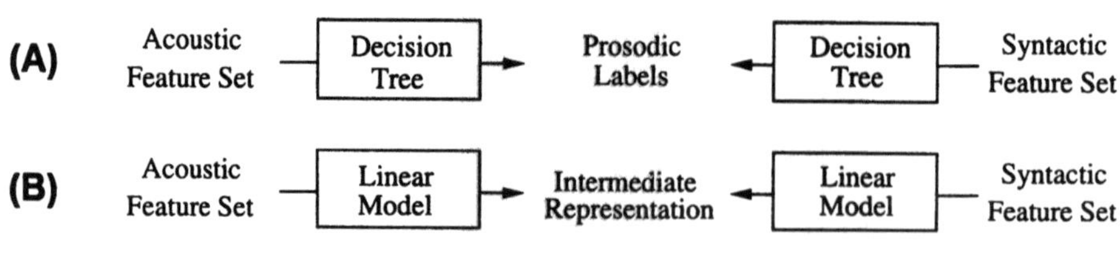

FIGURE 20.2. Two acoustic prosody-syntax model architectures.

20.3.2 Break Index Linear Regression Model

The goal of the new prosody-syntax models presented here is to identify a strong relationship between the sets of acoustic and syntactic features (at each word boundary) in such a way that the models can resolve syntactic ambiguities. The first model presented is the break index linear regression (BILR) model which is an adaptation of the architecture of Veilleux and colleagues. Sections 20.3.3 and 20.3.4 present the CCA and LDA models which further adapt the BILR modelling framework to support training without prosodic labels by using multivariate statistical techniques. The BILR model provides a benchmark for comparison of the CCA and LDA models so that the effect of prosodically unsupervised training can be determined.

The BILR model has a similar architecture to that used by Veilleux and colleagues, as shown in Figure 20.2(B). The two major differences are that linear regression models are substituted for decision trees and the intermediate representation of break indices is scalar instead of discrete.

Regression models were trained to predict break indices from both the acoustic and syntactic feature sets for both the corpora described in Sec. 20.2.1. The categorical features (phonetic identity of the syllabic nucleus and the link label) were incorporated into the regression model using contrasts [SPl93].

The regression model trained with the acoustic features for the radio news corpus was tested for its ability to label break indices on the ambiguous sentence corpus. It correctly labelled 50.4% of break indices and 90.3% of labels were accurate to within one break index level. Slightly higher accuracy was obtained for a model trained on the ambiguous sentence corpus and tested on the radio news corpus: 53.1% and 93.7% for the two measures. Comparable results were obtained for prediction with the syntactic feature set. In a closed-set test, the accuracies for the radio news corpus were 57.8% and 92.1%, respectively. For the ambiguous sentence corpus the closed-set accuracies were 52.0% and 93.3%.

These results show that both the acoustic and syntactic feature sets are reasonable predictors of break indices. The accuracies are comparable to (human) inter-labeller consistency [PBH94] for the ± 1 break index measure but lower for exact agreement. The accuracies are also comparable to other work on automatic break index labelling (e.g., [WO94]), however, close comparisons are not possible because of differences in the training and test data.

A recognition model can be obtained by comparison of the break index predictions from the acoustic and syntactic domains. Let $A = A_1, \ldots, A_q$ be the sets of acoustic features at word boundaries for an utterance with q boundaries. Let $A_i = a_{i1}, \ldots, a_{im}$ be the set of m acoustic features at each word boundary (as specified in Sec. 20.2.2). Let $\overline{A_i}$ be the prediction of break indices at the ith word boundary using weights w_j^a obtained using linear regression training with break index labels for the training data as described above:

$$\overline{A_i} = \sum_{j=1}^{m} w_j^a a_{ij} \tag{20.1}$$

Similarly, let $S = S_1, \ldots, S_q$ be the sets of syntactic features for an utterance, let $S_i = s_{i1}, \ldots, s_{in}$ be the set of n syntactic features (as specified in Sec. 20.2.3), and let $\overline{S_i}$ be the prediction of break indices at the ith word boundary using weights w_j^s obtained by linear regression training:

$$\overline{S_i} = \sum_{j=1}^{n} w_j^s s_{ij}. \tag{20.2}$$

The probability of the set of syntactic features for an utterance, S, given the acoustic features, A, can be estimated as follows. First assume that observations at word boundaries are independent:

$$p(S|A) = \prod_{i=1}^{q} p(S_i|A_i). \tag{20.3}$$

We can estimate the conditional probability on the right-hand side from the difference in the break index predictions of Eqs. (20.1) and (20.2) by assuming a Gaussian distribution of the error. The standard error was estimated separately for each link label on the training data. Thus,

$$p(S|A) = \prod_{i=1}^{q} p\left(\overline{A_i} - \overline{S_i}|e_{l_i}\right)^{1/q}, \tag{20.4}$$

where e_{l_i} is the standard error for the link label at the jth boundary and

$$\overline{A_i} - \overline{S_i} \sim N(0, e_{l_i}). \tag{20.5}$$

The exponent term in Equation 20.4 normalizes for sentence length so that longer sentences are not penalized.

It is worth pointing out that the framework described above is applicable to any intermediate prosodic representation which can be predicted by linear modelling. For example, break indices could be replaced by prominence labels. Moreover, multiple intermediate representations could be combined (with weightings) as Veilleux and Ostendorf did with break indices and prominence [VO93a].

20.3.3 CCA Model

The CCA model adapts the BILR model to allow training without using break indices as an intermediate representation. This is achieved by estimating the weights, w_j^s and w_j^a, using the multi-variate statistical technique canonical correlation analysis [And84]. Given the sets of acoustic and syntactic features for all boundaries in the training corpus, CCA determines the weights so that the linear combinations $\overline{A_i}$ and $\overline{S_i}$ are maximally correlated. All other aspects of the model described above remain unchanged.

CCA can provide multiple pairs of weights which provide decreasing correlations between the two feature sets. In previous work, these extra weights were used as additional intermediate vectors in the calculation of the probabilities [Hun94a, Hun95a]. It was found, however, that for the relatively small databases used in the current work, this lead to overtraining and hence reduced the accuracy and the robustness of the CCA recognition model. In the results presented here, only a single intermediate vector is used. A second difference between the current and previous work on the CCA model is that the categorical features are now modelled using contrasts instead of being estimated iteratively [Hun94a].

20.3.4 LDA Model

Like the CCA model, the LDA model adapts the BILR model to allow training without break indices, but uses another multi-variate statistical technique method, linear discriminant analysis (LDA) [And84], to estimate the weights w_j^s and w_j^a. Given a set of features which come from observations of different classes, LDA determines the linear weightings of the features that maximally separate the classes. In the LDA model, the linear weights for the acoustic features, w_j^a, were estimated to separate the observations of link labels, in other words to discriminate between the link labels. The link label was used as the predicted feature because earlier work consistently found that it was the most important of the eight syntactic features [Hun93, Hun95a, Hun95b].

Once the acoustic weights were obtained, linear regression was used to predict the intermediate acoustic vector, $\overline{A_i}$, from the set of syntactic features.

Linear discriminate analysis produces multiple sets of weights which provide decreasing discrimination. In the current work, only the first (and most accurate) discriminant vector is used. In previous work, the additional weights were also used in the calculation of the probabilities [Hun94c, Hun95a] but this lead to overtraining. The other difference between the current and previous work on the LDA model is that the weights of the syntactic features, w_i^s, are common for all link labels. In the previous work an interaction between link label and the weights was used. This substantially increased the number of weights and also lead to overtraining.

20.4 Results and Analysis

Three aspects of the three prosody-syntax recognition models were investigated: (1) the accuracy in resolving ambiguities in the ambiguous sentence corpus, (2) the correlation of the intermediate acoustic and syntactic representations, and (3) the correspondence of the internal model characteristics to expectations from previous work in the field. The relevance of the three criteria and results are presented below.

20.4.1 Criterion 1: Resolving Syntactic Ambiguities

The long-term goal of this research is to develop prosody-syntax models which can be applied to automatic speech understanding. The ability to resolve ambiguities in the ambiguous sentence corpus provides some indication of the recognition capability of the models. The ambiguous sentence corpus provides suitably controlled data and there are benchmark results for human subjects and other prosody recognition models.

Table 20.1 shows the recognition accuracies for the three recognition models when resolving syntactic ambiguities using the acoustic features. Results are presented for two test conditions: (1) recognition accuracy on the ambiguous sentence corpus for a model trained on the radio news corpus, (2) recognition accuracy for revolving training and testing on the ambiguous sentence corpus. In the revolving test, a model is trained on three of the speakers and tested on the fourth. This test is repeated with each speaker as the test speaker and the results averaged.

All the accuracies are significantly above chance level (50%; $p < 0.001$) but are significantly below human capabilities on the same task (84%, $p < 0.001$). The accuracy was reasonably consistent across speakers and syntactic forms. There is no significant difference in accuracy for the two test conditions for any of the models.

The result that the accuracies for the BILR and CCA models on the radio news corpus exceeded that for the revolving testing was unexpected. Lower

TABLE 20.1. Comparison of recognition model accuracies.

Model	Revolving test	Radio news
BILR model	74.6%	76.3%
CCA model	72.8%	73.7%
LDA model	71.1%	65.1%
[OWV93]	69%	—
[VO93a]	73%	—

accuracy was expected for the models trained on the radio news corpus because of the substantial differences in the syntactic forms of the two corpora and because the radio news corpus is a single speaker corpus and is thus not suitable for training speaker-independent models. Nevertheless, the result is very encouraging as it indicates that the CCA and BILR models generalize well across syntactic forms and across speakers.

Comparison of the results for the BILR model with the those for the CCA and LDA models indicates the extent to which using prosodically unsupervised training of the intermediate representation affects performance. The small decrease in accuracy from the BILR model to the CCA model (around 2.2%) indicates that unsupervised training is possible without substantial loss in accuracy. However, the more substantial decreases for the LDA model indicate that the method of unsupervised training is critical.

Table 20.1 also presents the recognition accuracy for previous work by Veilleux and colleagues. Close comparisons are difficult because of differences in the experimental conditions. In particular, the use of the ambiguous sentence data differs because not all of the ambiguous sentences were available for testing in the current work. The most direct comparison can be made between the BILR model, with an accuracy of 76.3% when trained on the radio news corpus, and the decision tree model which used only break indices and achieved 69% accuracy [OWV93]. The higher accuracy of the BILR model may be due to experimental differences, but may also be due to differences in the designs of the models such as (1) the use of the link grammar, (2) the use of linear regression and the scalar intermediate representation of break indices, or (3) the use of different acoustic features. It is an open question whether the linear framework of the BILR model could be improved by the addition of prominence to the intermediate representation as was achieved by Veilleux and Ostendorf (cf. [OWV93] with 69% accuracy to [VO93a] with 73% accuracy).

The CCA model achieves comparable accuracy to the decision tree-based models despite being trained without any hand-labelled prosodic features.

20.4.2 Criterion 2: Correlation of Acoustic and Syntactic Domains

The correlation between the intermediate acoustic and syntactic representations, $\overline{A_i}$ and $\overline{S_i}$, indicates the strength of the relationship that a model finds between the domains. As the correlation increases, a model should show greater discriminatory capability and should therefore be more effective for speech recognition. It is also of theoretical interest to know the extent to which acoustics and syntax are related.

Table 20.2 shows the correlations of the intermediate acoustic and syntactic representations for the three recognition models and for both corpora. All correlations in the table are statistically significant ($p <$

TABLE 20.2. Correlations of the intermediate representations.

Model	Radio news corpus	Amb. sent. corpus
BILR model	0.717	0.642
CCA model	0.778	0.805
LDA model	0.763	0.801

0.001). The results show that all three models can identify a strong linear relationship between the low level acoustic features and the higher level syntactic features. Moreover, this relationship applies across a wide range of syntactic forms and across a wide range of prosodic boundaries, from clitic boundaries to major phrase boundaries.

Not surprisingly, the CCA and LDA models show higher correlations than the BILR model. This is expected because their training methods explicitly maximize the correlations of their intermediate representations.

It is interesting to note that the substantial increase in intermediate correlation obtained by replacing break indices by a learned intermediate representation occurs along with a slight decrease in recognition accuracy. Also, the correlations for the CCA and LDA models are close, but the CCA model is substantially better in resolving ambiguities. Thus, the expectation that a higher correlation should indicate better recognition is not in fact supported.

20.4.3 Criterion 3: Internal Model Characteristics

It is of theoretical interest to know whether the internal characteristics of the CCA and LDA recognition models match expectations from previous research on the prosody-syntax relationship because, unlike the BILR model, they do not have linguistically motivated intermediate representations. From a practical viewpoint, an unsupervised model with "sensible" internal characteristics may be applicable to a wider range of speech technology applications, for example, the generation of prosody in speech synthesis.

Table 20.3 shows the correlations of the intermediate vectors to break indices for the three recognition models when trained without the two stress features in the acoustic feature set (described in Sec. 20.2.2). The correlations for the BILR model indicate the maximum obtainable correlations between the acoustic and syntactic intermediate representations and break indices for the two corpora. The correlations for the CCA and LDA models are between 3% and 22% lower. This suggests that the learned intermediate representations are reasonably close to break indices despite there being no explicit representation of break indices in the input features.

TABLE 20.3. Correlation of intermediate representations to break indices.

Model	Radio news corpus		Amb. sent. corpus	
	Acoustic	Syntax	Acoustic	Syntax
BILR model	0.842	0.796	0.737	0.770
CCA model	0.817	0.747	0.650	0.623
LDA model	0.754	0.704	0.614	0.598

With the two stress features included, the correlations to break indices drop substantially, but recognition accuracy improves. This result can be explained as follows. It has been suggested that prominence is relevant to syntactic disambiguation [POSHF91] and it has been found that including prominence can improve the accuracy of an automatic prosody-syntax recognition model [VO93a]. Since the stress features are correlated with phrasal prominence, it is possible that the intermediate representations using these features have some correlation to prominence placement and therefore lower correlation to the break indices alone. Furthermore, this could improve disambiguation accuracy.

The roles of the syntactic features in the CCA and LDA models were in agreement with theoretical predictions outlined in Sec. 20.2.3. The roles of the acoustic features in the models were in agreement with previous research. For example, as other researchers have found [WSOP92], the pause and rhyme durations were the most important of the acoustic features.

Thus, there is some evidence that despite the prosodically unsupervised training of the CCA and LDA models, many of their internal characteristics are in accord with previous research on prosodic modelling.

20.5 Discussion

The goal of the research presented here is the development of prosody-syntax recognition models which can be trained on large corpora for which there are no prosodic labels. The major contribution is the investigation of two prosody-syntax models which utilize multi-variate statistical techniques to provide training without prosodic labels. Despite being trained without prosodic labels, the CCA model achieved state-of-the-art accuracy for automatically resolving syntactic ambiguities using acoustic features. These accuracies are, however, slightly below that of the BILR model which has the same statistical framework but is trained with break index labels.

This suggests that training without prosodic labels can be effective but may be slightly less accurate than training with prosodic labels.

The recognition performance of the CCA model is clearly better than that of the LDA model. The most reasonable explanation for this is that the CCA training simultaneously maximizes the correlation between the complete sets of acoustic and syntactic features. In contrast, the LDA model first trains the discrimination of link labels using the acoustic features and then introduces the remaining syntactic features.

Close comparison of the three models with the previous work of Veilleux and colleagues is difficult because of the many experimental differences. Nevertheless, it is encouraging that similar recognition accuracies were obtained with the CCA model without hand-labelled prosodic features as were obtained for decision tree models trained with break index and prominence labels.

The CCA and LDA models can be integrated easily with other speech recognition system components because they produce likelihood scores. Veilleux and Ostendorf [VO93b] have already shown that prosody-syntax models can improve the accuracy of a speech understanding system. In that work on the ATIS task, they also found that dealing with disfluencies is also an important issue for prosodic models; this is an issue that has not been addressed for the CCA and LDA models.

Another issue requiring further consideration is that of automatic parsing. The current work used hand-corrected parse diagrams from the link parser. It is unclear what the effect of using a fully automatic parser would be. An interesting candidate parser, which was not available at the time this research was carried out, is the robust link parser [GLS95]. Initial tests suggest that it has many of the advantages of the older link parser used for this research but is capable of handling a much wider range of text input.

Further work on the CCA and LDA models could improve a number of areas of the models. Enhancements to the acoustic feature set are possible; for example, the introduction of segmental-normalized features and the introduction of features derived from pitch. Training on larger speech corpora is required to investigate the problems of overtraining that occurred when multi-dimensional intermediate represenations were used. Also, training on non-professional speech data is required to determine the robustness of the models to speech style. Finally, more work is required to determine the comparative effectiveness of the link grammar and more conventional Treebank analyses which have been used by other researchers.

Conclusion

Three prosody recognition models have been presented which can reliably resolve a range of syntactic ambiguities in professionally read speech with up to 76% accuracy. A novel characteristic of two of the models is that they can be trained without prosodic labels. The advantage of this prosodically unsupervised training is that the models are potentially applicable to very large corpora for which hand-labelling is prohibitively expensive and slow. Despite this novel training, the recognition accuracy is close to a comparable model trained with break index labels and to previous prosody-syntax recognition models using decision trees. Also, the models have internal characteristics which concur with the findings of previous research on the prosodic correlates of syntax. The application of the models to spoken language systems and the advantages and limitations of the new modelling approach were discussed.

Acknowledgments

I am grateful to Professor Mari Ostendorf for her very helpful comments on the draft of this paper and for providing the two speech corpora used in the research.

References

[And84] T. W. Anderson. *An Introduction to Multivariate Statistical Analysis*: 2nd ed. New York: Wiley, 1984.

[Cam93a] W. N. Campbell. Automatic detection of prosodic boundaries in speech. *Speech Communication*, 13:343–354, 1993.

[CH90] T. H. Crystal and A. S. House. Articulation rate and the duration of syllables and stress groups in connected speech. *J. Acoust. Soc. Am.*, 88:101–112, 1990.

[GLS95] D. Grinberg, J. Lafferty, and D. Sleator. A robust parsing algorithm for link grammars. In *Proceedings of the Fourth International Workshop on Parsing Technologies*, Prague, 1995.

[Hun93] A. J. Hunt. Utilising prosody to perform syntactic disambiguation. In *Proceedings of the European Conference on Speech Communication and Technology, Berlin, Germany*, pp. 1339–1342, 1993.

[Hun94a] A. J. Hunt. A generalised model for utilising prosodic information in continuous speech recognition. In *Proceedings of the International Conference on Acoustics, Speech, and Signal Processes*, pp. 169–172, 1994.

[Hun94b] A. J. Hunt. Improving speech understanding through integration of prosody and syntax. In *Proceedings of the 7th Aust. Joint Conference on Artificial Intelligence*, pp. 442–449, Armidale, Australia, 1994.

[Hun94c] A. J. Hunt. A prosodic recognition module based on linear discriminant analysis. In *Proceedings of the International Conference on Spoken Language Processing, Yokohama, Japan*, pp. 1119–1122, 1994.

[Hun95a] A. J. Hunt. Models of Prosody and Syntax and their Application to Automatic Speech Recognition. Ph.D. thesis, University of Sydney, 1995.

[Hun95b] A. J. Hunt. Syntactic influence on prosodic phrasing in the framework of the link grammar. In *Proceedings of the European Conference on Speech Communication and Technology, Madrid, Spain*, 1995.

[Kla75] D. H. Klatt. Vowel lengthening is syntactically determined in a connected discourse. *Journal of Phonetics*, 3:129–140, 1975.

[Mel88] I. A. Melcuk. *Dependency Syntax: Theory and Practice.* Albany: State University of New York Press, 1988.

[OPSH95b] M. Ostendorf, P. J. Price, and S. Shattuck-Hufnagel. The Boston University Radio News Corpus. Technical Report ECS-95-001, Boston University ECS Dept., 1995.

[OWV93] M. Ostendorf, C. W. Wightman, and N. M. Veilleux. Parse scoring with prosodic information: An analysis-by-synthesis approach. *Computer Speech and Language*, 7:193–210, 1993.

[PBH94] J. Pitrelli, M. E. Beckman, and J. Hirschberg. Evaluation of prosodic transcription labelling reliability in the ToBI framework. In *Proceedings of the International Conference on Spoken Language Processing, Yokohama, Japan*, Vol. 1, pp. 123–126, 1994.

[POSHF91] P. J. Price, M. Ostendorf, S. Shattuck-Hufnagel, and C. Fong. The use of prosody in syntactic disambiguation. *J. Acoust. Soc. Am.*, 90:2956–2970, 1991.

[SBP+92] K. Silverman, M. Beckman, J. Pitrelli, M. Ostendorf, C. Wightman, P. Price, J. Pierrehumbert, and J. Hirschberg. ToBI: a standard for labelling English prosody. In *Proceedings of the International Conference on Spoken Language Processing, Banff, Canada*, Vol. 2, pp. 867–870, 1992.

[SPl93] SPlus. *Guide to Statistical and Mathematical Analysis*. Seattle: StatSci, 1993.

[ST91] D. Sleator and D. Temperley. Parsing English with a link grammar. Technical report, CMU-CS-91-196, School of Computer Science, Carnegie Mellon University, 1991.

[VO93a] N. Veilleux and M. Ostendorf. Probabilistic parse scoring with prosodic information. In *Proceedings of the International Conference on Acoustics, Speech, and Signal Processes*, Vol. II, pp. 51–54, 1993.

[VO93b] N. M. Veilleux and M. Ostendorf. Prosody/parse scoring and its application in ATIS. In *Proceedings of the DARPA Workshop on Speech and Natural Language Processing*, 1993.

[VOW92] N. M. Veilleux, M. Ostendorf, and C. W. Wightman. Parse scoring with prosodic information. In *Proceedings of the International Conference on Spoken Language Processing, Banff, Canada*, pp. 1605–1608, 1992.

[WO94] C. W. Wightman and M. Ostendorf. Automatic labelling of prosodic patterns. *IEEE Trans. on Speech and Audio Processes*, 2:469–481, 1994.

[WSOP92] C. W. Wightman, S. Shattuck-Hufnagel, M. Ostendorf, and P. J. Price. Segmental durations in the vicinity of prosodic phrase boundaries. *J. Acoust. Soc. Am.*, 91:1707–1717, 1992.

21
Disambiguating Recognition Results by Prosodic Features

Keikichi Hirose

ABSTRACT For the purpose of realizing an effective use of prosodic features in automatic speech recognition, a method was proposed to check the suitability of a recognition candidate through its fundamental frequency contour. In this method, a fundamental frequency contour is generated for each recognition candidate and compared with the observed contour. The generation of fundamental frequency contours is conducted based on prosodic rules formerly developed for text-to-speech conversion, and the comparison is performed only on the portion with recognition ambiguity, by a newly developed scheme denominated partial analysis-by-synthesis. The candidate giving the contour that best matches the observed contour is selected as the final recognition result. The method was shown to be valid for detecting recognition errors accompanied by changes in accent types and/or syntactic boundaries, and was also evaluated as to its performance for detecting phrase boundaries. The results indicated that it can detect boundaries correctly or at least with a location error of one mora.

21.1 Introduction

Prosodic features of speech are known to be closely related with various linguistic and non-linguistic features, such as word meaning, syntactic structure, discourse structure, speaker's intention and emotion, and so on. In human speech communication, therefore, they play an important role in the transmission of information. In current speech recognition systems, however, their use is rather limited even in the linguistic aspect. Although the hidden Markov modelling has been successfully introduced in speech recognition and yields rather good results just with segmental features, prosodic features also need to be incorporated for further improvement.

However, different from the case of segmental features, the use of prosodic features should be supplementary in speech recognition. Since prosodic and linguistic features belong to two different aspects of language, respectively, spoken and written language, they do not bear a tight relationship. For instance, a major syntactic boundary (in written language) does not necessarily correspond to a major prosodic boundary (in spoken language).

Therefore, the incorporation of prosodic features in the speech recognition process should take this factor into consideration. One possibility is to assign a different (increased) likelihood to a recognition candidate when its expected prosodic features agree with the actual ones observed in the input speech.

In continuous speech recognition, unless the perplexity of the recognition task is small, much of computation is required for the searching process in the linguistic level, still not yielding a good result. Information on syntactic structures is thought to be effective in order to improve recognition performance, when utilized as the constraints for the searching process. From this point of view, several methods have already been reported to detect syntactic boundaries of spoken sentences from their prosodic features [KOI88],[OKI89],[WO91],[OEKS93],[G93],[BBBKNB94],[NS94]. We have pointed out that the simultaneous use of microscopic and macroscopic features of fundamental frequency contours (henceforth, F_0 contours) could decrease deletion errors in boundary detection, and have developed a method for the accurate detection of syntactic boundaries, with which 96% of manually detectable boundaries were correctly extracted for the ATR continuous read speech corpus on conference registration [HSK94].

Although syntactic boundaries can be detected quite well as mentioned above, the results depend highly on the speakers and speaking styles. The methods can be made more robust by introducing statistical techniques, but their effect on speech recognition is still limited. This is because, in these methods, syntactic boundaries are detected only by the prosodic features, without referring to the recognition results obtainable from segmental features. (In Ref. [WO91], the recognition results were utilized, but their use was limited to segmental boundaries.) From this point of view, we have proposed a method where an F_0 contour is generated for each recognition candidate using prosodic rules for speech synthesis and matched against the observed F_0 contour by a scheme of partial analysis-by-synthesis [HSK94]. The candidate giving the minimum distance is supposed to be the final recognition result. Of course, instead of using heuristic rules of prosody and the analysis-by-synthesis scheme, a statistical method, such as one based on hidden Markov modelling of prosodic features, could be used for the purpose. Although this method seems promising, it was not adopted because it requires a large amount of training data, and, moreover, cannot nicely include the model constraints.

The proposed method can be considered to be valid to detect recognition errors, to ensure recognition results, and thus to realize an effective search in continuous speech recognition. The method can also be used in conjunction with the above methods for the detection of syntactic boundaries: to check if an extracted boundary is correct or to identify the type of boundary.

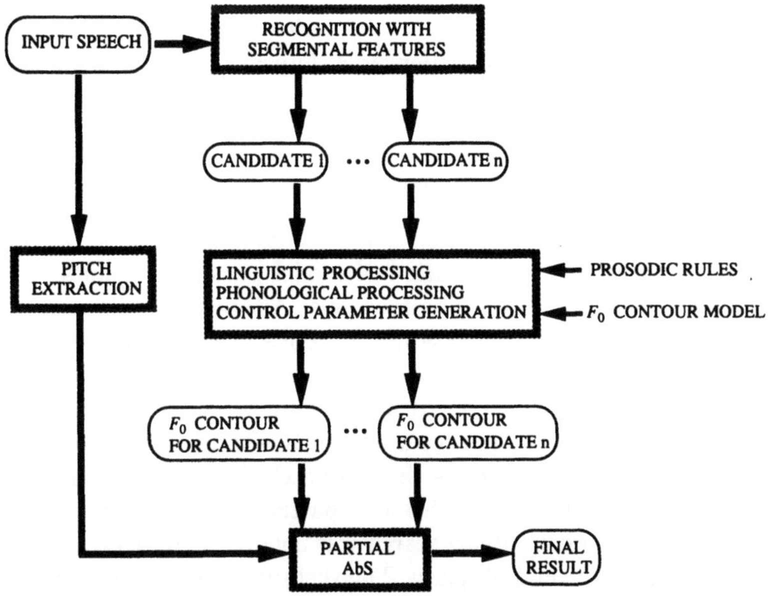

FIGURE 21.1. Total configuration of the method for finding correct recognition result from several candidates.

21.2 Outline of the Method

Figure 21.1 schematically shows the total configuration of the proposed method. Prosodic rules used for the generation of F_0 contours are those formerly constructed for a text-to-speech conversion system [HF93]. Because of considerably large utterance-to-utterance or speaker-to-speaker variations in the observed F_0 contours, a mere comparison between the generated contour and the observed contour could yield a large distance even for the correct recognition candidate. Therefore, before calculating the distance, the generated contour is adjusted in a limited extent to the observed contour by the partial analysis-by-synthesis scheme.

21.2.1 Model for the F_0 Contour Generation

The prosodic rules are based on a functional model for F_0 contour generation, originally proposed by H. Fujisaki [FS71a] and then slightly modified to the current formulation with the author [HF82]. This model represents a sentence F_0 contour in logarithmic frequency scale as a sum of phrase components, accent components, and a baseline component. Phrase components and accent components are, respectively, considered to be generated from impulse-like phrase commands and stepwise accent commands, which are known to have good correspondence, respectively, with syntactic structure and lexical accents. The generation processes of phrase and accent components from their corresponding commands are represented by critically damped second order linear systems. Details of the model can be found in the paper by H. Fujisaki, which is also included in this book.

21.2.2 Partial Analysis-by-synthesis

The method of analysis-by-synthesis based on the "hill-climbing" algorithm is widely used to find a combination of model parameter values yielding the contour that best fits the observed one. Although, in the case of F_0 contours, the best fitting search is usually done over the entire utterance unit (prosodic sentence) delimited by respiratory pauses [FH84], for the current purpose of evaluating recognition candidates, this procedure may be difficult if the unit includes several portions with recognition ambiguity. Even if possible, it would obscure the mismatch due to recognition errors. From this point of view, a new scheme of partial analysis-by-synthesis was developed, where the best fitting search was conducted only on the limited portion with recognition ambiguity. The distance between generated and observed contours is given as the following analysis-by-synthesis error per frame averaged over the voiced part of the portion:

$$E_r = \frac{\sum_{i=m}^{n}\{\ln(\frac{F_0(t_i)}{\hat{F}_0(t_i)})\}^2}{n-m+1}, \tag{21.1}$$

where $F_0(t_i)$ and $\hat{F}_0(t_i)$, respectively, denote the observed and model-generated fundamental frequencies at t_i, with t_i defined as the center of frame i. Frames from m to n are assumed to be included in the portion.

In order to start the analysis-by-synthesis process, a set of initial values is required for the parameters of the model; they are given by the prosodic rules for speech synthesis, as already mentioned. Although these rules generate three types of prosodic symbols (pause, phrase, and accent symbols) at appropriate syllabic boundaries using the linguistic information of the input sentence, pause symbols, representing pause lengths, are not necessary for the proposed method. This is because, unlike phrase and accent symbols, which represent magnitudes/amplitudes and timings of commands, pause symbols only carry timing information, which is easily obtainable from the input speech. In other words, phoneme boundaries are given by the segmental-based recognition process and no durational information need to be given by the rules. Table 21.1 shows the command values assigned to the phrase and accent symbols, which serve as the initial values for the analysis-by-synthesis process [HF93]. Each one of the phrase symbols, P1, P2, and P3, indicates a phrase command, with which a phrase component is generated, while the phrase symbol P0 indicates a symbol to reset the component sharply to zero. The symbol P0 is usually assigned before a respiratory pause in a sentence, or between sentences. As for accent symbols, each one of the symbols FH to DL in the table indicates

TABLE 21.1. Command magnitudes/amplitudes and positions assigned to the phrase and accent symbols in the prosodic rules. These will serve as the initial parameter values for the process of analysis-by-synthesis.

Type	Symbol	Command magnitude/ amplitude	Position with respect to voice onset (ms)
Phrase Symbol	P1	0.35	−210
	P2	0.25	−80
	P3	0.15	−80
	P0	(reset)	−80
Accent Symbol	FH	0.50	−70
	FM	0.25	−70
	FL	0.10	−70
	DH	0.50	−70
	DM	0.35	−70
	DL	0.15	−70
	A0	(reset)	−70

the onset of an accent command, with the counterpart A0 representing the end. Since, in standard Japanese, the word accent of type 0 (without rapid downfall in the F_0 contour) shows rather different features in prosodic rules when compared to other accent types with rapid downfall, different accent symbols have been prepared for type 0 accent (FH, FM and FL) and others (DH, DM and DL). According to the prosodic rules, the accent symbol FM or DH is assigned if a word is uttered in isolation. The other symbols, FH, FL, DM, and DL, appear in continuous speech due to "accent sandhi" [FHT93].

The initial positions of the commands with respect to the voice onset of the corresponding syllable are shown in Table 21.1. The initial values for the natural angular frequencies of the phrase control mechanism and accent control mechanism are set, respectively, to 3.0 and 20.0 s. The value of the baseline component was determined in such a way that the model-generated contour had the same average (on logarithmic frequency) as the observed contour.

Although, in the scheme of partial analysis-by-synthesis, the best fitting search is conducted only on a limited portion, it may possibly be affected by the phrase components generated prior to the portion. Therefore, proper assignment of the preceding phrase components is important for the performance of the method. According to the prosodic rules, the symbol P1 is usually placed at the beginning of a sentence. However, when a prosodic sentence starts with a conjunction word, such as "ippoo" (on the other hand), the symbol P1 is replaced by the symbol P2 with an additional symbol P3 after the word. The symbols P2 and P3 are placed at the syntactic boundaries of a sentence as shown in the following example:

"P1 kantookinkaiwa P3 namiga yaya takaku P2 enganbudewa P3 koikirino tame P3 mitooshiga waruku natteimasu node P2 funewa chuuishite kudasai P0." (Because the waves are rather high at the inshore sea in Kanto and heavy mist causes low visibility at the coast, careful navigation is recommended for ships in the area.)

To avoid complexity in the explanation, pause and accent symbols are not shown in the example above. Although, in the original prosodic rules, P2 or P3 are selected with the information on the depth of the syntactic boundary, in the proposed scheme, only the number of morae from the adjacent phrase command was taken into consideration. In concrete terms, P2 is selected if the number exceeds 5, and P3 is selected otherwise. If more than two phrase commands are assigned before the portion subject to the partial analysis-by-synthesis, they cannot be searched separately by the scheme. Therefore, in the proposed scheme, only the closest command to the portion is included in the searching process and the other commands are left unchanged. Since a phrase component decreases to almost zero in several morae due to its declining feature, the effect on the result caused by this simplification can be considered small.

In the conventional analysis-by-synthesis method, the search of parameter values is conducted within a wider range of the parameter space. This process may possibly yield similar contours for different recognition candidates and, therefore, may give the best fitting even for a wrong candidate. To cope with this problem, the searching space need to be limited to a smaller range. For the current scheme, the following constraints were put on the model parameters during the analysis-by-synthesis process:

T_0 (position of phrase command): ± 20 ms;

T_1 (onset of accent command): ± 20 ms;

T_2 (end of accent command): ± 20 ms;

A_p (magnitude of phrase command): ±20%;
A_a (amplitude of accent command): ±20%;
α (natural angular frequency for the phrase control mechanism): ±20%;
β (natural angular frequency for the accent control mechanism): ±20%.

21.3 Experiments on the Detection of Recognition Errors

The proposed method is considered to be valid for the detection of recognition errors causing changes in the accent types and/or syntactic boundaries. In order to show this point, several experiments have been conducted, after extracting the fundamental frequency from speech samples at every 10 ms, viz., with 10 ms frame shift. The pitch extraction was performed with few errors, using a method based on the correlation of the LPC residual, with frame length proportional to the time lag [HFS92]. No manual correction was made on the pitch extraction results before the experiment.

As for accent type changes, utterances of four short sentences were recorded for each of the following cases:

(1) Case 1: recognition error changing the word accent type from type N to type 0;

(2) Case 2: recognition error changing the word accent type from type 1 to type 0;

(3) Case 3: recognition error changing the word accent type from type 0 to type N;

(4) Case 4: recognition error changing the word accent type from type 0 to type 1.

Here, types 1 and N, respectively, denote accent types with a rapid downfall in the F_0 contour at the end of the first mora and with a rapid downfall at the end of the second or one of the following morae. The term "type N" was defined temporarily in this paper to denote accent types other than types 0 and 1. For each of cases 1 to 4, sentences U1, U2, U3, and U4 in Table 21.2 were adopted in the experiment. For each of these sentences, a phoneme recognition error was assumed in one of the consonants of the underlined prosodic word, producing a differenct sentence (such as U1', U2', U3' or U4'), and making the accent type of the word change according to one of cases 1 to 4. For instance, "ookuno GA'IKOTSUO mita" (U1 of case 2) was assumed to be wrongly recognized as "ookuno GAIKOKUO mita" (U1' of case 2) with the accent type changing from type 1 to type 0. The partial analysis-by-synthesis was conducted on the capitalized portions.

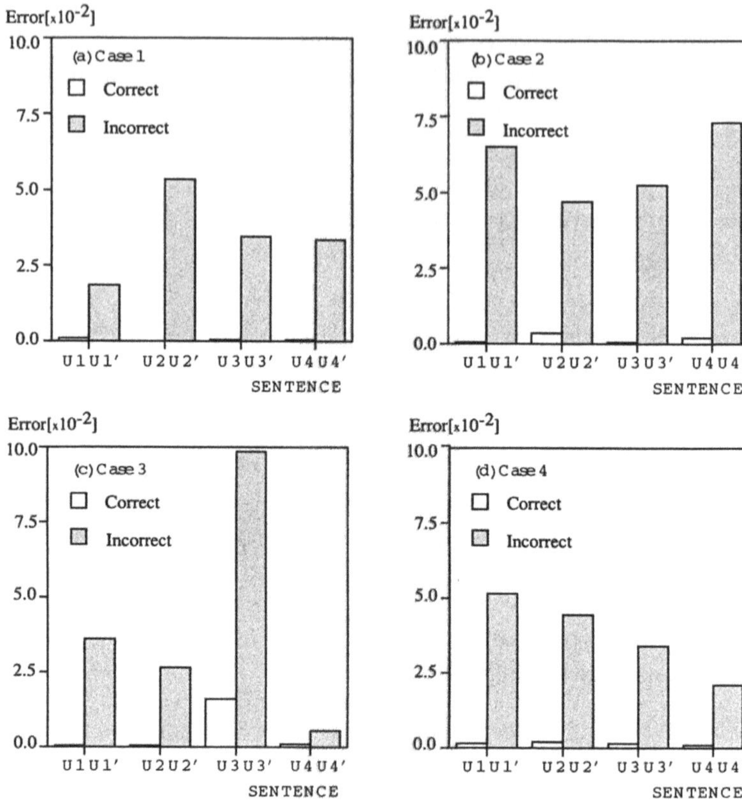

FIGURE 21.2. Partial analysis-by-synthesis errors for the utterances of cases 1 to 4 with correct and incorrect hypotheses on the accent types.

According to the prosodic rules, the accent symbols were assigned to the capitalized portion of U1 as "DH ga A0 i ko tsu o," and to the corresponding portion of U1' as "ga FM i ko ku o." Figure 21.2 shows the results of the experiment for utterences of a male speaker of the Tokyo dialect. For every utterance, a smaller error was obtained for the correct result, indicating the validity of the proposed method. However, the error was rather large for the correct result U3 of case 3, and, conversely, rather small for several wrong results, such as U4' of case 3. Fine alignment in the restrictions on the model parameters should be necessary.

As for syntactic boundary changes, an experiment was conducted for the following two speech samples:

(1) S1: "umigameno maeni hirogaru." (Stretching in front of a turtle.)

(2) S2: "kessekishita kuninno tamedesu." (It is for the nine who were absent.)

TABLE 21.2. Sentences used for the experiment on the detection of recognition errors accompanied by the changes in accent type. For each case, the speech samples were uttered as U14, but were supposed to be wrongly recognized as U1'4'. The capitalized parts indicate the portions for partial analysis-by-synthesis. The symbol " ' " indicates the expected position of the rapid downfall in the F_0 contour, in the Tokyo dialect. Two semantically incorrect sentences are marked with an asterisk.

Case 1	**U1 higa TOPPU'RI kureta (The sun set completely.)**
	U1'* higa TOKKURI kureta (The sun set 'tokkuri'.)
	U2 ishani KAKA'TTE iru (I'm under a doctor's care.)
	U2' ishani KATATTE iru (I'm talking to a doctor.)
	U3 anokowa UCHI'WAO motteita (She had a fan.)
	U3' anokowa UKIWAO motteita (She had a swim ring.)
	U4 sorewa FUKO'ODATO omou (I think it is unhappy.)
	U4' sorewa FUTOODATO omou (I think it is unfair.)
Case 2	**U1 ookuno GA'IKOTSUO mita (I saw many skeletons.)**
	U1' ookuno GAIKOKUO mita (I saw many foreign countries.)
	U2 kareo KA'NKOKUNI maneita (I invited him to Korea.)
	U2' kareo KANTOKUNI maneita (I invited him as a supervisor.)
	U3 tookuni GO'ORUGA mieta (I saw the goal far away.)
	U3' tookuni BOORUGA mieta (I saw the ball far away.)
	U4 ichiban KO'KUNA yarikatada (It is the most cruel way.)
	U4'* ichiban KOTSUNA yarikatada (It is the most 'kotsuna' way.)
Case 3	**U1 ishani KATATTE iru (I'm talking to a doctor.)**
	U1' ishani KAKA'TTE iru (I'm under a doctor's care.)
	U2 anokowa UKIWAO motteita (She had a swim ring.)
	U2' anokowa UCHI'WAO motteita (She had a fan.)
	U3 sorewa FUTOODATO omou (I think it is unfair.)
	U3' sorewa FUKO'ODATO omou (I think it is unfair.)
	U4 hisokani KITAIO yoseru (To expect secretly.)
	U4' hisokani KIKA'IO yoseru (To bring a machine closer in secret.)
Case 4	**U1 ookuno GAIKOKUO mita (I saw many foreign countries.)**
	U1' ookuno GA'IKOTSUO mita (I saw many skeletons.)
	U2 kareo KANTOKUNI maneita (I invited him as a supervisor.)
	U2' kareo KA'NKOKUNI maneita (I invited him to Korea.)
	U3 kanojono KOPPUNI tsugu (To pour into her cup.)
	U3' kanojono TO'PPUNI tsugu (To be second to her.)
	U4 tookuni BOORUGA mieta (I saw the ball far away.)
	U4' tookuni GO'ORUGA mieta (I saw the goal far away.)

Due to an error in detecting morpheme boundaries (S1) or a phoneme recognition error /ta/ $\Longrightarrow$ /ka/ (S2), these utterances can be wrongly recognized as follows:

(1) S1': "umiga menomaeni hirogaru." (The sea stretches before our eyes.)

(2) S2': "kessekishi kakuninno tamedesu." (Being absent. This is for the confirmation.)

The portion subject to partial analysis-by-synthesis was chosen so as to begin at the earliest syntactic boundary in question, ending 5 morae later. In this case, the portion "menomaeni" of S1 and the portion "takuninno" of S2 were selected. According to the prosodic rules, additional phrase components (phrase components generated by the symbols P2 or P3) occur at F_0 contours corresponding to major syntactic boundaries. In the experiment, the following three cases were assumed as the possible hypotheses for the additional phrase component:

(1) H1: an onset of additional phrase component (additional phrase command) immediately before the portion,

(2) H2: no additional phrase command around the portion,

(3) H3: an additional phrase command inside of the portion, viz., between "umigameno" and "maeni" for S1 and between "kessekishita" and "kuninno" for S2.

The hypothesis H1 corresponds to the results S1' and S2' of the incorrect recognition. Although both hypotheses H2 and H3 were assumed as the F_0 contours for the correct recognition, the hypothesis H2 agreed with the prosodic rules for S1, while hypothesis H3 agreed with those for S2. Namely, prosodic symbols were assigned to the portions of partial analysis-by-synthesis as follows:
S1: "(P1 u DH mi ga) me no ma A0 e ni";
S1': "P3 me DH no ma A0 e ni";
S2: "ta A0 P2 ku DH ni A0 n no";
S2': "P3 ka FM ku ni n no".
Distances between observed contours and model-generated contours are shown as errors of the partial analysis-by-synthesis in Fig. 21.3. In both samples, smaller distances were observed for the correct recognition, viz., hypothesis H2 for S1 and hypothesis H3 for S2, indicating that the final recognition results can be correctly selected from several candidates using prosodic features.

21.4 Performance in the Detection of Phrase Boundaries

Although the proposed method should be evaluated after being incorporated in segmental-based recognition systems, its performance was tested in the detection of phrase boundaries. This is because information on phrase boundaries is very useful as the constraints in the recognition process, but

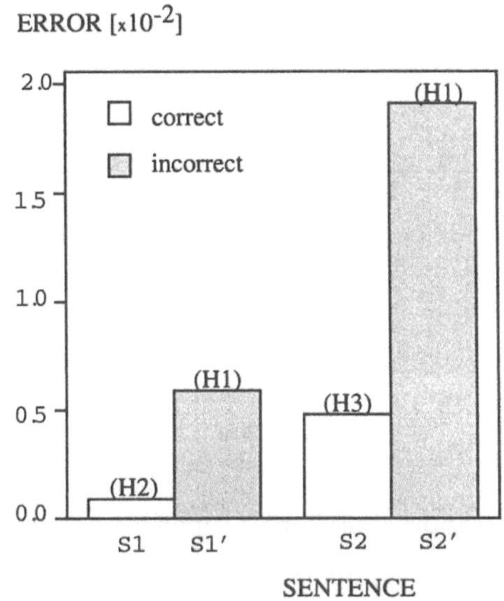

FIGURE 21.3. Partial analysis-by-synthesis errors for samples S1 and S2 with hypotheses of correct and incorrect recognition.

their correct detection is sometimes quite difficult using conventional methods based only on prosodic features. Assuming that phrase boundary positions had been shifted by one or two morae due to recognition errors (one of the hardest conditions for those methods) the proposed method was evaluated as to whether it could detect such shifts [HS96]. The evaluation was conducted using the ATR continuous speech corpus on conference registration. The speech samples used were uttered by the male speaker MAU with an approximate speech rate of 10 morae/s. First, major syntactic boundaries were selected manually from the written text of the corpus, and, then, for each selected boundary, the existence of a phrase command was checked for the observed F_0 contours using the conventional analysis-by-synthesis method. The experiment was conducted for phrase boundaries actually accompanied by phrase commands. We excluded phrase boundaries with long pauses, viz., those corresponding to phrase commands of level P1, because these boundaries can be easily detected.

Unlike the previous section, the portion to be subject to the partial analysis-by-synthesis was automatically set as the period of 1 s with the initial position of the command for correct recognition at the center. Figure 21.4 shows the positions for two speech samples Q1 and Q2, whose contents are noted later. Besides the case of correct recognition, the partial analysis-by-synthesis was conducted after shifting the initial position of the phrase command backward and forward, by one and two morae. When shifting the phrase command forward, we had to note some peculiarities of Japanese

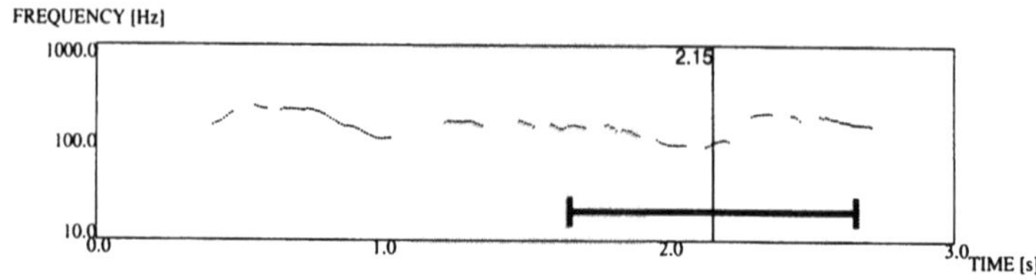

Q1: "koozabangooo shiteeshiteitadakereba / jidootekini hikiotosaremasu."

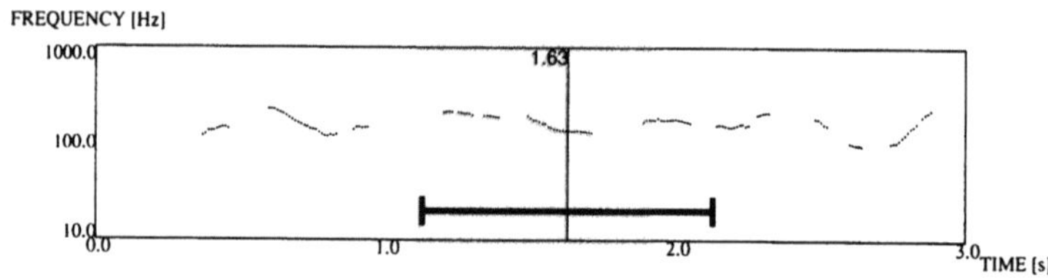

Q2: "mochiron happyoonotokimo / nippongode yoroshiinodesune."

FIGURE 21.4. Portions of partial analysis-by-synthesis for two sentence speech samples Q1 (upper) and Q2 (lower), indicated by thick horizontal bars. The vertical lines in the F_0 contour, as well as the slashes '/' in the Roman-letter descriptions indicate the locations of phrase boundaries.

accentuation. In standard Japanese, an n-mora word is uttered only in one out of $n + 1$ accent types, although the F_0 contour could be formed by 2^n combinations of high and low constituent morae. As a result, if the first mora of a word has a high F_0 contour, the following morae should have a low F_0 contour. This accent type is denoted by type 1, as already mentioned in the previous section. On the contrary, if the first mora has a low F_0 contour, the second mora must have a high F_0 contour. Therefore, if the accent type of the first word in a phrase is originally non-type 1, after a one-mora forward shift of the phrase command, it can still remain non-type 1 with a one-mora forward shift of the onset of the accent command (case 1) or it should be transformed into type 1 with no shift in the onset of the accent command (case 2). When the original accent type was of type 1, it was left unchanged, with a one-mora forward shift on both the onset and end of the accent command.

Figure 21.5 shows the results of the partial analysis-by-synthesis for the following two speech samples:

(1) Q1: "koozabangooo shiteeshiteitadakereba / jidootekini hikiotosaremasu."

(If the banking account is specified, the charge will be automatically subtracted.)

(2) Q2: "mochiron happjoonotokimo / nippongode yoroshiinodesune?"

(Naturally, we can make the presentation also in Japanese, can't we?)

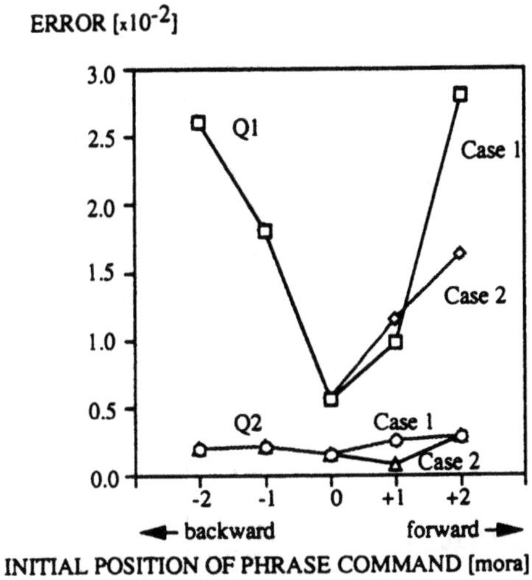

FIGURE 21.5. Partial analysis-by-synthesis errors for the sentence speech samples Q1 and Q2 as functions of the initial position of phrase command. Two hypotheses were considered when the phrase boundary was shifted forwards: Case 1 and Case 2.

The slash "/" indicates the original position of the phrase command searched by the experiment. The horizontal axis of the figure indicates the positions of assumed phrase boundaries represented by the number of morae with respect to the correct boundary location. The results for these two samples indicate two extreme cases: the first one, when the boundary is detected correctly at the right position and the second one, when the correct detection is quite difficult. A close inspection of these two and other examples indicated that the exact detection of phrase boundaries became difficult when the portion of partial analysis-by-synthesis included long voiceless parts and/or the magnitude of the phrase command was small. In all, 38 phrase boundaries were analysed in this way, and the results showed that about 95% of the phrase command positions could be determined with the maximum deviation of 1 mora, and about 40% with no deviation.

Because of microprosodic undulations in F_0 contours, sample-to-sample variations could sometimes be large in terms of distances between the observed contours and the generated contours for correct recognition. A large variation makes it difficult to set a proper threshold for the correct/incorrect decision of phrase boundaries. To cope with this problem, a smoothing process was further introduced on the observed F_0 contour before the process of partial analysis-by-synthesis. In concrete terms, the F_0 contour was treated as a waveform expressed as a function of time and was filtered by a 10 Hz low-pass filter. With this additional process, the mean and the standard deviation of the distance for the correct recognition were reduced by around 20%.

Conclusion

A method was proposed for the selection of the correct recognition result out of several candidates. Although the experiments showed that the method is valid for the detection of recognition errors causing changes in accent types or syntactic boundaries, the following studies are necessary: (1) to increase the performance of the scheme of partial analysis-by-synthesis; (2) to construct a criterion to relate the partial analysis-by-synthesis errors and the boundary likelihood; (3) to combine the method with other prosody-based methods; and (4) to incorporate the method in recognition systems.

Acknowledgment

I would like to express my appreciation to Atsuhiro Sakurai, a graduate student of the author's laboratory, who offered a great help in preparing this paper.

References

[BBBKNB94] G. Bakenecker, U. Block, A. Batliner, R. Kompe, E. Noth and P. Regel–Brietzmann, "Improving parsing by incorporating 'prosodic clause boundaries' into a grammer," In *Proceedings of the International Conference on Spoken Language Processing, Yokohama*, Vol. 3 pp. 1115–1118, 1994.

[FH84] H. Fujisaki and K. Hirose. Analysis of voice fundamental frequency contours for declarative sentences of Japanese. *J. Acoust. Soc. Japan (E)*, 5:233–242, 1984.

[FHT93] H. Fujisaki, K. Hirose, and N. Takahashi. Manifestation of linguistic information in the voice fundamental frequency contours of spoken Japanese. *IEICE Trans. Fundamentals of Electronics, Communications and Computer Sciences*, E76-A:1919–1926, 1993.

[FS71a] H. Fujisaki and H. Sudo. A generative model for the prosody of connected speech in Japanese. *Annual Report of Engineering Research Institute 30*, pp. 75–80, 1971.

[G93] E. Geoffrois, "A pitch contour analysis guided by prosodic event detection," In *Proceedings of the European Conference on Speech Communication and Technology, Berlin*, pp. 793–797, 1993.

[HF82] K. Hirose and H. Fujisaki. Analysis and synthesis of voice fundamental frequency contours of spoken sentences. In *Proceedings of the International Conference on Acoustics, Speech, and Signal Processes*, pp. 950–953, 1982.

[HF93] K. Hirose and H. Fujisaki. A system for the synthesis of high-quality speech from texts on general weather conditions. *IEICE Trans. Fundamentals of Electronics, Communications, and Computer Sciences*, E76-A, 1971–1980, 1993.

[HFS92] K. Hirose, H. Fujisaki and N. Seto, "A scheme for pitch extraction of speech using autocorrelation function with frame length proportional to the time lag." In *Proceedings of the International Conference on Acoustics, Speech, and Signal Processing, San Francisco*, Vol. 1, pp. 149–152, 1992.

[HS96] K. Hirose and A. Sakurai. Detection of syntactic boundaries by partial analysis-by-synthesis of fundamental frequency contours. *Proceedings of the International Conference on Acoustics, Speech, and Signal Processes*, Atlanta, Vol. 4, pp. 809–812, 1996.

[HSK94] K. Hirose, A. Sakurai, and H. Konno. Use of prosodic features in the recognition of continuous speech. In *Proceedings of the International Conference on Spoken Language Processing, Yokohama*, Vol. 3, pp. 1123–1126, 1994.

[KOI88] A. Komatsu, E. Oohira, and A. Ichikawa. Conversational speech understanding based on sentence structure inference using prosodics, and word spotting. *Trans. IEICE (D)*, J71–D:1218–1228, 1988.

[NS94] M. Nakai and H. Shimodaira. Accent phrase segmentation by finding N-best sequences of pitch pattern templates. In *Proceedings of the International Conference on Spoken Language Processing, Yokohama, Japan*, Vol. 1, pp. 347–350, 1994.

[OEKS93] S. Okawa, T. Endo, T. Kobayashi, and K. Shirai, "Phrase recognition in conversational speech using prosodic and phonemic information," *IEICE Trans. Information and Systems*, Vol. E76–D, No. 1 pp. 44–50, 1993.

[OKI89] E. Oohira, A. Komatsu, and A. Ichikawa, "Structure inference algorithm of conversational speech sentence using prosodic information," *Trans. IEICE (A)*, Vol. 72-A, No. 1, pp. 23-31, 1989. (in Japanese).

[WO91] C. W. Wightman and M. Ostendorf, "Automatic recognition of prosodic phrases," In *Proceedings of the International Conference on Acoustics, Speech, and Signal Processing, Toronto*, Vol. 1, pp. 321–324, 1991.

22

Accent Phrase Segmentation by F_0 Clustering Using Superpositional Modelling

Mitsuru Nakai
Harald Singer
Yoshinori Sagisaka
Hiroshi Shimodaira

ABSTRACT We propose an automatic method for detecting minor phrase boundaries in Japanese continuous speech by using F_0 information. In the training phase, F_0 contours of hand labelled minor phrases are parameterized according to a superpositional model proposed by Fujisaki and Hirose, and assigned to some clusters by a clustering method, in which model parameter of reference templates are calculated as an approximation of each cluster's centroid. In the segmentation phase, automatic N-best extraction of boundaries is performed by one-stage Dynamic Programming (DP) matching between the reference templates and the target F_0 contour. About 90% of minor phrase boundaries were correctly detected in speaker independent experiments with the ATR (Advanced Telecommunications Research Institute International) Japanese continuous speech database.

22.1 Introduction

To realize a more natural conversation between machines and human beings, speech recognition has become an important technique. But continuous speech is a difficult task for recognition or understanding and it is costly in terms of CPU time and memory. So, it is thought that phrase boundary information is useful for raising the recognition accuracy and reducing processing time or memory [LMS75, KOI88]. Therefore, the extraction of phrase boundaries from the input speech has become an important problem.

Since the Japanese minor phrase appears in the F_0 (fundamental frequency) contour as a rise–fall pattern, most of the studies are based on prosodic structure. For example, a method for detecting the minor phrase boundaries directly from the local features of the F_0 contour have been proposed[UNS80, SSS89]. Analysis by synthesis, based on the F_0 generation model[FHL92], and methods for utilizing the duration time of phonemes or pauses without using F_0 information[WO91] have also been proposed.

On the other hand, we have proposed an indirect method for the detection of the minor phrase boundaries [SKS90, NS94]. This method is based on the assumption that all F_0 contours of minor phrases can be expressed by a limited number of typical patterns (templates) and that a whole F_0 contour can be approximated by a connection of these patterns. We can thus reformulate the problem of phrase boundary extraction as a recognition of minor phrase patterns. We have implemented this approach as a one-stage DP matching of whole input F_0 contour against a sequence of templates.

In our previous research, templates were constructed by clustering the plain F_0 contours of minor phrases without using a parametric model for minor phrase patterns. By contrast, our new segmentation method mentioned in this chapter is based on a superpositional model. This structured expression enables us to use stochastic modelling of the correlation between adjacent prosodic phrases, and has achieved higher performance than the previous extraction scheme using plain F_0 clustering. By using the F_0 generation model, the minor phrase patterns can be expressed by very few parameters, so that templates can be constructed using comparatively little training data. Furthermore, a major benefit of using an accent model is that constraints on the path in the one-stage DP matching can be applied by the F_0 generation function, and the calculation cost can be considerably reduced.

22.2 Outline of Prosodic Segmentation System

As is shown in Figure 22.1, our segmentation system has two phases: template training and automatic segmentation. In both phases, pause regions are detected at first and they are excluded from the following

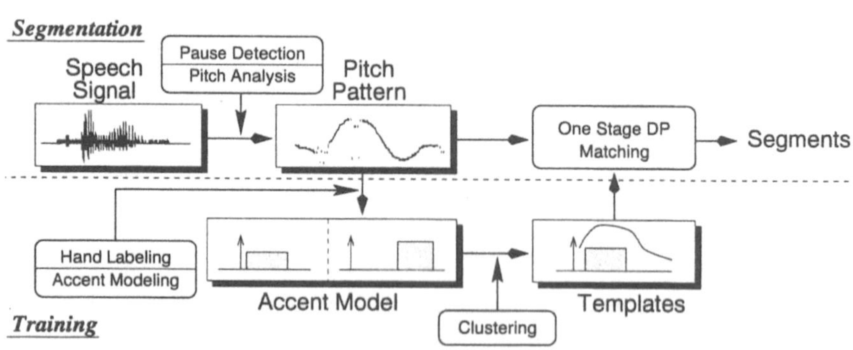

FIGURE 22.1. Block diagram of the prosodic segmentation system.

analysis. The fundamental frequency is analysed by the lag-window method[SF78] and the maximum value of the auto-correlation obtained during the F_0 period estimation is used as a reliability indicator for the estimated F_0 value.

During the training process, F_0 contours of minor phrases are determined by hand-labelled boundaries and modelled semi-automatically by using parameters of the F_0 generation function. Then, reference templates (F_0 templates) are automatically produced by clustering model patterns of minor phrases.

On the other hand, the prosodic segmentation is performed by one-stage DP matching between the F_0 contour of continuous input speech and multiple reference templates. We can search the N-best sequences of F_0 templates using the criterion of the N least squared errors. Then, the connection frames of template sequences are considered to be the minor phrase boundaries of input speech. The point to note is that the input pattern is the observed plain F_0 contour extracted by F_0 analysis and the model parameters are not used in the segmentation phase. The model parameters are used only in the training phase. Therefore, an automatic algorithm for the estimation of F_0 function parameters is unnecessary in this system.

22.3 Training of F_0 Templates

22.3.1 Modelling of Minor Phrase Patterns

To express F_0 contours of minor phrases with a small number of parameters, we use the F_0 generation model proposed by Fujisaki and Hirose[FH84]. In this model, the fundamental frequency F_0 as a function of time t is given by

$$\begin{aligned} \ln F_0(t) &= \ln F_{\min} + \sum_{i=1}^{I} A_{p_i} G_{p_i}(t - T_{pi}) \\ &+ \sum_{j=1}^{J} A_{a_j} \left\{ G_{a_j}(t - T_{aj}) - G_{a_j}(t - (T_{aj} + \tau_{aj})) \right\}, \quad (22.1) \end{aligned}$$

where

$$G_{p_i}(t) = \begin{cases} \alpha_i^2 t e^{-\alpha_i t}, & (t \geq 0) \\ 0, & (\text{otherwise}) \end{cases} \quad (22.2)$$

indicates the impulse response function of the phrase control mechanism and

$$G_{a_j}(t) = \begin{cases} \min[1-(1+\beta_j t)e^{-\beta_j t}, \theta_j], & (t \geq 0) \\ 0, & (\text{otherwise}) \end{cases} \quad (22.3)$$

indicates the step response function of the accent control mechanism. The symbols in the above equations indicate

$F_{\min}$: bias level;
I, J: number of phrase and accent commands;
A_{p_i}: magnitude of the i-th phrase command;
A_{a_j}: amplitude of the j-th accent command;
T_{p_i}: instant of occurrence of the i-th phrase command;
T_{a_j}, τ_{a_j}: onset and duration of the j-th accent command;
α_i, β_j: natural angular frequency of the phrase and accent control mechanisms;
θ_j: ceiling level of the accent component.

Among these parameters we decided to keep α_i, β_j, θ_j fixed because there is no large variation of these parameters between different speakers or speaking styles[Fuj83, HHS94]. Thus the k-th minor phrase pattern occurring at time T_k is represented by the six parameters

$$\mathcal{M}_k = \{T_p^{\mathcal{M}_k}, A_p^{\mathcal{M}_k}, T_a^{\mathcal{M}_k}, \tau_a^{\mathcal{M}_k}, A_a^{\mathcal{M}_k}, \tau_b^{\mathcal{M}_k}\} \quad (22.4)$$

shown in Figure 22.2.

For example, the accent component of k-th minor phrase model $\mathcal{M}_k$ are defined by

$$T_a^{\mathcal{M}_k} = T_{a_k} - T_k, \quad (22.5)$$
$$A_a^{\mathcal{M}_k} = A_{a_k}, \quad (22.6)$$
$$\tau_a^{\mathcal{M}_k} = \tau_{a_k}, \quad (22.7)$$

where the number of minor phrases corresponds with the number of accent commands. Previous accent components which occur before the k-th minor phrase are not contained in this model because each accent component appears as a relatively rapid rise–fall pattern and it does not have an influence on the F_0 contour of succeeding minor phrases. On the other hand, a phrase command given by the impulse response function generates a reduction slope in the F_0 contour for a few succeeding minor phrases, therefore it is necessary to sum up all previous phrase components and to represent them by one phrase command. Then the occurrence instant of

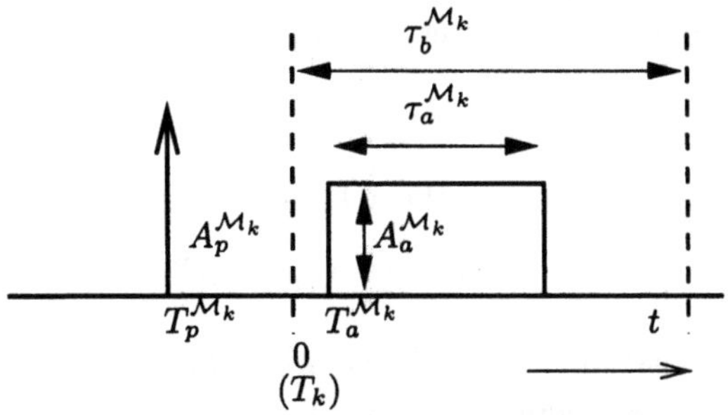

$\tau_b^{\mathcal{M}_k}$:	duration of accent pattern
$A_a^{\mathcal{M}_k}$:	amplitude of accent command
$T_a^{\mathcal{M}_k}$:	onset of accent command
$\tau_a^{\mathcal{M}_k}$:	duration of accent command
$A_p^{\mathcal{M}_k}$:	magnitude of phrase command
$T_p^{\mathcal{M}_k}$:	occurrence of phrase command

FIGURE 22.2. Model parameter set for the minor phrase.

the phrase command and the magnitude of phrase command are defined by

$$T_p^{\mathcal{M}_k} = \frac{\sum_{i=1}^{k'} A_{p_i}(T_{p_i} - T_k)e^{\alpha(T_{p_i}-T_k)}}{\sum_{i=1}^{k'} A_{p_i}e^{\alpha(T_{p_i}-T_k)}}, \tag{22.8}$$

$$A_p^{\mathcal{M}_k} = \frac{\sum_{i=1}^{k'} A_{p_i}e^{\alpha(T_{p_i}-T_k)}}{e^{\alpha T_p^{\mathcal{M}_k}}}, \tag{22.9}$$

where $k'(\leq k)$ is the number of phrase commands occurring before the k-th minor phrase.

22.3.2 Clustering of Minor Phrase Patterns

From the parameterized patterns $\mathcal{M}_j$, a new set of F_0 contours $\hat{P}_j$ is regenerated, where

$$\hat{P}_j = (\hat{p}_{j1}, \ldots, \hat{p}_{ji}, \ldots, \hat{p}_{jL}) \tag{22.10}$$

with $\hat{p}_{ji}$ the logarithmic F_0 value of the i-th frame for the j-th minor phrase and L a fixed length in common for all patterns. Then, the distance between a pair of patterns, $\hat{P}_j$ and $\hat{P}_k$; can be defined by Euclidean distance:

$$D(\hat{P}_j, \hat{P}_k) = \sum_{i=1}^{L}(\hat{p}_{ji} - \hat{p}_{ki})^2. \tag{22.11}$$

After the LBG clustering[LBG80] operation, the model parameters for each cluster are calculated and a set of templates,

$$\mathcal{R} = \{\mathcal{R}_0, \mathcal{R}_1, \ldots, \mathcal{R}_{K-1}\}, \tag{22.12}$$

is constructed. The parameters of the k-th reference template $\mathcal{R}_k$,

$$\mathcal{R}_k = \{T_p^{\mathcal{R}_k}, A_p^{\mathcal{R}_k}, T_a^{\mathcal{R}_k}, \tau_a^{\mathcal{R}_k}, A_a^{\mathcal{R}_k}\}, \tag{22.13}$$

are derived from the parameters of all the minor phrase patterns belonging to the k-th cluster C_k as follows:

$$T_p^{\mathcal{R}_k} = \frac{\sum_{i \in C_k} A_p^{\mathcal{M}_i} T_p^{\mathcal{M}_i} e^{\alpha T_p^{\mathcal{M}_i}}}{\sum_{i=1}^{N_k} A_p^{\mathcal{M}_i} e^{\alpha T_p^{\mathcal{M}_i}}}, \tag{22.14}$$

$$A_p^{\mathcal{R}_k} = \frac{\sum_{i \in C_k} A_p^{\mathcal{M}_i} e^{\alpha T_p^{\mathcal{M}_i}}}{N_k e^{\alpha T_p^{\mathcal{R}_k}}}, \tag{22.15}$$

$$T_a^{\mathcal{R}_k} = \frac{\sum_{i \in C_k} T_a^{\mathcal{M}_i}}{N_k}, \tag{22.16}$$

$$\tau_a^{\mathcal{R}_k} = \frac{\sum_{i \in C_k} \tau_a^{\mathcal{M}_i}}{N_k}, \tag{22.17}$$

$$A_a^{\mathcal{R}_k} = \frac{\int_{T_a^{\mathcal{R}_k}}^{T_a^{\mathcal{R}_k} + \tau_a^{\mathcal{R}_k}} \sum_{i \in C_k} f_i(t)dt}{N_k \tau_a^{\mathcal{R}_k}}, \tag{22.18}$$

$$f_i(t) = \begin{cases} A_a^{\mathcal{M}_i} & T_a^{\mathcal{M}_i} \le t \le T_a^{\mathcal{M}_i} + \tau_a^{\mathcal{M}_i} \\ 0 & \text{otherwise} \end{cases}, \tag{22.19}$$

where N_k is the number of minor phrase patterns in the k-th cluster.

Figure 22.3 shows the reference templates in the case of $K = 8$. The F_0 templates and their model parameters are shown on the left-hand side. But there is a problem with using F_0 templates in that we have to estimate the $F_{\min}$ value for unknown speakers. So, we trained ΔF_0 templates by clustering delta patterns of logarithmic F_0 contours which are generated from minor phrase patterns. ΔF_0 templates are shown on the right-hand side of Figure 22.3.

22.4 Prosodic Phrase Segmentation

22.4.1 One-Stage DP Matching under a Constraint of the F_0 Generation Model

Automatic segmentation is performed by one-stage DP[Ney84] matching between the reference templates and the target F_0 contour. The matching

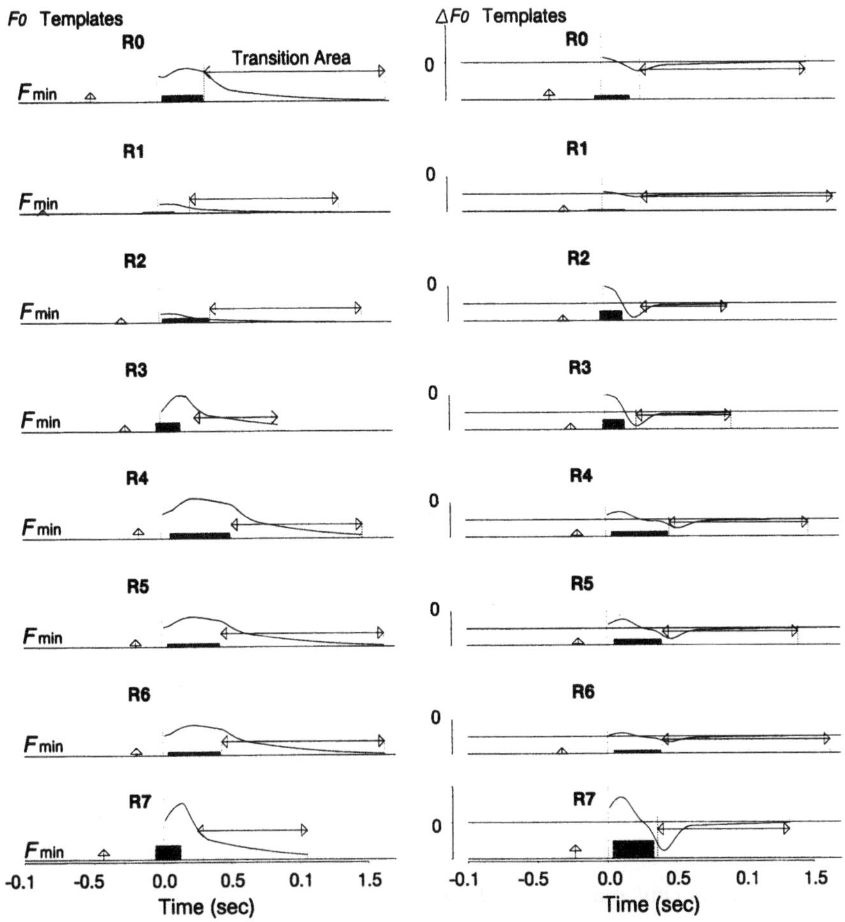

FIGURE 22.3. F_0 contour, ΔF_0 contour, and the corresponding parameters for each minor phrase cluster.

path can be constrained to 45 degrees as shown in Fig. 22.4. In other words, the DP grid $g(i, j, k)$, which is a cross point of input i-th frame and reference j-th frame of the k-th template, does not succeed except grid $g(i-1, j-1, k)$. This is because in the superpositional model with fixed angular frequency (α, β), any F_0 value in a minor phrase is completely defined in terms of time from the onset of commands. In order to allow flexible time warping despite the rigid path constraint, some transition area to another template is defined for each template. The end point of the area is set to $\max \tau_b$, which represents the maximum duration of τ_b of the minor phrase patterns belonging to the cluster for the template. The beginning point of the area is set to the maximum value among the following three values:

(1) the minimal length ($\min \tau_b$) of all minor phrase patterns in the cluster for this template;

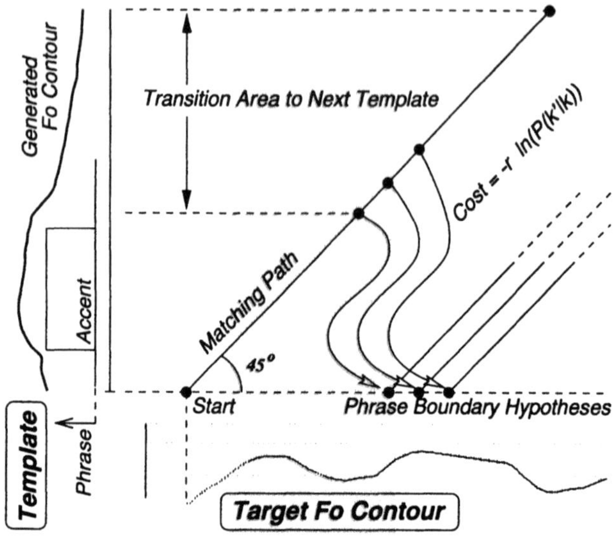

FIGURE 22.4. Matching path between templates and target F_0 contour.

(2) half of the average minor phrase pattern length ($\bar{\tau}_b$);

(3) end of accent command ($T_a + \tau_a$).

Before calculating the distance at each grid, the bias $\ln F_{\min}$, which varies among the speakers and is difficult to estimate, must be added to the logarithmic F_0 value of the template in advance. The ΔF_0 templates shown in the previous section can be used for this problem. In the case of using ΔF_0 templates, it is unnecessary to modify the one-stage DP matching algorithm, but only to fix a variable offset value of templates to zero.

As there is a strong correlation between adjacent templates, we use this additional information by introducing bigram probabilities of minor phrases as a template connection cost defined by

$$C(k^*, k) = -\gamma * \ln\left(P(k \mid k^*)\right), \qquad (22.20)$$

where $P(k \mid k^*)$ is a transition probability from the k^*-th template to the k-th template, and γ is strength factor of bigram constraints.

Algorithm (Case: F_0 templates, 1-best)

Frame number of input pattern	:	$i = 0, \cdots, I-1;$
F_0 template number	:	$k = 0, \cdots, K-1;$
Frame number of k-th F_0 template	:	$j = 0, \cdots, J_k - 1;$
ln F_0 value of input pattern	:	$P(i);$
ln F_0 value of k-th F_0 template	:	$T_k(j);$
F_0 reliability (auto-correlation of F_0 period)	:	$r(i);$
Frame distance on DP grid $g(i,j,k)$	:	$d(i,j,k);$

$$d(i,j,k) = r(i)(P(i) - (T_k(j) + \ln F_{\min}))^2 \;;$$

Cumulative distance of DP grid $g(i,j,k)$	:	$D(i,j,k);$
Transition area of k-th F_0 template	:	$E_k;$
Transition cost from F_0 template k^* to k	:	$C(k^*,k).$

Step 1 Initialize ($i := 0$)
for $k := 0$ to $K-1$ do
$D(0,0,k) = C(pause,k) + d(0,0,k)$
for $j := 1$ to $J_k - 1$ do
$D(0,j,k) = \infty.$

Step 2 (a) for $i := 1$ to $I-1$ do steps (b)–(e)
(b) for $k := 0$ to $K-1$ do steps (c)–(e)
(c) Candidate selection on start frame of templates ($j := 0$)
$(j^*, k^*) = \arg\min_{j' \in E_{k'}, k'} [D(i-1, j', k') + C(k', k)]$
$D(i,0,k) = D(i-1, j^*, k^*) + d(i,0,k) + C(k^*, k)$
(d) for $j := 1$ to $J_k - 1$ do step (e)
(e) shift along a linear matching path
$D(i,j,k) = D(i-1, j-1, k) + d(i,j,k).$

Step 3 Boundary detection by tracing back the path of the optimum template sequence.

22.4.2 N-best Search

The above algorithm is a special case of 1-best search and it sometimes fails to detect whole minor phrase boundaries. To achieve high segmentation accuracy, the technique of N-best search[SC90] is very useful. The basic idea of the N-best method is to keep top N different candidate sub-sequences of templates at every DP grid $g(i,j,k)$. Then, on the final input frame we can find the top N sequences among all possible combinations of templates. Furthermore, the benefit of model based templates is that it is quicker than

TABLE 22.1. Prosodic feature of four speakers.

speaker		50 sentences used for segmentation			
		F_0 (av.)	Phrase length (av.)	No.pauses	No.boundaries
Male	MYI	136.3 Hz	546.3 ms	70	206
	MHO	120.2 Hz	536.4 ms	125	151
Female	FKN	217.6 Hz	652.0 ms	93	183
	FKS	177.4 Hz	599.8 ms	60	216

conventional plain F_0 templates, because the rank of top N candidates on the grid $g(i, j, k)$ is not changed from grid $g(i-1, j-1, k)$ in the case of linear matching path. The N-best selection is necessary only on the starting grid $g(i, 0, k)$ of the k-th template.

22.5 Evaluation of Segmentation System

22.5.1 Experimental Condition

The speech database used in this evaluation test consists of a continuous speech database of phoneme balanced 503 Japanese sentences uttered by 5 male speakers and 2 female speakers[STA+90].

For a total of 565 sentences from 3 speakers (MHT, MSH, MTK), model parameters were semi-automatically extracted[HIV+93]. Then 8 F_0 templates were constructed and bigram probabilities between templates were estimated. Automatic phrase segmentation was performed with 50 sentences from different speakers (MYI, MHO, FKN, FKS) shown in Table 22.1, which are also different in contents from the training sentences, and the 10 best candidates of template sequences were retained. The detail of F_0 analysis parameters is shown in Table 22.2.

Detected boundaries located within 100 ms from the hand labelled boundaries are treated as correct. Correct rate (R_c) and insertion rate (R_i) for each candidate are defined by

$$R_c = \frac{\#\text{ correct detected boundaries}}{\#\text{ hand labelled boundaries}}, \tag{22.21}$$

$$R_i = \frac{\#\text{ incorrect detected boundaries}}{\#\text{ hand labelled boundaries}}. \tag{22.22}$$

Figure 22.5 is an example of the segmentation result in which the eight templates are matched against the F_0 contour in [c] or ΔF_0 contour in

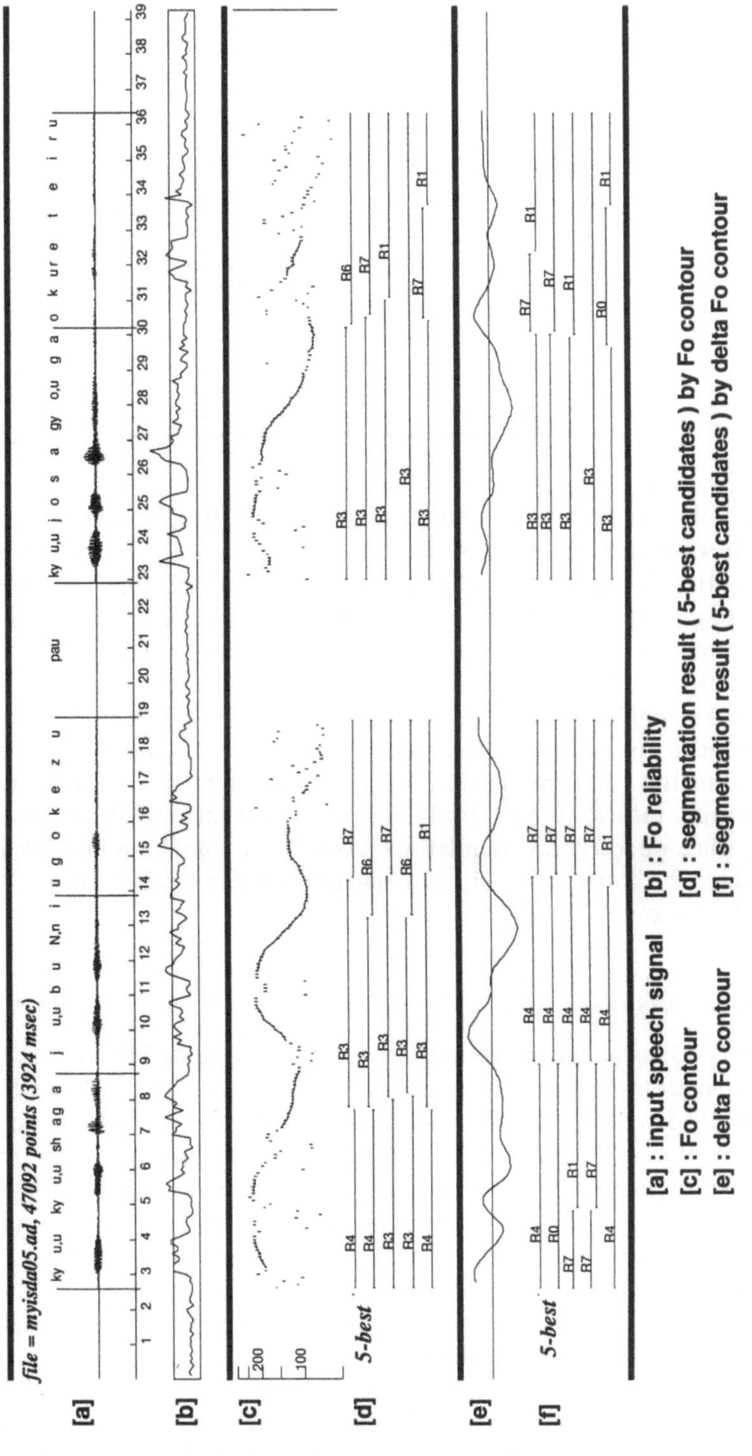

FIGURE 22.5. Example of segmentation result.

TABLE 22.2. Experimental condition.

F_0 extraction	
Window length	512 point (42.7 ms)
Analysis interval	120 point (10.0 ms)
F_0 search	50 – –500 Hz
Extraction method	lag-window method
Automatic segmentation	
No. template	8
No. candidate	10-best

[e] and the five best results are given in [d] and [f]. [a] displays the input speech wave and the vertical lines show the hand-labelled minor phrase boundaries. [b] shows the reliability of the F_0 values, which is used as a weighting coefficient for the squared error between reference template and F_0 contour. The labels on top of each minor phrase candidate in [d] and [f] refer to the templates given in Figure 22.3.

In the example of [d] in Fig. 22.5, the number of hand labelled boundaries in the second part of the sentence after the pause is one, and the correct rate R_c of the first candidate is 100% (1/1). On the average over 5 candidates, the correct rate $\bar{R}_c$ is 80% (4/5), and the insertion rate $\bar{R}_i$ is 20% (1/5). Also, when we merge all boundaries on the 5 best candidates into 1 sequence together, the correct rate of the sequence, which we call "5 best" correct rate R_c^5, becomes 100% (1/1).

22.5.2 *Results*

Figure 22.6 shows the segmentation accuracy of speaker MYI when varying the strength of bigram constraints γ. As γ increases from 0.0 to 1.0, both the averaged correct rate $\bar{R}_c$ and the averaged insertion rate $\bar{R}_i$ decrease, but the "10 best" correct rate R_c^{10} does not decrease so rapidly because undetected boundaries for higher ranking candidates can be detected in lower ranking candidates. Varying γ between 0.0 and 0.05, we notice a reduction of the insertion errors $\bar{R}_i$ from 85.68% to 46.99% while R_c^{10} remains about 92%. Thus the template bigram is a useful constraint for insertion error control. From these results, we fixed γ to 0.05 in the following experiments of multiple speakers.

Figure 22.7 shows the segmentation accuracy of speaker MYI with a variable $F_{\min}$ value in the case of $\gamma = 0.05$. We found that if we set the $F_{\min}$ value incorrectly, the averaged insertion rate $\bar{R}_i$ becomes very large,

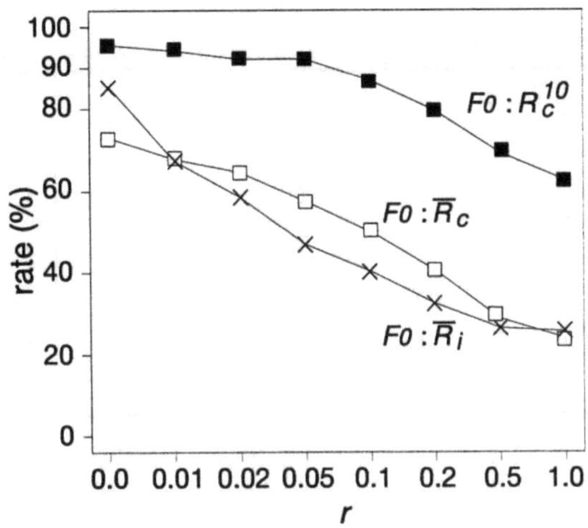

FIGURE 22.6. Segmentation accuracy with a variable strength of bigram constraints γ (MYI).

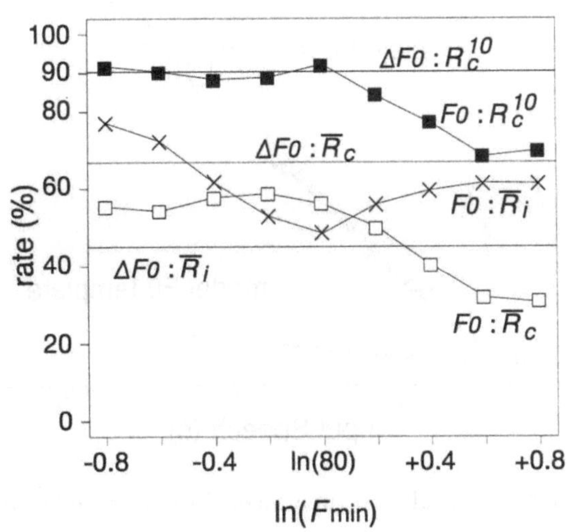

FIGURE 22.7. Segmentation accuracy of F_0 templates with a variable $F_{\min}$ value and segmentation accuracy of ΔF_0 templates (MYI).

and if we set $F_{\min}$ at a high value, the 10 best correct rate R_c^{10} begins to decrease. These results show that the accuracy of the phrase segmentation using F_0 templates is dependent on the accuracy of Fmin estimation. Also, Figure 22.7 shows the segmentation accuracy of ΔF_0 templates in the case of $\gamma = 0.05$. We can see that ΔF_0 templates can achieve high segmentation accuracy as well as the F_0 templates achieved with the desirable $F_{\min}$ value.

TABLE 22.3. Segmentation accuracy by speakers.

Speaker	Plain template		Model based template				
	F_0		F_0			ΔF_0	
	R_c^{10}	$\bar{R}_i$	$F_{\min}$	R_c^{10}	$\bar{R}_i$	R_c^{10}	$\bar{R}_i$
MYI	89.8	52.7	80 Hz	92.2	46.9	90.2	44.1
MHO	85.6	80.9	50 Hz	90.0	75.9	88.1	77.4
FKN	82.5	110.5	120 Hz	85.3	73.7	83.1	88.1
FKS	83.5	69.3	110 Hz	90.8	69.9	87.9	46.9

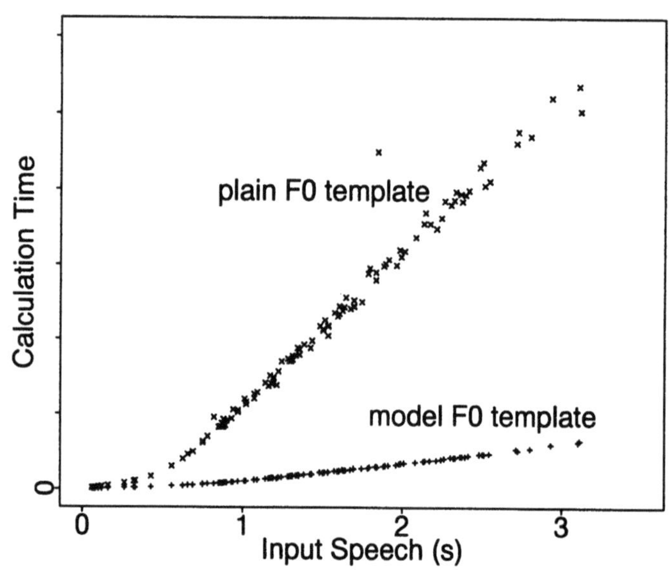

FIGURE 22.8. Calculation time taken for one-stage DP matching.

Similarly, the optimum $F_{\min}$ value for each speaker has been chosen so as to perform high segmentation accuracy, and its results are listed in Table 22.3. The comparison in processing time between using plain F_0 templates with the dynamic time warping (DTW) path and using model F_0 templates with linear matching path is shown in Figure 22.8. As a result, the characteristics of each templates can be described as follows.

Plain F_0 templates: (1) Segmentation accuracy is high with desirable input F_0 contour (such as MYI), but under an influence of many

F_0 extraction errors or a large difference of averaged F_0 value (such as FKN), incorrected phrase boundaries insert frequently.

(2) Processing of N-best sort on DTW path takes large cost in terms of CPU time and memory.

(3) Template training is very easy because it is unnecessary to estimate parameters of the F_0 generation model.

Model F_0 templates: (1) Regardless of F_0 extraction errors, segmentation accuracy is higher than using plain F_0 templates.

(2) The Segmentation process is very fast because of the linear matching path.

(3) It is hard to establish the $F_{\min}$ value for unknown speakers in the automatic segmentation phase.

(4) It is necessary to estimate the parameters of the F_0 generation model in the templates training phase.

Model ΔF_0 templates: (1) Since the ΔF_0 contour is heavily influenced by the errors of F_0 extraction, segmentation accuracy is slightly inferior than when using model F_0 templates, but the accuracy is stable because of the unnecessity of $F_{\min}$ estimation.

(2) Similarly with model F_0 templates, templates training is not easy and the segmentation process is very fast.

Conclusion

We have proposed a segmentation scheme using structured expressions of F_0 contours based on superpositional modelling. These structured expressions enable stochastic modelling of the correlation between adjacent prosodic phrases and permit higher performance than the previous extraction scheme using plain F_0 clustering.

Another interesting aspect of our method is that we do not rely on automatic extraction of parameters for the superpositional model during automatic segmentation. These parameters are used only during training and can thus be hand-corrected.

As a second step, we are now developing an algorithm for a continuous speech recognition system which will use this phrase boundary information effectively.

References

[FH84] H. Fujisaki and K. Hirose. Analysis of voice fundamental frequency contours for declarative sentences of Japanese. *J. Acoust. Soc. Japan (E)*, 5:233–242, 1984.

[FHL92] H. Fujisaki, K. Hirose, and H. Lei. Prosody and syntax in spoken sentences of Standard Chinese. In *Proceedings of the International Conference on Spoken Language Processing, Banff, Canada*, pp. 433–43692, 1992.

[Fuj83] H. Fujisaki. Dynamic characteristics of voice fundamental frequency in speech and singing. In P. MacNeilage, editor, *The Production of Speech*, pp. 39–55. Berlin: Springer-Verlag, 1983.

[HHS94] N. Higuchi, T. Hirai, and Y. Sagisaka. Effect of speaking style on parameters of voice fundamental frequency generation model. In *Proceedings of the Conference IEICE*, Vol. SA–5–3, pp. 488–489, 1994.

[HIV+93] T. Hirai, N. Iwahashi, H. Valbert, N. Higuchi, and Y. Sagisaka. Fundamental frequency contour modelling using statistical analysis. In *Proceedings of the Acoust. Soc. Jpn. Autumn 93*, pp. 225–226, 1993.

[KOI88] A. Komatsu, E. Oohira, and A. Ichikawa. Conversational speech understanding based on sentence structure inference using prosodics, and word spotting. *Trans. IEICE, (D)*, J71–D:1218–1228, 1988.

[LBG80] Y. Linde, A. Buzo, and R. M. Gray. An algorithm for vector quantizer design. *IEEE Trans. Commun.*, COM-28:84–95, 1980.

[LMS75] W. A. Lea, M. F. Medress, and T. E. Skinner. A prosodically guided speech understanding strategy. *IEEE Trans. Acoust., Speech, Signal Processing*, ASSP-23:30–38, 1975.

[Ney84] H. Ney. The use of a one-stage dynamic programming algorithm for connected word recognition. *IEEE Trans. Acoust., Speech, Signal Processing*, ASSP-32:263–271, 1984.

[NS94] M. Nakai and H. Shimodaira. Accent phrase segmentation by finding N-best sequences of pitch pattern templates. In *Proceedings of the International Conference on Spoken Language Processing, Yokohama, Japan*, Vol. 1, pp. 347–350, 1994.

[SC90] R. Schwartz and Y. L. Chow. The N-best algorithm: an efficient and extract procedure for finding the N most likely sentence hypotheses. In *Proceedings of the International Conference on*

Acoustics, Speech, and Signal Processes, Vol. S2. 12, pp. 81–84, 1990.

[SF78] S. Sagayama and S. Furui. A technique for pitch extraction by lag-window method. In *Proceedings of the Conference IEICE*, 1235, 1978.

[SKS90] H. Shimodaira, M. Kimura, and S. Sagayama. Phrase segmentation of continuous speech by pitch contour DP matching. In *Papers of Technical Group on Speech*, Vol. SP90-72. IEICE, 1990.

[SSS89] Y. Suzuki, Y. Sekiguchi, and M. Shigenaga. Detection of phrase boundaries using prosodics for continuous speech recognition. *Trans. IEICE, (D-II)*, J72–D-II:1606–1617, 1989.

[STA+90] Y. Sagisaka, K. Takeda, M. Abe, S. Katagiri, T. Umeda, and H. Kuwabara. A large-scale Japanese speech database. In *Proceedings of the International Conference on Spoken Language Processing, Kobe, Japan*, pp. 1089–1092, 1990.

[UNS80] T. Ukita, S. Nakagawa, and T. Sakai. A use of pitch contour in recognizing spoken Japanese arithmetic expressions. *Trans. IEICE, (D)*, J63–D:954–961, 1980.

[WO91] C. W. Wightman and M. Ostendorf. Automatic recognition of prosodic phrases. In *Proceedings of the International Conference on Acoust., Speech, and Signal Processes*, pp. 321–324, 1991.

23
Prosodic Modules for Speech Recognition and Understanding in VERBMOBIL

Wolfgang Hess[1]
Anton Batliner
Andreas Kiessling
Ralf Kompe
Elmar Nöth
Anja Petzold
Matthias Reyelt
Volker Strom

ABSTRACT Within VERBMOBIL, a large project on spoken language research in Germany, two modules for detecting and recognizing prosodic events have been developed. One module operates on speech signal parameters and the word hypothesis graph, whereas the other module, designed for a novel, highly interactive architecture, only uses speech signal parameters as its input. Phrase boundaries, sentence modality, and accents are detected. The recognition rates in spontaneous dialogs are for accents up to 82.5%, for phrase boundaries up to 91.7%.

In this paper we present an overview about ongoing research on prosody and its role in speech recognition and understanding in the framework of the German spoken language project VERBMOBIL. In Sec. 23.1 some general aspects of the role of prosody in speech understanding will be discussed. Section 23.2 will give some information about the VERBMOBIL project, which deals with automatic speech-to-speech translation. In Secs. 23.3 and 23.4 we then present more details about the prosodic modules currently under development.

---

[1] W. Hess, A. Petzold, and V. Strom are with the Institut für Kommunikationsforschung und Phonetik (IKP), Universität Bonn, Germany; A. Batliner is with the Institut für Deutsche Philologie, Universität München, Germany; A. Kiessling and R. Kompe are with the Lehrstuhl für Mustererkennung, Universität Erlangen-Nürnberg, Germany, and M. Reyelt is with the Institut für Nachrichtentechnik, Technische Universität Braunschweig.

23.1 What Can Prosody Do for Automatic Speech Recognition and Understanding?

The usefulness of prosodic information for speech recognition has been known for a rather long time and emphasized in numerous papers (for a survey see Lea [Lea80], Waibel [Wai88], Vaissière [Vai88], or Nöth [Nö91]). Nevertheless, only very few speech recognition systems did actually make use of prosodic knowledge. In recent years, however, with the growing importance of automatic recognition of spontaneous speech, an increasing interest in questions of prosody and its incorporation in speech recognition systems can be registered.

The role of prosody in speech recognition is that of supplying side information. In principle, a speech recognition system can do its main task without requiring or processing prosodic information. However, as Vaissière [Vai88] pointed out, prosodic information can (and does) support automatic speech recognition on all levels. Following Vaissière [Vai88] as well as Nöth and Batliner [NB95], these are mainly the following.

(1) Prosodic information disambiguates. On almost any level of processing, from morphology over the word level to semantics and pragmatics there are ambiguities that can be resolved (or at least reduced) by prosodic information. As prosody may be regarded as the most individual footprint of a language, the domain in which prosodic information can help depends strongly on the language investigated. For instance, in many languages there are prosodic minimal pairs, i.e., homographs and homophones with different meaning or different syntactic function that are distinguished only by word accent. This is a rather big issue for Russian with its free lexical accent which may occur on almost any syllable. In English there are many noun–verb or noun–adjective pairs where a change of the word accent indicates a change of the word category. In German, the language on which our investigations concentrate, such prosodic minimal pairs exist[2] but play a minor role because they are not too numerous. This holds for single words; yet if continuous speech is looked at, this issue becomes more important in German due to the almost unlimited possibilities to construct compounds. Since word boundaries are usually not indicated by acoustic events and must thus be hypothesized during speech recognition, prosodic information may prove crucial for determining whether a sequence of syllables forms a compound or two separate words [for instance, "*Zwei*räder" (with the accent on the first syllable)—"bicycles" vs "zwei *Rä*der"—"two wheels"]. (Note, however, that "*zwei* Räder" with a contrastive accent on "zwei" cannot be told apart from the compound.)

---

[2] For instance, "ein Hindernis umfahren" would mean "to run down an obstacle" when the verb "umfahren" is accented on the first syllable as opposed to "to drive around an obstacle" when the verb is accented on the second syllable.

(2) On the word level, prosodic information helps limiting the number of word hypotheses. In languages like English or German where lexical accent plays a major role, the information which syllables are accented supports scoring the likelihood of word hypotheses in the speech recognizer. At almost any time during processing of an utterance, several competing word hypotheses are simultaneously active in the word hypothesis graph of the speech recognizer. Matching the predicted lexical stress of these word hypotheses with the information about realized word accents in the speech signal helps enhancing those hypotheses where predicted lexical stress and realized accent coincide, and helps suppressing such hypotheses where they are in conflict (cf., e.g., Nöth and Kompe [NK88]). When we compute the probability of a subsequent boundary for each word hypothesis and add this information into the word hypothesis graph, the syntactic module can exploit this prosodic information by rescoring the partial parses during the search for the correct/best parse (cf. Bakenecker *et al.* [BBB+94], Kompe *et al.* [KKN+95b]). This results in a disambiguation between different competitive parses and in a reduction of the overall computational effort.

(3) On the sentence and higher levels, prosody is likely—and sometimes the only means—to supply "the punctuation marks" to a word hypothesis graph. Phrase and sentence boundaries are for instance marked by pauses, intonation contour resets, or final lengthening. In addition prosody is often the only way to determine sentence modality, i.e., to discriminate, e.g., between statements and (echo) questions (cf. Kiessling *et al.* [KKN+93] or Kompe *et al.* [KBK+94], [KNK+94]). In spontaneous speech we cannot expect that one contiguous utterance or one single dialog turn will consist of one and only one sentence. Hence prosodic information is needed to determine where a sentence begins or ends during the turn. Kompe *et al.* [KKN+95b] supply a practical example from one of the VERBMOBIL time scheduling dialogs. Consider the output of the word hypothesis graph to be the following (correctly recognized) sequence: "ja zur Not geht's auch am Samstag". Depending on where prosodic boundaries are, two of more than 40 (!) meaningful versions[3] possible would read as (1) "Ja, zur Not geht's auch am Samstag." (yes, if necessary it will also be possible on Saturday) or (2) "Ja, zur Not. Geht's auch am Samstag?" (yes, if necessary. Will it also be possible on Saturday?). In contrast to read speech, spontaneous speech is prone to making deliberate use of prosodic marking of phrases so that a stronger dependence on prosody may result from this change in style.

---

[3]"Meaningful" here, means there exists more than 40 different versions (different on the syntactic level including sentence modality) of this utterance all of which are syntactically correct and semantically meaningful. The number of possible different interpretations of the utterance is of course much lower.

Prosodic information is mostly associated to discrete events which come with certain syllables or words, such as accented syllables or syllables followed by a phrase boundary. These prosodic events are highly biased, i.e., syllables or words marked with such events are much less frequent than unmarked syllables or words. In our data, only about 28% of all syllables in continuous speech are accented, and strong phrase boundaries (cf. Sec. 23.3.1) occur only after about 15% of all syllables (which is about 19% of all word boundaries). This requires special cost functions in pattern recognition algorithms to be applied for recognizing and detecting prosodic events. Moreover, as the prosodic information serves as side information to the mainstream of the recognition process, a false alarm is likely to cause more damage to the system performance than a miss, and so it is appropriate to design the pertinent pattern recognition algorithms in such a way that false alarms (i.e., the indication of a prosodic event in the signal when none is there) are avoided as much as possible. We can also get around this problem when the prosodic module passes probabilities or likelihoods, i.e., scores rather than hard decisions to the following modules which, in turn, must then be able to cope with such information.

23.2 A Few Words About VERBMOBIL

VERBMOBIL [Wah93] is a multidisciplinary research project on spoken language in Germany. Its goal is to develop a tool for machine translation of spoken language from German to English and (in a later stage) also from Japanese to English. This tool (which is also called VERBMOBIL) is designed for face-to-face appointment scheduling dialogs between two partners of different nationalities (in particular, German and Japanese). Each partner is supposed to have good passive yet limited active knowledge of English. Correspondingly, the major part of a dialog will be carried out in English without intervention by VERBMOBIL. However, when one of the partners is temporarily unable to continue in English, he (or she) presses a button and starts speaking to VERBMOBIL in his/her native language. The button is released when the turn is finished. VERBMOBIL is then intended to recognize the utterance, to translate it into English, and to synthesize it as a spoken English utterance. A first demonstrator was built in early 1995, and the second milestone, the so-called research prototype, is due in late 1996. Twenty-nine institutions from industry and universities participate in this project.

It was specified that any speech recognition component of VERBMOBIL should include a prosody module.

The architecture of the 1995 demonstrator is mostly sequential. If the speaker invokes VERBMOBIL, the spoken utterance is first processed by the speech recognition module for German. From this module, word hypotheses are passed to the syntactic analysis module and on to the translation path with the modules of *semantic construction*, *transfer*, *generation* (English), and *speech synthesis* (English). The flow of data and hypotheses is controlled by the semantic analysis and dialog processing modules. If an utterance is not or not completely recognized or translated, the dialog processing module invokes a generation module for German whose task is to create queries for clarification dialogs or requests to the speaker (for instance, to talk louder or more clearly). Such utterances are then synthesized in German.

During the dialog parts which are carried out in English, a word spotter (for English) is intended to supply the necessary domain knowledge for the dialog processing module to be able to "follow" the dialog. As the input is "controlled spontaneous" speech, German utterances to be translated may be elliptic so that such knowledge is needed to resolve ambiguities. (The word spotter is likely to be replaced with a complete speech recognizer for English in a later stage.)

The scope of the prosodic analysis module (for German) currently under development for the VERBMOBIL research prototype is shown in Figure

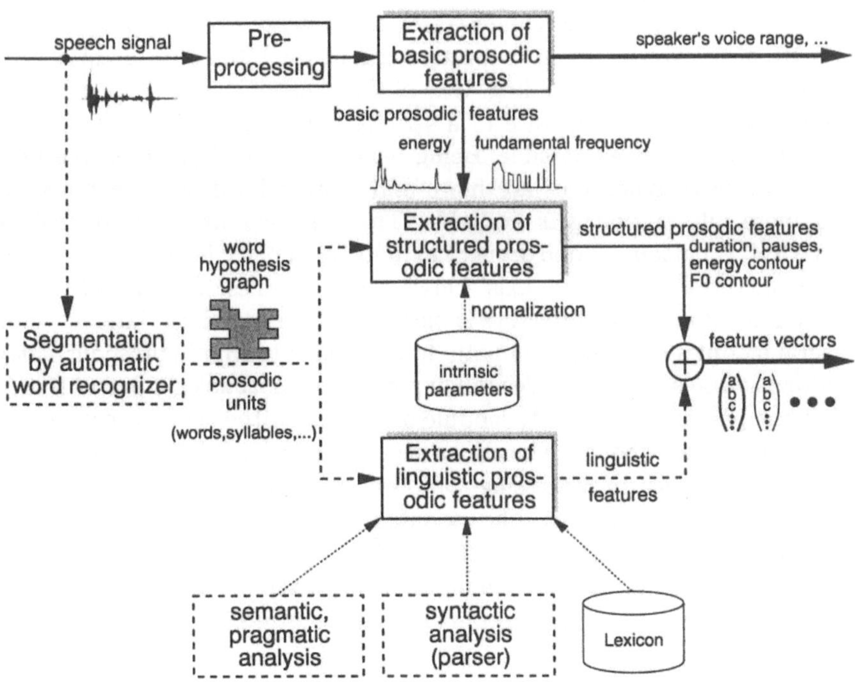

FIGURE 23.1. Prosodic analysis module for the VERBMOBIL research prototype. For more details, see the text. Figure provided by Nöth et al. (personal communication).

23.1. In the present implementation, the module operates on the speech signal and the word hypothesis graph (as supplied by the speech recognition module). From the speech signal basic prosodic features and parameters [KKN+92], such as energy or fundamental frequency, are extracted, whereas the word hypothesis graph carries information about word and syllable boundaries. Interaction with and feedback from higher information levels (such as syntax and semantics) and the pertinent modules are envisaged. The output of the module consists of information about the speaker (voice range etc.) to be used for speaker adaptation (this cannot be discussed here due to lack of space), and the feature vectors which are used as input to the boundary and accent classifiers. The module is described in Sec. 23.3.

For training and test a large database of (elicited) spontaneous speech has been collected [HKT95]. The data consist of appointment scheduling dialogs in German; they have been recorded at four university institutes in different regions of Germany; the speakers were mostly students. To obtain utterances that are as realistic (with respect to the VERBMOBIL environment) as possible, each speaker has to press a button when speaking and keep it pressed during his/her whole turn. The whole database was transcribed into an orthographic representation, and part of it was also labelled prosodically (cf. Sec. 23.3.2).

Besides developing the demonstrator and research prototypes, VERBMOBIL also investigates an innovative and highly interactive architecture model for speech understanding. One of the goals of this activity is to develop algorithms that operate in a strictly incremental way and provide hypotheses as early as possible. Being rather crude and global in the first moment, these hypotheses are more and more refined as time proceeds and more information gets available. The pertinent architecture (called INTARC) is bottom-up and sequential in its main path; however, top-down and transversal connections exist between the modules. The prosody module contained in this architecture is placed separately and can interact with several modules from the main path; it is intended to supply prosodic (side) information to several modules ranging from the morphologic parser to the semantic parser. The prosody module only exploits the acoustic signal and some information about the locations of syllabic nuclei as bottom-up inputs; however, it is open to processing top-down information such as prediction of sentence mode or accent. The module is described in Sec. 23.4.

As work on these modules is still ongoing, this paper will be a progress report. Most results will thus be preliminary or still incomplete. For more details the reader is referred to the original literature.

23.3 Prosody Module for the VERBMOBIL Research Prototype

This section discusses the module developed in Erlangen and Munich (cf. Kompe *et al.* [KKN+95b] and earlier publications by the same authors) which was originally trained on read speech. In read speech and the pertinent train inquiry the recognition rates were rather high: 90.3% for primary accents, and 94.3% for the phrase boundaries. This module was adapted to the VERBMOBIL spontaneous speech environment. First results show that the recognition rates are considerably lower than for read speech, but that the presence of the module positively contributes to the overall performance of the speech understanding system.

23.3.1 Work on Read Speech

According to the three application areas mentioned in Sec. 23.1, prosodic analysis algorithms were developed for (1) recognition of accents, (2) detection of boundaries, and (3) detection of sentence modality. A large corpus of read sentences was available for this task. The so-called Erlanger Bahnanfragen; Erlangen train inquiries (ERBA) corpus contains a set of 10,000 unique sentences generated by a stochastic sentence generator (which was based on a context-free grammar and 38 sentence templates). It was read by 100 naive speakers (with 100 sentences per speaker). Out of these 100 speakers, 69 were used for training, 21 for test, and the utterances of the remaining 10 speakers were used for perceptual tests and for evaluating parts of the classifiers.

Syntactic boundaries were marked in the grammar and included in the sentence generation process with some context-sensitive processing [KNK+94]. Listening tests [BKBN95] showed a high agreement (92%) between these automatically generated labels and the listeners' judgments.

Four types of boundaries are distinguished (with the notation close to that applied in the prosodic description system ToBI [SBP+92]).

- Boundaries B3—full prosodic phrase boundaries (between clauses); such boundaries. are expected to be prosodically well marked.

- Boundaries B2—boundaries between constituents in the same phrase or intermediate (secondary) phrase boundaries; such boundaries tend to carry a weak prosodic marking.

- Boundaries B1—boundaries that syntactically pertain to the B2 category but are likely to be prosodically unmarked because the pertinent constituent is integrated with the preceding or following constituent to form a larger prosodic phrase.

- Boundaries B0—any other word boundary. It was assumed that there is no difference between the categories B0 and B1 in the speech signal so that these two categories were treated as one category in the recognition experiments.

An example is given in Fig. 23.3 (cf. Sec. 23.4.2).

In addition different accent types were defined [KKB+94]: primary accents A3 (one per B3 boundary), secondary accents A2 (one per B2 phrase), other accents A1, and the category A0 for non-accented syllables.

Computation of the acoustic features is based on a time alignment of the words on the phoneme level as obtained during word recognition. For each syllable to be classified and for the six immediately preceding and following syllables a feature vector is computed which contains features such as normalized duration of the syllabic nucleus; F_0 minimum, maximum, onset, and offset, maximum energy and the position of the pertinent frames relative to the position of the actual syllable; mean energy and F_0, and information about whether this syllable carries a lexical accent. In total 242 features per syllable are extracted and calculated.

For the experiments using ERBA all these 242 features were fed into a multi-layer perceptron (MLP) with two hidden layers and one output node per category [KKN+95a]. The output categories of the MLP are six combinations of boundaries and accents: (1) A0/B0-1, (2) A0/B2, (3) A0/B3, (4) A1-3/B0-1, (5) A1-3/B2, and (6) A1-3/B3. To obtain accent and boundary classification separately, the categories were regrouped; in each case the pertinent MLP output values were added appropriately. The most recent results [KKN+95b] showed recognition rates for boundary recognition of 90.6% for B3, 92.2% for B2, and 89.8% for B0/1 boundaries; the average recognition rate was 90.3%. Primary accents were recognized with an accuracy of 94.9%.

As an alternative a polygram classifier was used. As Kompe *et al.* [KNK+94] had shown, the combination of an acoustic-prosodic classifier with a stochastic language model improves the recognition rate. To start with, a modified n-gram word chain model was used which was specifically designed for application in the prosody module. First of all, the n-gram model was considerably simplified by grouping the words into a few rather crude categories whose members are likely to behave prosodically in a similar way (for ERBA these were: names of train stations, days of the week, month names, ordinal numbers, cardinal numbers, and anything else). This enabled us to train rather robust models on the ERBA corpus. Prosodic information, i.e., boundaries (B2/3) and accents (A2/3), was incorporated in much the same way as ordinary words. For instance, let

$$v_i \in V \ \ (= \{\neg\text{B3},\ \text{B3}\})$$

be a label for a prosodic boundary attached to the i-th word in the word chain $(w_1, \ldots, w_m)$. As the prosodic labels pertaining to the other words in the chain are not known, the a priori probability for v_i is determined from

$$P(w_1 \ \ldots \ w_i \ v_i \ w_{i+1} \ \ldots \ w_m) \ .$$

The MLP classifier, on the other hand, provides a probability or likelihood

$$P(v_i|\mathbf{c}_i),$$

where $\mathbf{c}_i$ represents the acoustic feature vector at word w_i. The two probabilities are then combined to

$$Q(v_i) \ = \ P(v_i|\mathbf{c}_i) \ P^{\xi}(w_1 \ \ldots \ w_i \ v_i \ w_{i+1} \ \ldots \ w_m) \ ;$$

ξ is an appropriate heuristic weight. The final estimate $v_i^\star$ is then given by

$$v_i^\star \ = \ \text{argmax} \ Q(v_i) \ ; \quad v_i \in V \ .$$

To enable the polygram classifier to be used in conjunction with word hypothesis graphs, the language model had to be further modified. In a word hypothesis graph, as is supplied by the speech recognizer, each edge contains a word hypothesis. This word hypothesis usually can be chained with the acoustically best immediate neighbors (i.e., the best word hypotheses pertaining to the edges immediately preceding and following the one under investigation) to form a word chain which can then be processed using the language model as described before. In addition to the word identity each hypothesis contains its acoustic probability or likelihood, the numbers of the first and last frame, and a time alignment of the underlying phoneme sequence. This information from the word hypothesis graph is needed by the prosodic classifier as part of its input features. In turn the prosodic classifier computes the probability of a prosodic boundary to occur after each word of the graph, and provides a prosodic score which is added to the acoustic score of the word (after appropriate weighing) and can be used by the higher-level modules.

As expected, the polygram classifier works better than the MLP alone for the ERBA data, yielding recognition rates of up to 99% for the three-class boundary detection task. Kompe *et al.* [KKN+95b], however, state that this high recognition rate is at least partly due to the rather restricted syntax of the ERBA data.

23.3.2 Work on Spontaneous Speech

The prosodic module described in Sec. 23.3.1 was adapted to spontaneous speech data and integrated in the VERBMOBIL demonstrator. For spontaneous speech it goes almost without saying that it is no longer possible to generate training and test data in such a systematic way as was done for the read speech data of the ERBA corpus. To adapt the prosodic module to the spontaneous-speech VERBMOBIL scenario, real training data had to be available, i.e., prosodically labelled original utterances from the VERBMOBIL-PHONDAT corpus. A three-level labelling system containing one functional and two perceptual levels was developed for this purpose [Rey93], [Rey95]. The labels on the functional level comprise sentence modality and accents. On the first perceptual level (perceived) prosodic boundaries are labelled. These are (cf. Sec. 23.3.1): full prosodic phrase boundaries (B3), intermediate (secondary) phrase boundaries (B2), and any other (word) boundaries (B0). (Note that the boundaries carry the same labels for the spontaneous VERBMOBIL data and for the read speech of ERBA; since the boundaries in the spontaneous data are perceptually labelled rather than syntactically predicted, their meaning may be somewhat different.) To cope with hesitations and repair phenomena as they occur in spontaneous speech, an additional category "irregular boundary" (B9) was introduced. On the second perceptual level intonation is labelled using a descriptive system which is rather close to ToBI [SBP+92]. At present the prosodically labelled corpus contains about 670 utterances from 36 speakers (about 9500 words or 75 min of speech); this corpus is of course much smaller than ERBA, although it is continuously enlarged.

In principle, Kompe *et al.* [KKN+95b] used the same classifier configuration for the spontaneous data. Since the neural network used for the ERBA database proved too large for the smaller corpus of training data, separate nets, each using only a subset of the 242 input features, were established for the different classification tasks. One network distinguishes between the accents A0 and A1/2/4 (A4 meaning emphasis or contrast accent; A3 accents were not labelled for this corpus), the second one discriminates between the two categories B3 and B0/2/9 (i.e., any other boundary), and the third one classifies all categories of boundaries (B0, B2, B3, and B9) separately. The language model for the polygram classifier comprises a word list of 1186 words which were grouped into 150 word categories.

For each word in the word hypothesis graph the prosodic classification results were added together with their scores [NP94].

First results show that the recognition performance goes down considerably when compared to the read-speech scenario. This is not surprising because there is much less training data and because the variability between speakers and utterances is much larger. The most recent results [KKN+95b] (referring to word chains) are displayed in Table 23.1.

The main difference between the results of the multi-layer perceptron (without language model) and the polygram classifier is the recognition rate for the B0, i.e., the non-boundary category. Since the B0 category is much more frequent than any of the others, a poor recognition rate for B0 results in a many false alarms which strongly degrade the results. The improvement for B0 resulting from the language model goes mostly at the expense of weak (B2) and irregular (B9) boundaries, and even the recognition rate for B3 boundaries goes down although the overall recognition rate mounts by more than 20% points.

TABLE 23.1. Prosodic module by Kompe *et al.* [KKN+95b]: recognition results for boundaries (all numbers in percent). (MLP) Multi-layer perceptron classifier; (LM3) polygram classifier with a three-word left and right context. In all experiments the training data were different from the test data.

	Overall	B0	B2	B3	B9
MLP	60.6	59.1	48.3	71.9	68.5
LM3	82.1	95.9	11.4	59.6	28.1

In the current VERBMOBIL implementation the syntactic, semantic, and dialog modules are most interested in obtaining estimates of B3 boundaries. For this purpose the above-mentioned two-class (B0/2/9 vs B3) boundary recognition algorithm was implemented and trained. In contrast to the four-class recognizer (B0, B2, B3, and B9) where many of the confusions occurred between B0 and B2/B9, the overall recognition rate was much improved. For the neural network without language model, the best results were 78.4% for B0/2/9 vs 66.2% for B3, and in a combination of the neural network and a polygram classifier, where a two-word context was used for the language model, the recognition rates amounted to 90.5% for B0/2/9 vs 54.1% for B3. Note that again for the polygram classifier the number of false B3 alarms was greatly reduced at the expense of a drop in the B3 boundary recognition rate. When using the word chain instead of the word hypothesis graph, better results (up to 91.7% for B0/2/9 vs B3) could be achieved.

Even though the results are still to be improved, Bakenecker *et al.* [BBB+94] as well as Kompe *et al.* [KKN+95b] report that the presence of prosodic information considerably reduced the number of parse trees in the syntactic and semantic modules and thus decreased the overall search complexity.

As to the recognition of accented vs non-accented syllables on the same database, 78.4% were achieved for word graphs and 83.5% for word chains. First results concerning the exploitation of prosodically marked accents in the semantic module are described in (Bos *et al.* [BBK95]).

23.4 Interactive Incremental Module

The prosody modules developed in Bonn by Strom [Str95a] and Petzold [Pet95] for the INTARC architecture (cf. Sec. 23.2) work in an incremental way. Eleven features suitable for direct classification are derived from the F_0 contour and the energy curve of the speech signal for consecutive 10 ms frames (Sec. 23.4.1). Further processing is carried out in three steps (Secs. 23.4.2 and 23.4.3). For word accent detection, a statistical classifier is applied. Another Gaussian classifier works on phrase boundaries and sentence mode detection. Finally a special module deals with focus detection when the focus of an utterance is marked by prosody.

23.4.1 F_0 Interpolation and Decomposition

All the input features used in the prosody module are (1) short-time energy and the F_0 contour of the speech signal, and (2) information about the locations of the syllabic nuclei. No further input information is needed for the basic processing.

From Fujisaki's well known intonation model [Fuj83] we adopted the principle of linear decomposition of the F_0 contour into several subbands. In Fujisaki's model an F_0 contour is generated by superposition of the output signals of two critically damped linear second-order systems with different damping constants. One of these systems generates the representation of word accents in the F_0 contour and uses a sequence of rectangular time functions, the so-called accent commands, as its input. The second system, the so-called phrase accent system, is responsible for the global slope of the F_0 contour within a prosodic phrase; it is driven by the pulse-shaped phrase commands. It has been shown that this model is able to approximate almost any F_0 contour very accurately (cf. Möbius *et al.* [MPH93], Mixdorff and Fujisaki [MF94]) and thus proves to be an excellent tool, e.g., for speech synthesis. For recognition purposes an algorithm for automatic parametrization of F_0 contours using this model had been developed earlier [MPH93] which yielded good results for several categories of one-phrase and two-phrase sentences. In the present application, however, where F_0 contours of sentences of arbitrary phrase structure have to be processed in an incremental way it proved more appropriate to use features which are closer to the original F_0 contour than the phrase and accent commands of Fujisaki's model. As the phrase and accent components have different damping constants, their output signals, which are added together in the model to yield the (synthesized) F_0 contour, occupy different frequency bands; hence the decomposition of the F_0 contour into frequency bands that roughly correspond to the damping constants of the phrase and accent commands in Fujisaki's model will provide features that correspond to the accent and phrase components and are sufficiently robust for automatic processing under adverse conditions at the same time.

This decomposition of the F_0 contours, however, is still a non-trivial task. Since fundamental frequency does not exist during unvoiced segments (i.e., pauses and voiceless sounds), an interpolation of the F_0 contour is required for these frames so that jumps and discontinuities introduced by assigning arbitrary "F_0" values are smoothed out prior to the decomposition into several frequency bands. To obtain an interpolation which is band limited in the frequency domain, an iterative procedure is applied (Fig. 23.2). Per

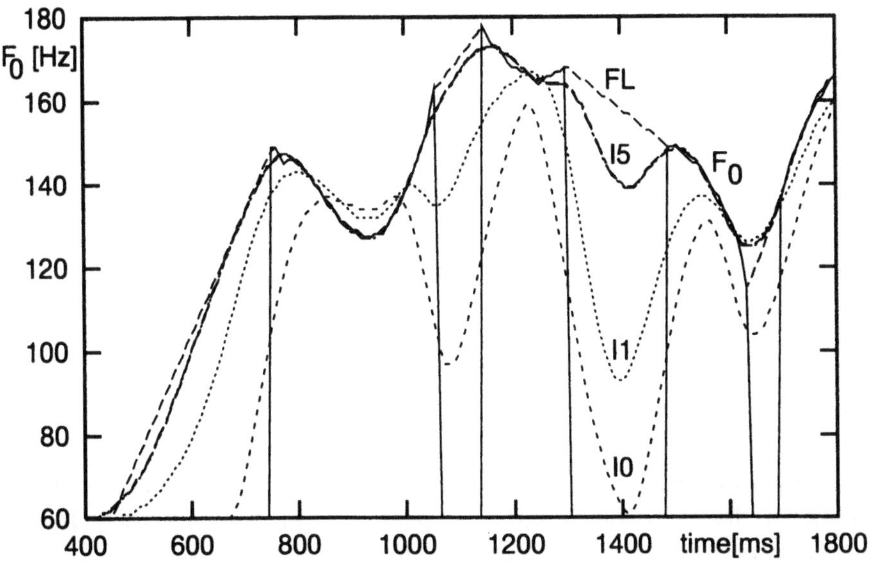

FIGURE 23.2. Interpolation of F_0 through unvoiced segments by iterative filtering. After Strom [Str95b], [Str95a]. (FL) Linear interpolation of the F_0 contour through unvoiced segments; (I0) contour after low-pass filtering; (I1) contour after first iteration; (I5) contour after fifth iteration.

definition, a low, constant value (greater than zero) is assigned to unvoiced frames within the utterance. Moreover, the F_0 contour is defined to descend linearly toward this value before the first and after the last voiced frame of the utterance. The contour is then low-pass filtered using a Butterworth filter with almost linear-phase behavior. As the output of the low-pass filter strongly deviates from the original contour, all voiced frames are restored to their original F_0 values, and, finally, continuity between the original contour and the output of the low-pass filter at the beginning and end of an unvoiced segment is enforced by weighting the difference between the output of the low-pass filter and a linear interpolation of the F_0 contour across the unvoiced segment. These three steps (low-pass filtering, restoring the original F_0 values in voiced frames, and enforcing continuity) are then repeated until, after five iterations, the interpolated "F_0" values in unvoiced frames match well with the original parts of the contour in the voiced frames. Since this procedure only uses digital filters (including a moving average for weighting) and local decisions it is compatible with the requirement of incrementality.

The next step is the decomposition of the interpolated F_0 contour into three subbands. These subbands, ranging from 0 to about 0.5 Hz, from 0.5 to about 1.5 Hz, and from 1.5 to about 2.5 Hz, roughly correspond to the accent and phrase components of Fujisaki's model; the exact values of the edge frequencies were optimized with respect to the recognition rate of the word accent classifier. Digital Butterworth filters with negligible phase distortions are used to perform this task. The three subbands and the original F_0 contour (after interpolation) together yield four F_0 features. The time derivatives of these four features, approximated by regression lines over 200 ms, yield four ΔF_0 features. In addition three energy features, as proposed by Nöth [Nö91], are calculated for three frequency bands of the speech signal (50–300 Hz, 300–2300 Hz, and 2300–6000 Hz); these features are derived from the power spectrum of the signal followed by a time-domain median smoothing.

23.4.2 Detecting Accents and Phrase Boundaries, and Determining Sentence Mode

For accent detection based on the 11 features from Sec. 23.4.1 a modified Gaussian classifier [Nie83] with a special cost function was used. In the training phase every frame was grouped into one of five classes: (1) no vowel; (2) vowel in non-accented syllable; (3) vowel with primary accent; (4) vowel with secondary accent; and (5) vowel with emphasis. These classes were recombined to the categories "accented vowel yes/no", followed by a filter that suppresses segments marked as accented when they are shorter than six consecutive frames.[4] Figure 23.3 shows the output of the accent detector for a sample utterance together with the F_0 contour, the interpolated F_0 contour, the three subband contours, and the three energy measures. A syllable was regarded as accented when at least one frame within that syllable was marked accented by the classifier. Table 23.2 shows the results for a corpus of utterances consisting of a total of 9887 syllables. The total recognition rate was 74.0%, whereas the average recognition rate was 71.5%. The ratio between non-accented and accented syllables was about 3:1.

The boundary detector processes a moving window of four consecutive syllables, where the output refers to the boundary between the second and the third syllable. A Gaussian classifier was trained to distinguishing between all combinations of the four types of boundaries (B3, B2, B0, and B9) and the three sentence modes (question, statement, progredient). These classes were then remapped onto the four boundary types on the one hand, and onto the sentence modes question, statement, and progredient when a B3 boundary was detected, and zero (as the dummy category) otherwise. With the corpus of 9887 syllables from the prosodically labelled VERBMOBIL database, the total recognition rate for the boundaries was

[4]With framewise classification there are much more training data available than with a syllable-based classification scheme. For this reason a frame-by-frame classification strategy was applied in the present version. As the prosodically labelled corpus is continuously enlarged, we intend to classify accents on a syllable-based scheme in future versions of the accent detector.

TABLE 23.2. Confusion matrix of the accent detector (after Strom [Str95a]). All numbers in percent. (RFO) Relative frequency of occurrence; (A) accented; (NA) non-accented.

	Classified as		
Accenting	A	NA	RFO
A	66.53	33.47	25.39
NA	23.45	76.55	74.61

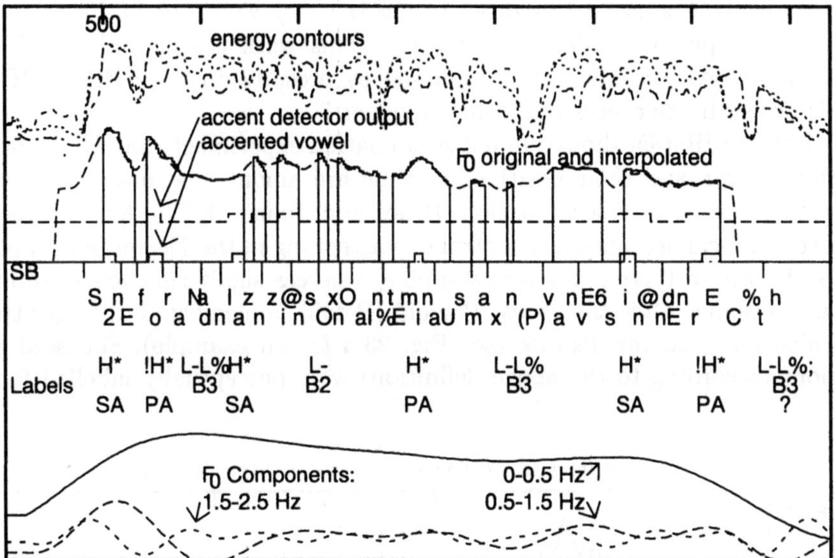

FIGURE 23.3. Accent detection by decomposition of the F_0 contour and subsequent classification (after Strom [Str95a]). Utterance "schön hervorragend, dann lassen Sie uns doch noch ein' Termin ausmachen (P). wann wär's Ihnen denn recht." Phonetic transcription (in SAM-PA notation; word boundaries marked by spaces for better legibility): "S2n EforaN dan lazn zi @ns Ox nO aIn %tEmin aUsmaxn (P) van vE6s in@n dEn rEC%t". In the figure the phonetic transcription had to be displayed in two rows for reason of space. (P) Pause; (SB) Syllable boundaries (word boundaries marked by longer lines); (labels) prosodic labelling. Upper line: tone labelling; middle line: boundaries; lower line: accents (PA—primary accent; SA—secondary accent).

80.8%, and the average recognition rate was 58.8%. This drop is due to the bad score of the B2 and B9 boundaries where only 32.9% and 47.6% were correctly recognized. These two boundary types together, on the other hand, only occur in 7.3% of all syllables. For sentence modality the total recognition rate amounts to 85.5% and the average recognition rate to 61.9%. This difference stems from the fact that only those 16% of the syllables which are associated with B3 boundaries carry a sentence mode label, and that the classification errors with respect to the boundary type influence the results of the sentence mode classifier as well.

23.4.3 Strategies for Focal Accent Detection

In this investigation [Pet95] focus is defined as the semantically most important part of an utterance which is in general marked by prosodic features. If it is marked by other means (e.g., by word order), its prosody no longer provides salient information. This work is thus only confined to those focal accents that are marked by prosody. In the VERBMOBIL dialogs such utterances are rather frequent.

Batliner [Bat89] showed in a discrimination experiment that F_0 maxima and minima and their positions in time are among the most significant features for focus discrimination. Bruce and Touati [BT90] found that in Swedish focal accents often control downstepping in the F_0 contour: in prefocal position there is no downstepping, whereas significant downstepping can be found after the focus. Petzold [Pet95] implemented an algorithm which relies on this feature (see Fig. 23.4 for an example). Focussed regions (according to the above definition) were perceptually labelled for 7

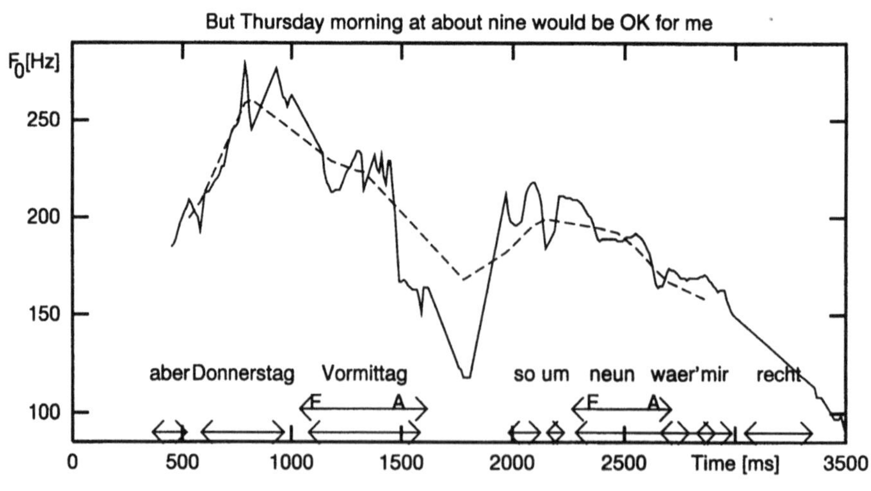

FIGURE 23.4. Utterance from a dialog with labelled focus (after Peizold [Pet95]).

dialogs of the VERBMOBIL data (154 turns, 247 focal accents found, but only about 20% of all frames pertain to focussed regions). To detect significant downsteps in the F_0 contour, Petzold's algorithm first eliminates such frames where F_0 determination errors are likely, or where the influence of microprosody is rather strong (for instance at voiced obstruents). The remaining frames of the F_0 contour are then processed using a moving window of 90 ms length; if a significant maximum (with at least a two-point fall on either side) is found within the window, its amplitude and position are retained; the same holds for significant minima. By connecting these points a simplified F_0 contour is created. To serve as a candidate for a focal accent, a fall must extend over a segment of at least 200 ms in the simplified F_0 contour. If such a significant downstep is detected, the nearest F_0 maximum (of the original F_0 contour) is taken as the place of the focus.

First results, based on these seven dialogs, are not too good yet but in no way disappointing. As only a minority of the frames fall within focussed regions, and as particularly in focus detection false alarms may do more damage than a focus that remains undetected, the recognition rates for focus areas are lower than for nonfocus areas. Table 23.3 displays a synopsis of the results for all dialogs.

Experiments are under way to incorporate knowledge about phrase boundaries and sentence mode. Batliner [Bat89] showed that in questions with a final rising contour focus cannot be determined in the same way as in declarative sentences; we could therefore expect an increase in recognition rate from separating questions and non-questions. Phrase boundaries could help us to restrict focus detection to single phrases and therefore to split the recognition task.

TABLE 23.3. First results for detection of focussed regions in seven spontaneous dialogs [Pet95]. The figures for the "best" and "worst" lines are not necessarily taken from the same dialog. All numbers are given in percent.

	Focussed	Recognition Rate		Recognition for	
	part	Global	Average	Focus	Non-focus
Average	18.4	78.6	66.7	45.8	87.5
Best		88.2	80.0	63.0	97.5
Worst		74.5	55.8	20.5	78.8

Concluding Remarks

Vaissière ([Vai88], p. 96) stated that "it is often said that prosody is complex, too complex for straightforward integration into an ASR system. Complex systems are indeed required for full use of prosodic information. [...] Experiments have clearly shown that it is not easy to integrate prosodic information into an already existing system [...]. It is necessary therefore to build an architecture flexible enough to test 'on-line' integration of information arriving in parallel from different knowledge sources [...]." The concept of VERBMOBIL has enabled prosodic knowledge to be incorporated from the beginning on and has given prosody the chance to contribute to automatic speech understanding. Although our results are still preliminary and most of the work is still ahead, it is shown that prosodic knowledge favorably contributes to the overall performance of speech recognition. Even if the incorporation of a prosodic module does not significantly increase word accuracy, it decreases the number of word hypotheses to be processed and thus reduces the overall complexity.

Our prosodic modules developed so far rely on acoustic features that are classically associated with prosody, i.e., fundamental frequency, energy, duration, and rhythm. With these features and classical pattern recognition methods, such as statistical classifiers or neural networks, typical detection rates for phrase boundaries or word accents range from 55% to 75% for spontaneous speech like that in the VERBMOBIL dialogs. We are sure that these scores can be increased when more prosodically labelled training data become available. It is an open question, however, how much prosodic information is really contained in the acoustic features just mentioned, or, in other words, whether a 100% recognition of word accents, sentence mode or phrase boundaries is possible at all when it is based on these features alone without reference to the lexical information of the utterance. Both prosodic modules described in this paper make little use of such information. The module by Kompe, Nöth, Batliner *et al.* (Sec. 3) only exploits the word hypothesis graph to locate syllables that can bear an accent and can be followed by boundaries, and the module by Strom (Sec. 4) uses the same information in a more elementary way by applying a syllable nucleus detector. Perceptual experiments are now under way to investigate how well humans perform when they have to judge prosody only from these acoustic features [Str96]. In any case more interaction between the segmental and lexical levels on the one hand and the prosody module on the other will be needed for the benefit of both modules. This requires—as Vaissière [Vai88] postulated—a flexible architecture that allows for such interaction. As VERBMOBIL offers this kind of architecture, it will be an ideal platform for more interactive and sophisticated processing of prosodic information in the speech signal.

Acknowledgments

This work was funded by the German Federal Ministry for Education, Science, Research, and Technology (BMBF) in the framework of the VERBMOBIL project under Grants 01 IV 102 H/0, 01 IV 102 F/4, and 01 IV 101 D/8. The responsibility for the contents of the experiments lies with the authors. Only the first author should be blamed for the deficiencies of this presentation.

References

[Bat89] A. Batliner. Zur intonatorischen Indizierung des fokus im Deutschen. In H. Altmann and A. Batliner, editors, *Zur Intonation von Modus und Fokus im Deutschen*, pp. 21–70. Tübingen: Niemeyer, 1989.

[BBB+94] G. Bakenecker, U. Block, A. Batliner, R. Kompe, E. Nöth, and P. Regel-Brietzmann. Improving parsing by incorporating 'prosodic clause boundaries' into a grammar. In *Proceedings of the International Conference on Spoken Language Processing, Yokohama, Japan*, pp. 1115–1118, 1994.

[BBK95] J. Bos, A. Batliner, and R. Kompe. On the use of prosody for semantic disambiguation in VERBMOBIL (Heidelberg, Munich, Erlangen, VERBMOBIL Memo 82-95), 1995.

[BKBN95] A. Batliner, A. Kiessling, S. Burger, and E. Nöth. Filled pauses in spontaneous speech. In *Proceedings of the 13th International Congress of Phonetic Sciences, Stockholm, Sweden*, Vol. 3, pp. 472–475, 1995.

[BT90] G. Bruce and P. Touati. On the analysis of prosody in spontaneous dialogue. In *Working Papers 36*, pp. 37–55. Department of Linguistics, Lund University, 1990.

[Fuj83] H. Fujisaki. Dynamic characteristics of voice fundamental frequency in speech and singing. In P. MacNeilage, editor, *The Production of Speech*, pp. 39–55. Berlin: Springer-Verlag, 1983.

[HKT95] W. Hess, K. Kohler, and H. Tillmann. The Phondat-Verbmobil speech corpus. *Proceedings of the European Conference on Speech Communication and Technology, Madrid, Spain*, pp. 863–866, 1995.

[KBK+94] R. Kompe, A. Batliner, A. Kiessling, U. Kilian, H. Niemann, E. Nöth, and P. Regel-Brietzmann. Automatic classification of prosodically marked boundaries in German. *Proceedings of the International Conference on Acoustics, Speech, and Signal Processes*, 2:173–176, 1994.

[KKB+94] A. Kiessling, R. Kompe, A. Batliner, H. Niemann, and E. Nöth. Automatic labelling of phrase accents in German. In *Proceedings of the International Conference on Spoken Language Processing, Yokohama, Japan*, pp. 115–118, 1994.

[KKN+92] A. Kiessling, R. Kompe, H. Niemann, E. Nöth, and A. Batliner. DP-based determination of F_0 contours from speech signals. *Proceedings of the International Conference on Acoustics, Speech, and Signal Processes*, 2, II/17-II/20, 1992.

[KKN+93] A. Kiessling, R. Kompe, H. Niemann, E. Nöth, and A. Batliner. Roger, sorry, I'm still listening: Dialog guiding signals in information retrieval dialogs. *Working Papers 41, Proceedings of the ESCA Workshop on Prosody, Lund University, Sweden*, pp. 140–143, 1993.

[KKN+95a] A. Kiessling, R. Kompe, H. Niemann, E. Nöth, and A. Batliner. Detection of phrase boundaries and accents. *Progress and Prospects of Speech Research and Technology: Proc of the CRIM/ORWISS workshop, Sankt Augustin*, pp. 266–269, 1995.

[KKN+95b] R. Kompe, A. Kiessling, H. Niemann, E. Nöth, E. Schukat-Tala maz zini, A. Zottmann, and A. Batliner. Prosodic scoring of word hypotheses graphs. *Proceedings of the European Conference on Speech Communication and Technology, Madrid, Spain*, pp. 1333–1336, 1995.

[KNK+94] R. Kompe, E. Nöth, A. Kiessling, T. Kuhn, M. Mast, H. Niemann, K. Ott, and A. Batliner. Prosody takes over: Towards a prosodically guided dialog system. *Speech Commun.*, 15:155–167, 1994.

[Lea80] W. A. Lea. Prosodic aids to speech recognition. In W. A. Lea, editor, *Trends in Speech Recognition*, pp. 166–205. Englewood Cliffs: Prentice-Hall, 1980.

[MF94] H. Mixdorff and H. Fujisaki. Analysis of voice fundamental frequency contours of German utterances using a quantitative model. In *Proceedings of the International Conference on Spoken Language Processing, Yokohama, Japan*, pp. 2231-2234, 1994.

[MPH93] B. Möbius, M. Pätzold, and W. Hess. Analysis and synthesis of F_0 contours by means of Fujisaki's model. *Speech Commun.*, 13:53–61, 1993.

[Nö91] E. Nöth. *Prosodische Information in der Automatischen Spracherkennung.* Tübingen: Niemeyer, 1991.

[NB95] E. Nöth and A. Batliner. Prosody in speech recognition. *Lecture at the Symposium on Prosody, Stuttgart, Germany,* Feb. 1995.

[Nie83] H. Niemann. *Klassifikation von Mustern.* Berlin: Springer, 1983.

[NK88] E. Nöth and R. Kompe. Der Einsatz Prosodischer Information im Spracherkennungssystem EVAR. In H. Bunke *et al.*, editor, *Mustererkennung 1988 (10. DAGM Symposium)*, pp. 2–9. Berlin: Springer, 1988.

[NP94] E. Nöth and B. Plannerer. Schnittstellendefinition für den Worthypothesengraphen. *Erlangen, Munich, Verbmobil Memo*, pp. 2–94, 1994.

[Pet95] A. Petzold. Strategies for focal accent detection in spontaneous speech. In *Proceedings of the 13th International Congress of Phonetic Sciences, Stockholm, Sweden,* Vol. 3, pp. 672–675, 1995.

[Rey93] M. Reyelt. Experimental investigation on the perceptual consistency and the automatic recognition of prosodic units in spoken German. *Working Papers 41, Proceedings of the ESCA Workshop on Prosody, Lund University, Sweden,* pp. 238–241, 1993.

[Rey95] M. Reyelt. Ein System Prosodischer Etiketten zur Transkription von Spontansprache. *Studientexte zur Sprachkommunikation,* 12, (Techn. Univ. Dresden):167–174, 1995.

[SBP$^+$92] K. Silverman, M. Beckman, J. Pitrelli, M. Ostendorf, C. Wightman, P. Price, J. Pierrehumbert, and J. Hirschberg. ToBI: a standard for labelling English prosody. In *Proceedings of the International Conference on Spoken Language Processing, Banff, Canada,* Vol. 2, pp. 867–870, 1992.

[Str95a] V. Strom. Detection of accents, phrase boundaries, and sentence modality in German with prosodic features. *Proceedings of the European Conference on Speech Communication and Technology, Madrid, Spain,* pp. 2039–2041, 1995.

[Str95b] V. Strom. Die Prosodiekomponente in INTARC I. 3. Technical Report 33, Bonn, VERBMOBIL, 1995.

[Str96] V. Strom. What's in the pure prosody? *Forum Acusticum, Antwerp, Belgium,* 1996.

[Vai88] J. Vaissière. The use of prosodic parameters in automatic speech recognition. In H. Niemann *et al.*, editor, *Recent Advances in Speech Understanding and Dialog Systems 46, NATO-ASI Series F*, pp. 71–100. Berlin: Springer, 1988.

[Wah93] W. Wahlster. Verbmobil—Translation of face-to-face dialogs. *Proceedings of the European Conference on Speech Communication and Technology, Berlin, Germany*, pp. 29–38, 1993.

[Wai88] A. Waibel. *Prosody and speech recognition.* London: Pitman, 1988.

Author Index

Citation Index

Subject Index

MIX
Papier aus verantwortungsvollen Quellen
Paper from responsible sources
FSC® C105338

If you have any concerns about our products,
you can contact us on
ProductSafety@springernature.com

In case Publisher is established outside the EU,
the EU authorized representative is:
Springer Nature Customer Service Center GmbH
Europaplatz 3, 69115 Heidelberg, Germany

Printed by Libri Plureos GmbH
in Hamburg, Germany